Simplified Robust Adaptive Detection and Beamforming for Wireless Communications

Simplified Robust Adaptive Detection and Beamforming for Wireless Communications

Ayman Elnashar
Emirates Integrated Telecommunications Company (EITC)
Dubai
UAE

This edition first published 2018

Registered Offices
John Wiley & Sons, Inc., 111 River Street, Hoboken, NJ 07030, USA
John Wiley & Sons Ltd, The Atrium, Southern Gate, Chichester, West Sussex, PO19 8SQ, UK

Editorial Office
The Atrium, Southern Gate, Chichester, West Sussex, PO19 8SQ, UK

For details of our global editorial offices, customer services, and more information about Wiley products visit us at www.wiley.com.

Wiley also publishes its books in a variety of electronic formats and by print-on-demand. Some content that appears in standard print versions of this book may not be available in other formats.

Library of Congress Cataloging-in-Publication Data

Names: Elnashar, Ayman, author.
Title: Simplified robust adaptive detection and beamforming for wireless communications / by Ayman Elnashar.
Description: First edition. | Hoboken, NJ : John Wiley & Sons, 2018. | Includes index. |
Identifiers: LCCN 2017060724 (print) | LCCN 2018007145 (ebook) | ISBN 9781118938232 (pdf) | ISBN 9781118938225 (epub) | ISBN 9781118938249 (cloth) |
Subjects: LCSH: Adaptive signal processing. | Beamforming. | Wireless communication systems.
Classification: LCC TK5102.9 (ebook) | LCC TK5102.9 .E36 2018 (print) | DDC 621.382/4–dc23
LC record available at https://lccn.loc.gov/2017060724

Cover design: Wiley
Cover image: © scanrail/istockphoto;
© hakkiarslan/istockphoto

Set in 10/12pt WarnockPro by SPi Global, Chennai, India
Printed in Singapore by C.O.S. Printers Pte Ltd
10 9 8 7 6 5 4 3 2 1

This book is dedicated to the memory of my parents (God bless their souls). They gave me the strong foundation and unconditional love, which remains the source of motivation and is the guiding light of my life.

Also, this book is dedicated to my PhD supervisors, Prof. Said Elnoubi from Alexandria University and Prof. Hamdi Elmikati from Mansoura University. They have guided and encouraged me during my PhD thesis and inspired me to author this book.

To my dearest wife, your encouragement and patience has strengthened me always.

To my beloved children Noursin, Amira, Yousef, and Yasmina. You are the inspiration!

Finally, I acknowledge the contribution of Tamer Samir from mobily for chapter 8.

–Ayman Elnashar, PhD

Contents

About the Author

Ayman Elnashar, PhD, has 20+ years of experience in telecoms industry including 2G/3G/LTE/WiFi/IoT/5G. He was part of three major start-up telecom operators in MENA region (Orange/Egypt, Mobily/KSA, and du/UAE). Currently, he is Vice President and Head of Infrastructure Planning - ICT and Cloud with the Emirates Integrated Telecommunications Co. "du", UAE. He is the founder of the Terminal Innovation Lab and UAE 5G innovation Gate (U5GIG). Prior to this, he was Sr. Director – Wireless Networks, Terminals and IoT where he managed and directed the evolution, evaluation, and introduction of du wireless networks including LTE/LTE-A, HSPA+, WiFi, NB-IoT and currently working towards deploying 5G network in UAE. Prior to this, he was with Mobily, Saudi Arabia, from June 2005 to Jan 2008 as Head of Projects. He played key role in contributing to the success of the mobile broadband network of Mobily/KSA. From March 2000 to June 2005, he was with Orange Egypt.

He published 30+ papers in wireless communications arena in highly ranked journals and international conferences. He is the author of "Design, Deployment, and Performance of 4G-LTE Networks: A Practical Approach" published by Wiley & Sons, and "Practical Guide to LTE-A, VoLTE and IoT: Paving the way towards 5G" to be published in May 2018. His research interests include practical performance analysis, planning and optimization of wireless networks (3G/4G/WiFi/IoT/5G), digital signal processing for wireless communications, multiuser detection, smart antennas, massive MIMO, and robust adaptive detection and beamforming.

About the Companion Website

This book is accompanied by a companion website:

www.wiley.com/go/elnashar49

The website include:

- Matlab scripts

1 Introduction

1.1 Motivation

This book presents alternative and simplified approaches for the robust adaptive detection and beamforming in wireless communications. The book adopts several system models, including:

- DS/CDMA, with and without antenna array
- MIMO-OFDM with antenna array
- general smart antenna array model.

Recently developed detection and beamforming algorithms are presented and analyzed with an emphasis on robustness. In addition, simplified and efficient robust adaptive detection and beamforming techniques are developed and compared with existing techniques. The robust detectors and beamformers are implemented using well-known algorithms including, but not limited to:

- least-mean-square
- recursive least-squares (RLS)
- inverse QR decomposition RLS (IQRD-RLS)
- fast recursive steepest descent (RSD)
- block-Shanno constant modulus (BSCMA)
- conjugate gradient (CG)
- steepest descent (SD).

Simplified Robust Adaptive Detection and Beamforming for Wireless Communications, First Edition. Ayman Elnashar.

Companion website: www.wiley.com/go/elnashar49

The robust detection and beamforming methods are derived from existing detectors/beamformers including, but not limited to:

- the robust minimum output energy (MOE) detector
- partition linear interference canceller (PLIC) detector
- linearly constrained constant modulus (CM) algorithm (LCCMA),
- linearly constrained minimum variance (LCMV) beamforming with single constraint,
- minimum variance distortionless response (MVDR) beamformer with multiple constraint
- block Shanno constant modulus algorithm (BSCMA) based detector/beamformer
- adaptive minimum bit error rate (BER) based detectors.

The adopted cost functions include the mean square error (MSE), BER, CM, MV and the signal-to-noise or signal-to-interference-plus-noise ratios (SINR/SNR). The presented robust adaptive techniques include:

- quadratic inequality constraint (QIC)
- diagonal loading techniques
- single and multiple worst-case (WC) constraint(s)
- ellipsoidal constraint
- joint constraints

Detailed performance analysis in terms of MSE, SINR, BER, computational complexity, and robustness are conducted for all the presented detectors and beamformers. Practical examples based on the above system models are provided to exemplify the developed detectors and beamforming algorithms. Moreover, the developed techniques are implemented using Matlab and the relevant Matlab scripts are provided to allow the readers to develop and analyze the presented algorithms. The developed algorithms will be presented in the context of DS/CDMA, MIMO-OFDM, and smart antenna arrays, but they can be easily extended to other domains and other applications. Figure 1.1 provides a high-level description of the book.

Recently, robust adaptive detection/beamforming has become a hot topic. Researchers seek to provide robustness against uncertainty in the direction of arrival (DOA) or the

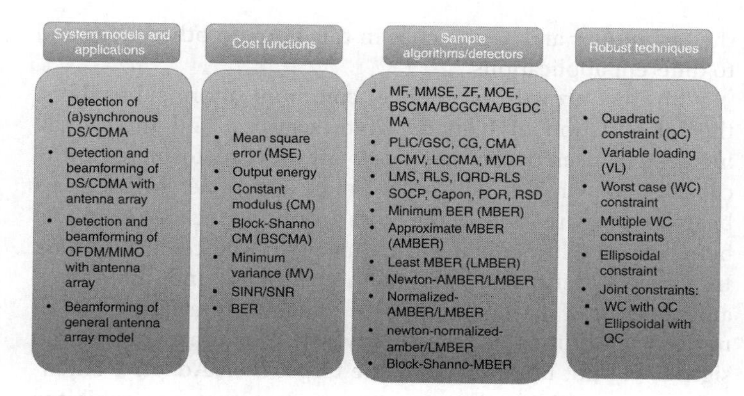

Figure 1.1 Summary of the book.

signature waveform, accuracy errors, calibration errors, small sample sizes, mutual coupling in antenna arrays, and so on. The major concern with the robust algorithms is the compromises involving robustness, complexity, and optimality. This book is aims to efficiently address this concern by presenting alternative and simplified approaches for robust adaptive detection and beamforming in wireless communications systems. The presented algorithms have low computational complexity while offering optimal or close-to-optimal performance and can be practically implemented. Wireless communication applications using DS/CDMA, MIMO-OFDM, and smart antenna systems are presented to demonstrate their robustness and to compare their complexity with established techniques and optimal detectors/beamformers.

The book presents and addresses current hot topics in adaptive signal processing: robustness and simplified adaptive implementation. It presents simplified approaches that add robustness to adaptive signal processing algorithms, with less computational complexity, while maintaining optimality. In addition, the presented algorithms are illustrated with practical examples and simulation results for major wireless communications systems, including DS/CDMA, MIMO-OFDM, and smart antenna systems. Moreover, Matlab scripts are provided for further analysis and development. The reader can easily extend the techniques and approaches in this book to other areas and to different applications.

With the growth of mobile communication subscribers, the introduction of high data-rate services, and the overall increase in user traffic, new ways are needed to increase the capacity of wireless networks. Smart antennae, MIMO and beamforming are some of the most promising technologies now being exploited to enhance the capacity of the cellular system. In wireless networks, the traditional omni and directional antennae of a base-station cause higher interference than necessary. Additionally, they are wasteful, as most transmitted signals will not be received by the target user. Adaptive antennas are a multidiscipline technology area that has exhibited growth steadily over the last four decades, primarily due to the impressive advances in the field of digital processing. Exploiting the spatial dimension using adaptive antennae promises impressive increases in system performance in terms of capacity, coverage, and signal quality. This will ultimately lead to increased spectral efficiency and extended coverage, especially for higher-frequency bands, such as millimetre waves (mmWave), that will be adopted for 5G evolution.

1.2 Book Overview

In Chapter 1, the mathematical models of DS/CDMA and MIMO-OFDM systems are presented. These form the foundation for the robust adaptive detection and beamforming algorithms that will be presented and/or developed in this book. DS/CDMA and OFDM are used in 3GPP 3G and 4G systems respectively. The 5G system under development by 3GPP will use evolved versions of MIMO-OFDM. The algorithms presented in this book may fit any of these systems and may also be extended to other systems. The focus of the 3G and 4G evolutions were on mobile broadband, as a result of widespread smartphone adoption. The internet of things (IoT) evolution will lead to billions of devices being connected to the internet and this has directed the 3GPP and mobile communications industry towards narrowband technologies. 3GPP has modified the LTE system to meet the IoT requirements by introducing NB-IoT. Other proprietary technologies, such as low-power wide-area networks, have used narrowband or ultra-narrowband technologies such as chirp spread spectrum. The focus of this book is not on certain technologies and readers will need to expend some effort in order to apply the detection and beamforming algorithms outlined here to specific systems. The focus of the book is the development and comparative analysis of robust adaptive detection and beamforming algorithms based on simplified system models. All the results in the book are simulated using Matlab and the developed scripts are provided along with the book. The reader may need to slightly modify the scripts depending on the Matlab version. In addition, some algorithms developed by other authors are provided as part of the software package with this book for the purpose of comparative analysis.

In Chapter 3, we will provide a survey of adaptive detection algorithms based on the DS/CDMA model. However, the adaptive techniques that are summarized in this survey can be easily extended to MIMO-OFDM and smart antenna arrays. The DS/CDMA model is the most complicated system model, because of its need for multiuser interference cancellation and since the channel is frequency selective, as explained in Chapter 2. Despite the various advantages of the DS/CDMA system, it is interference limited due to multiuser interference and it cannot be easily extended to ultra-broadband systems. A conventional DS/CDMA receiver treats each user separately as a signal, with other users considered as noise or multiple access interference (MAI). A major drawback of such conventional DS/CDMA systems is the near–far problem: degradation in performance due to the sensitivity to the power of the desired user against the power of the interference. A reliable demodulation is impossible unless tight power control algorithms are exercised. The near–far problem can significantly reduce the capacity. Multiuser detection (MUD) algorithms can give dramatically higher capacity than conventional single-user detection techniques. MUD considers signals from all users, which leads to joint detection. MUD reduces interference and hence leads to a capacity increase, alleviating the near–far problem. Power control algorithms can be used but are not necessary.

Linear receiver design by minimization of some inverse filtering criterion is explained in Chapter 4. Appropriate constraints are used to avoid the trivial all-zero solution. A well-known cost function for the constrained optimization problem is the variance or the power of the output signal. An MOE detector for multiuser detection is developed, based on the constrained optimization approach. In an additive white Gaussian environment with no multipath, this detector provides a blind solution with MMSE performance. In Chapter 4, linearly constrained IQRD-RLS algorithms with multiple constraints are developed and implemented for MUD in DS/CDMA systems. As explained above, the same algorithms can be extended to MVDR beamforming algorithms. Two approaches are considered, the first with a constant constrained vector and the other with an optimized constrained vector. Three IQRD-based detectors are developed as follows:

- a direct form MOE detector based on the IQRD update method with fixed constraints
- a MOE detector in the PLIC structure based also on the IQRD-RLS algorithm
- an optimal MOE algorithm built using the IQRD update method and a subspace tracking algorithm for tracking the channel vector.

The constrained vector (estimated channel vector) is obtained using the max/min approach with IQRD-RLS based subspace tracking algorithms that are analyzed and tested for channel vector tracking.

The recently developed subspace tracking algorithms are tested and analyzed for channel estimation in Chapter 4. These are the fast orthogonal projection approximation subspace tracking (OPAST) algorithm and the normalized orthogonal Oja (NOOja) algorithm. In addition, a fast subspace tracking algorithm based on the Lagrange multiplier methodology and the IQRD algorithm will be developed and adopted for channel vector estimation and tracking. Moreover, a new strategy for combining the max/min channel estimation technique with the robust quadratic constraint technique is proposed anchored in the direct form algorithm. Specifically, a robust MOE detector is developed, based on the max/min approach and QIC on the weight vector norm to overcome noise enhancement at low SNR. A direct form solution is introduced for the quadratically constraint detector with a variable loading (VL) technique employed to satisfy the QIC. Thus, the IQRD algorithm acts as a core to the proposed receivers, which facilitate real-time implementation through systolic implementation. However, the same algorithms can be easily implemented using fast and robust RLS-based algorithms.

A robust low-complexity blind detector is presented in Chapter 5. This is based on a recursive steepest descent (RSD) adaptive algorithm rather than the RLS algorithm and a QIC on the weight vector norm. The QIC is employed to manage the residual signal mismatch and other random perturbations errors. In addition, the QIC will make the noise constituent in the output SINR constant and hence overcome noise enhancement at low SNR. Quadratic constraints have been used in adaptive beamforming for a variety of purposes, such as improving robustness against mismatch and modeling errors, controlling mainlobe response, and enhancing interference cancellation capability. The quadratic constraint will be analyzed along with beamforming algorithms in Chapter 7.

Analogous to the recursive conjugate gradient (RCG) algorithm, a fast RSD algorithm is developed in Chapter 5. A low-computational complexity recursive update equation for the gradient vector is derived. Furthermore, a variable step-size approach is introduced for the step-size update of the RSD algorithm based on an optimum step-size calculation. The RSD algorithm is exploited to update the adaptive weight vector of the PLIC structure to suppress MAI. The same technique will be extended to MVDR beamforming in Chapter 7. From this similarity, the reader can easily extend the algorithms in this book to other systems and even beyond the realm of wireless communications. From this it can be seen that we have simplified the deployment of the robust techniques, such as quadratic constraints, uncertainty constraints, worst-case constraint optimization, and constrained optimization.

The drawbacks of diagonal loading techniques are tackled in Chapter 5. An alternative way of robust adaptive detection based on the RSD adaptive algorithm is presented. This involves an accurate technique for precisely computing the diagonal loading level without approximation or eigendecomposition. We combined the QIC with the RSD algorithm to produce a robust recursive implementation with $O(N^2)$ complexity. A new optimal VL technique is developed and integrated into the RSD adaptive algorithm. In addition, the diagonal loading term is optimally computed, with $O(N)$ complexity, using a simple quadratic equation. Geometrical interpretations of the scaled projection (SP) and VL techniques, along with RLS and RSD algorithms, are illustrated and analyzed. The performance of the robust detectors is compared with traditional detectors and the former are shown to be more accurate and more robust against signal mismatch and random perturbations. Finally, the presented approach can be reformulated to handle an uncertainty constraint – imposed on the signature waveform in MUD, or on a steering vector in beamforming – such as the ellipsoidal constraint. It can also be exploited with any of the robust approaches to produce a simple recursive implementation.

In Chapter 6, the quadratic inequality constraint is imposed on the weight vector norm of the LCCMA and BSCMA algorithms in order to enhance their performance. The weight norm constraint will control the gradient vector norm, meaning that there is no need to check the gradient vector norm increase in BSCMA. Additionally, the iteration inside the block can continue without affecting algorithm stability due to the weight vector norm constraint. We will investigate the effect of adding a quadratic inequality constraint on the LCCMA and BSCMA algorithms. The proposed VL technique in Chapter 5 is exploited to estimate the optimum diagonal loading value. The LCCMA and BSCMA algorithms are used to update the adaptive vector of the PLIC structure. The PLIC structure with multiple constraints is employed to identify the MAI and hence help in avoiding interference capture. Moreover, the different forms of BS-CMA algorithms – the block-conjugate gradient CMA algorithm (BCGCMA) and block gradient descent constant modulus algorithm (BGDCMA) – are investigated as well. The resistance of BSCMA-based algorithms against the near–far effect is discussed and evaluated.

In Chapter 7, we will present four approaches for robust adaptive beamforming as follows:

Improved recursive realization for robust LCMV beamforming We first develop an improved recursive realization for robust LCMV beamforming. This includes an ellipsoidal uncertainty constraint on the steering vector. The robust recursive implementation presented here is based on a combination of the ellipsoidal constraint formulation and the variable diagonal loading technique demonstrated in Chapter 5. As a consequence, an accurate technique for computing the diagonal loading level without eigendecomposition or SOCP is developed. The geometrical interpretation of the diagonal loading technique is demonstrated and compared with eigendecomposition approach. Note that this approach adopts a spherical constraint on the steering vector to optimize the beamformer output power. Unfortunately, the adaptive beamformer developed here is apt to noise enhancement at low SNR and an additional constraint is required to bolster the ellipsoidal constraint.

Joint constraint approach for a joint robustness beamformer The second approach is the development of a joint constraint approach for a joint robustness beamformer. A joint constraint approach is presented for joint robustness against steering vector mismatch and unstationarity of interferers. An alternative approach involves imposing an ellipsoidal uncertainty constraint and a quadratic constraint on the steering vector and the beamformer weights, respectively. We introduce a new simple approach to get the corresponding diagonal loading value. The quadratic constraint is invoked as a cooperative constraint to overcome noise enhancement at low SNR. The performance of the robust adaptive schemes developed and other robust approaches are demonstrated in scenarios with steering vector mismatch and several moving jammers.

Beamformer with a single WC constraint In the third approach, a robust MVDR beamformer with a single WC constraint is implemented using an iterative gradient minimization algorithm. This involves a simple technique to estimate the Lagrange multiplier instead of a Newton-like algorithm. This algorithm exhibits several merits, including simplicity, low computational load, and no need for either sample-matrix inversion or eigendecomposition. A geometric interpretation of the robust MVDR beamformer is demonstrated to supplement the theoretical analysis.

LCMV beamformer with MBWC constraints In the last approach, a robust LCMV beamformer with multiple-beam WC (MBWC) constraints is developed using a novel multiple-WC constraints formulation. The optimization problem entails solving a set of nonlinear equations. As a consequence, a Newton-like method is mandatory to solve the system of nonlinear equations, which yields a vector of Lagrange multipliers. The Lagrange method is used to give the solution.

The traditional MMSE detector is the most popular technique for beamforming. An adaptive implementation of the MMSE can be achieved by minimizing the MSE between the desired output and the actual array output. The LCMV and MVDR beamformers in Chapter 7 are different forms of MMSE detectors. For a practical communication system, it is the BER or block BER, not the MSE performance, that really matter. Ideally, the system design should be based directly on minimizing the BER rather than the MSE. For application in single-user channel equalization, multiuser detection, and beamforming, it has been shown that the MMSE solution can, in certain situations, be distinctly inferior to the minimum BER (MBER) solution. However, the BER cost function is not a linear function of the detector or the beamformer, making it difficult to minimize. Several adaptive MBER beamformer/detectors implementations are developed in the literature.

It must be stated here that the cost function of the MMSE criterion has a circular shape. This means that we have one global minimum. Hence, convergence can be easily achieved. In contrast, the cost function of the BER is highly nonlinear. This means that during minimization steps we may converge to a local minimum. The MMSE and MBER solutions lead to very different detector weight vectors. Clearly, the MBER design is more intelligent in utilizing the detector's resources. However, special attention is mandatory with the minimization algorithm in order to avoid convergence to a local minimum, and hence the algorithm diverging rather than converging.

Beamforming is a key technology in smart antenna systems, and can increase capacity and coverage and mitigate multipath propagation in mobile radio communication systems. The most popular criterion for linear beamforming is MMSE. However, the MSE cost function is not optimal in terms of the bit error probability performance of the system. In Chapter 8, a class of adaptive beamforming algorithms using direct minimization of the BER cost function is presented. Unfortunately, the popular least minimum BER stochastic beamforming algorithm suffers from low convergence speeds. Gradient Newton algorithms are presented as an alternative. These speed up the convergence rate and enhance performance but only at the expense of complexity. In Chapter 8, a block processing objective function for the MBER is formulated, and a nonlinear optimization strategy that produces the so-called 'block-Shanno MBER' is developed. A complete consideration of the complexity calculations of the proposed algorithm is given. Simulation scenarios are carried out in a multipath Rayleigh-fading DS-CDMA system to explore the performance of the proposed algorithm. Simulation results show that the proposed algorithm offers good performance in terms of convergence speed, steady-state performance, and even system capacity, compared to other MBER- and MSE-based algorithms.

Finally, we will extend the adaptive filtering algorithms using the concept of spatial multiuser detection in a MIMO-OFDM system model rather than beamforming in a DS-CDMA model. As stated above, a fundamental goal in any digital communications system is to directly minimize the BER. Wiener solution based algorithms indirectly minimize the BER by optimizing other cost functions (SNR, SINR, or MSE), which may result in suboptimal BER performance.

2

Wireless System Models

2.1 Introduction

In this chapter, mathematical models of DS/CDMA and orthogonal frequency division multiplexing (OFDM) systems are presented. These form the foundation of the robust adaptive detection algorithms that will be presented in this book. DS/CDMA and OFDM are used in 3GPP 3G and 4G systems respectively. The 5G systems under development by 3GPP will use evolved versions of OFDM. The algorithms presented in this book suit any of these systems and may also be extended to other systems.

The focuses of 3G and 4G were on mobile broadband thanks to smartphone adoption. The internet of things (IoT), with billions of devices expected to be connected to the internet, has directed the 3GPP and industry towards narrowband technologies. 3GPP has modified the LTE system to meet IoT requirements by introducing NB-IoT. Other proprietary technologies in unlicensed bands, known as low-power wide area (LPWA) networks, have used narrowband or ultra-narrowband approaches such as SigFox. Other LPWA networks have adopted a wideband CDMA approach based on chirp spread spectrum (CSS). One example is LoRa.

The focus of this book is not on particular commercial technologies, and readers will require some effort in order to match the detection and beamforming algorithms developed in this book to specific practical systems. The focus of the book is on the development and comparative analysis of robust adaptive detection and beamforming algorithms based on simplified

Simplified Robust Adaptive Detection and Beamforming for Wireless Communications, First Edition. Ayman Elnashar.

Companion website: www.wiley.com/go/elnashar49

system models. All results in the book are simulated using Matlab and the scripts used are provided along with the book. The reader may need to slightly modify the scripts depending on the Matlab version. Also, some algorithms developed by other authors are provided as part of the software package with this book for the purpose of comparative analysis.

Due to the complex nature of CDMA systems, there have been many different formulations of the DS/CDMA model. In this book, we consider a general DS-CDMA system model, which account for user asynchronism, multipath propagation, and frequency-selective fading propagation channels. The link from the base station to the mobile station is referred to as the "downlink" and is typically characterized by synchronous data transmission. More challenging for demodulation is the "uplink" from mobile to base station, where different user transmissions are typically asynchronous and of widely disparate power levels. Reliable modulation might also require mitigation of multipath interference, especially in wideband CDMA (WCDMA) schemes where multipath effects can be significant. The robust adaptive receivers/detectors presented and/or developed in this book are suitable for both uplink and downlink scenarios. Important performance measures are formulated for assessment and comparative analysis. DS/CDMA was adopted in 3G/HSPA+ mobile communication systems and IEEE 802.15.4. The main reasons for using CDMA techniques are as follows:

- resistance to unintended or intended jamming/interference
- sharing of a single channel among multiple users
- reduced signal/background-noise level hampering interception
- determination of relative timing between transmitter and receiver
- robustness against multipath propagation and frequency selective channels.

Second- and third-generation mobile systems are based on either TDMA or CDMA technologies. Although these technologies can theoretically be extended to next-generation wireless broadband systems, practical implementation issues and complexities limit their adoption. On the other hand,

OFDM offers an easier solution and simple implementation. However, OFDM is not without its issues.

Multipath signal propagation makes the channel response time dispersive; the amount of signal dispersion depends on the environment of operation. For example, the channel dispersion is about 5 μs in typical urban areas and 15–20 μs in rural and hilly terrain. The factor that affects the receiver is the number of resolvable channel taps over the channel dispersion interval. In a TDMA system, it is the ratio of the channel dispersion to signal symbol time. However, in a CDMA system, it is the number of channel taps with strong energy at chip-time resolution over the channel dispersion period. The channel time dispersion is viewed as frequency-selective or non-selective in the frequency domain. A frequency non-selective channel means the signal, over its entire bandwidth, will have the same effect as due to the multipath channel. This is also called a flat fading channel. In the time domain, the channel is not dispersive relative to its symbol time, and hence there is no intersymbol interference (ISI). In the frequency-selective channel, the signal will have independent effects over its bandwidth due to the channel, and it is time dispersive relative to its symbol time.

In narrowband TDMA systems, such as GSM, multipath propagation makes the channel frequency non-selective or less selective, making the receiver less complex. Extending TDMA techniques to broadband systems makes the receiver complexity unmanageable, as the channel becomes highly frequency selective. More specifically, GSM is a 200 kHz channel TDMA system, of 270.833 kHz symbol rate with either binary GMSK or 8-PSK modulations. The baseband signal uses partial response signalling, which spreads the symbol to three symbol periods. For a typical urban case with about 5 μs channel dispersion, the received signal can have a signal dispersion of about 5 symbol periods, including its partial response signalling. Therefore, a typical GSM receiver requires a 16-state MLSE (maximum likelihood sequence estimation) equalizer for Gaussian minimum shift keying (GMSK) modulation and an 8- or 64-state DFSE (decision feedback sequence equalizer) for an 8-PSK EDGE signal. Suppose we want to scale up this technique to a wideband or broadband system by factor of 100; that is, a 20 MHz channel bandwidth like LTE system with 27.0833 MHz symbol rate. For the same amount of channel dispersion, the received symbol will be spread over 200 symbol periods, which means a very frequency-selective channel. The receiver with an equalizer for 200 channel taps will be impractical to implement due to the very high complexity.

Similarly, WCDMA can also be extended to broadband systems, but its complexity increases, as it requires a larger number of rake receiver fingers. Complexity of a rake receiver, and often its gain, are based on the number of rake fingers the receiver can process. A typical WCDMA rake receiver requires about 5–8 rake fingers for a typical urban channel with channel spread of 5 μs. More advanced receivers, such as generalized rake receivers (G-rake), require even more fingers, placed around the desired signal, and often called "interference fingers". Extending WCDMA to a 20 MHz broadband system will require higher chip rates, meaning that it can resolve channel taps with finer resolution. This results in more fingers for the rake receiver with strong signal energy. Therefore, extension of WCDMA/HSPA+ systems to a 20 MHz broadband system requires expansion, by a similar factor, of the number of fingers in the rake receiver, and thus an increase in its detection complexity especially at the mobile side. In addition, the receiver design will be further complicated if we need to add multiple-input, multiple-output (MIMO) on top of such complicated receivers; the gain of MIMO will be minimized due to the frequency selective nature and the high number of detector taps. This is the main reason behind the delay in deploying MIMO with 3G/HSPA+ mobile systems while MIMO 2 × 2 has been used with 4G/LTE OFDM based systems from day one and currently MIMO 4 × 4 and MIMO 8 × 8 systems are in development. Moreover, 5G evolution will take MIMO and beamforming to the next level by introducing evolved node B (eNB), with 64 antenna elements. The 3GPP has introduced other ways of extending HSPA+ systems to broadband, based on multicarrier HSPA. Most commercial 3G networks adopt DC-HSPA+ and have evolved to DC/2C/4C-HSPA+ by combining 2, 3 and 4 carriers. The three/four carriers provide 63 Mbps/84 Mbps in the DL, respectively. However, it is still challenging to go beyond the dual carrier approach due to the network optimization complexity with each carrier considered as one cell. In addition, DC is mainly deployed in the downlink and its adoption in the uplink is still limited. More particularly, the multicarrier approach in the UL is still not widely deployed. With 3C-HSPA+ modems, dual carriers in the UL will be supported. The complexity in deploying multicarrier HSPA+ systems has been a motivation to develop and deploy the LTE system.

OFDM has become a most favored technique for broadband wireless systems due to the susceptibility to signal spread under multipath conditions. OFDM can also be viewed as a multicarrier narrowband system, where the whole system bandwidth is split into multiple smaller subcarriers with simultaneous transmission. Simultaneous data transmission and reception over these subcarriers are handled almost independently. Each subcarrier is usually narrow enough that the multipath channel response is flat over the individual subcarrier frequency range; that is, there is a frequency non-selective fading channel. Therefore, the OFDM symbol time is much larger than the typical channel dispersion. Hence OFDM is inherently susceptible to channel dispersion due to multipath propagation.

The fading channels are illustrated in Figure 2.1. The fading can be mainly classified into two different types: large-scale fading and small-scale fading. Large-scale fading occurs as the mobile moves over a large distance, say of the order of the cell size [52]. It is caused by the path loss of the signal as a function of distance, and shadowing by large objects such as buildings, intervening terrain, and vegetation. Shadowing is a slow-fading process, characterized by variation of median path loss between the transmitter and receiver in fixed locations. In other words,

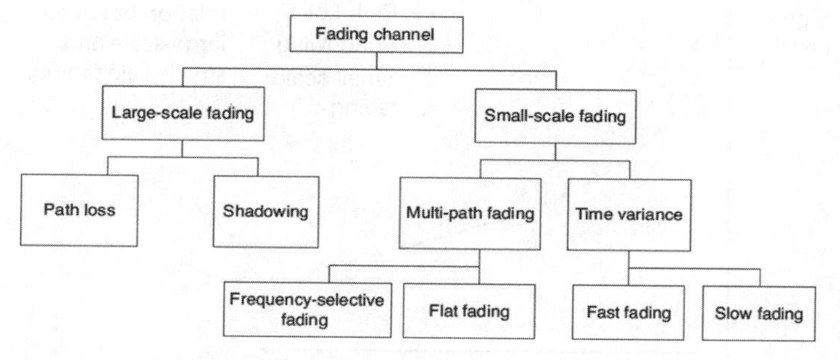

Figure 2.1 Classification of fading channels.

large-scale fading is characterized by average path loss and shadowing. On the other hand, small-scale fading involves the rapid variation of signal levels due to the constructive and destructive interference of multiple signal paths (multipaths) when the mobile station moves over short distances. Depending on the relative extent of a multipath, frequency selectivity of a channel is characterized (say, as frequency-selective or frequency-flat) for small-scaling fading. Meanwhile, depending on the time variation in a channel due to mobile speed (characterized by the Doppler spread), short-term fading can be classified as either fast fading or slow fading. Large-scale fading and time-variance fading are addressed in the link budget of the system. The path loss and shadowing identify the cell radius in the mobile system. Multipath fading is addressed in the receiver design and the multiple access technique deployed. The link budget usually assumes slow-fading channels. For fast-fading channels, such as with users inside a high-speed train, a fast-fading compensation factor is added to the link budget [52].

The relationship between large-scale fading and small-scale fading is illustrated in Figure 2.2 [52]. Large-scale fading is manifested by the mean path loss, which decreases with distance, and shadowing, which varies along the mean path loss. The received signal strength may be different at two different places, even at the same distance from a transmitter, due to shadowing caused by obstacles on the path. Furthermore, the scattering components incur small-scale fading, which ultimately gives a short-term variation of the signal, which has already

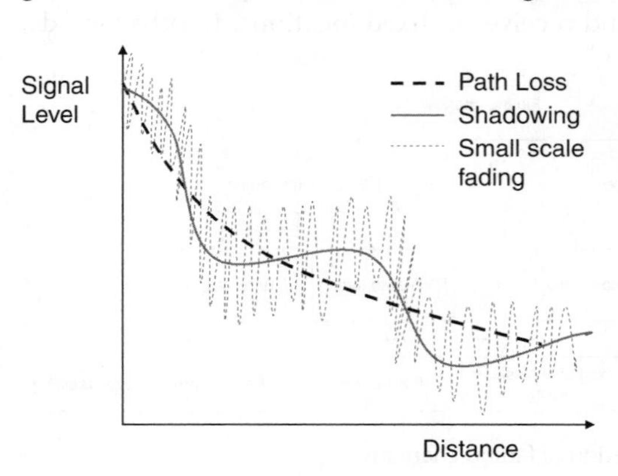

Figure 2.2 The relation between large-scale and small-scale fading.

experienced shadowing. A detailed practical link budget for an LTE system is provided in the literature [53]. This illustrates how all of these fading factors are addressed in the link budget.

In traditional systems (TDMA- or CDMA-based systems), symbol detection is on the samples at either the symbol or chip rate, and the carrier-to-interference level only matters at the sampling points. However, OFDM symbol detection requires that the entire symbol duration be free of interference from previous symbols, thus preventing ISI. Even though the OFDM symbol duration is much larger than the channel dispersion, even a small amount of channel dispersion causes some spilling of each OFDM symbol to the next symbol, thus causing some ISI. However, this ISI spillover is limited to only the initial part of the neighboring symbol. Hence this ISI spill-over at the beginning of each symbol can easily be tackled by adding a cyclic prefix (CP) to each transmit symbol. This prefix extends each symbol by duplicating a portion of the signal at the symbol ends. The prefix is removed at the receiver. The amount of symbol extension – that is, the length of the CPs – is a system design parameter, and is based on the expected signal dispersion in the environment of system operation. For example, the LTE system uses an OFDM symbol of 66 μs plus 5 μs of CP. This means it is susceptible to a maximum signal dispersion of 5 μs due to multipath channel propagation.

The LTE FDD frame structure is shown in Figure 2.3 [53] for normal CP. Each LTE FDD radio frame is $T_{\mathrm{d}} = 307200\ T_{\mathrm{s}} = 10$ ms long and consists of 20 slots of length $T_{\mathrm{slot}} = 15360\ \mathrm{T_s} = 0.5$ ms, numbered from 0 to 19. The number of UL SC-FDMA symbols in a slot depends on the CP length configured by the higher layer parameter (Table 2.1). The number of DL OFDM symbols in a slot depends on the CP length and the subcarrier spacing configured (Table 2.2). The CP length $N_{\mathrm{CP},l}$ that is to be used is provided in Table 2.2. Note that different OFDM symbols within a slot in some cases have different CP lengths.

During radio network planning, it is important to identify the CP. Radio network planning usually involves a normal CP of 4.7 μs, as this sufficiently combats multipath fading delay spread. In 3GPP LTE standard fading channel models, the maximum delay spread is 2.5 μs, which is much shorter than a normal

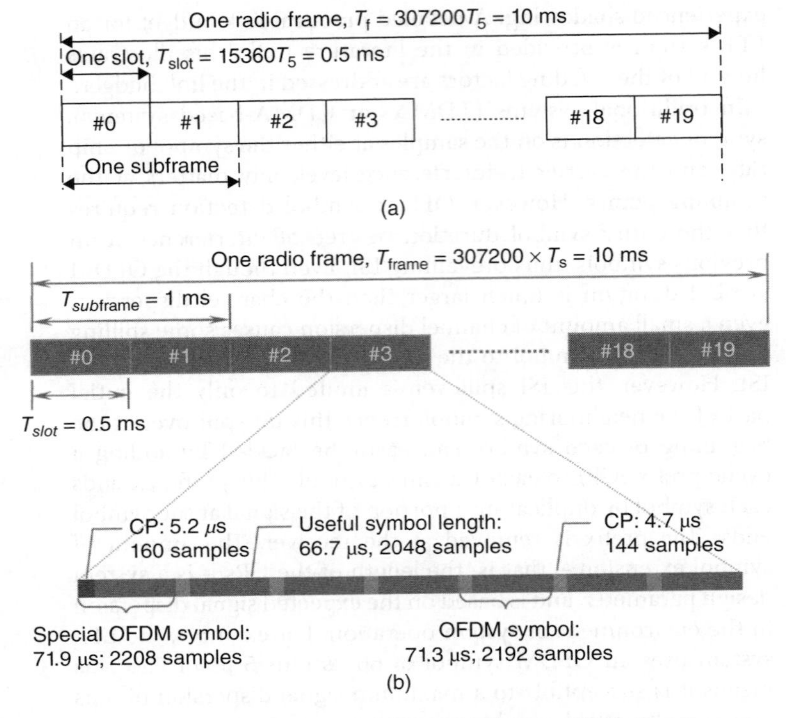

Figure 2.3 LTE FDD frame and slot structure for normal CP.

Table 2.1 CP type and relevant resource block parameters in LTE UL with SC-FDMA.

Configuration	N_{sc}^{RB}	$N_{symb}^{UL/DL}$
Normal CP	12	7
Extended CP	12	6

CP. Moreover, a normal CP consumes fewer overheads in the total OFDM symbol than extended CP, and therefore provides higher throughput performance. Extended CP is used for two applications:

- for an extended cell radius where the maximum delay spread is larger than 4.7 μs

Table 2.2 CP and physical RB parameters in LTE DL with OFDM.

Configuration	Δf (kHz)	OFDM symbol N^{DL}_{symb}	Sub-carrier N^{RB}_{sc}	Cyclic prefix length in samples $N_{CP,l}$	Cyclic prefix length (μs)
Normal CP	15	7	12	160 for $l = 0$ 144 for $l = 1,2,\ldots,6$	5.2 for 1st symbol 4.7 for other symbols
Extended CP	15	6		512 for $l = 0,1,\ldots,5$	16.7
Extended CP	7.5	3	24	1024 for $l = 0, 1,2$	33.3

- with evolved multimedia broadcast multicast service (eMBMS), which is required for broadcasting TV channels on LTE networks.

SC-FDMA is a promising technique for high data-rate uplink communication and has been adopted by 3GPP for the 4G LTE system. SC-FDMA is a modified form of OFDM, with similar throughput performance and complexity. It is often viewed as DFT-coded OFDM, where time-domain data symbols are transformed to the frequency domain via a discrete Fourier transform (DFT) before going through the standard OFDM modulation. Thus, SC-FDMA inherits all the advantages of OFDM over other well-known techniques, such as TDMA and CDMA. The major problem in extending GSM TDMA and wideband CDMA to broadband systems is the increase in complexity with the multipath signal reception. The main advantage of OFDM, as for SC-FDMA, is its robustness against multipath signal propagation, which makes it suitable for broadband systems. SC-FDMA brings the additional benefit of low peak-to-average power ratios compared to OFDM, making it suitable for uplink transmissions by user terminals.

A typical OFDM transmitter and receiver structure is shown in Figure 2.4. A transmitter includes a baseband modulator, subcarrier mapping, inverse Fourier transform, CP addition, parallel-serial conversion, and a digital-to-analog converter followed by an I-Q RF modulator. Unlike other modulation techniques that operate symbol by symbol, OFDM transmits a block of data symbols simultaneously over one OFDM symbol. An OFDM symbol is the time used to transmit all the subcarriers that are modulated by the block of input data symbols. The baseband modulator transforms the input binary bits into a set of multilevel complex numbers that correspond to different modulation formats, such as BPSK, QPSK, 16QAM, 64QAM or 256QAM. The type of modulation format used often depends on the signal-to-noise level of the received signal and the receiver's ability to decode them correctly. These modulated symbols are then mapped to subcarriers. An inverse-FFT (IFFT) is used to transform the modulated subcarriers in the frequency domain to time-domain samples [41].

In general, the same modulation format is used in all the subcarriers to keep the control-information overhead small. However, it is possible to have different modulation formats over multiple subcarriers, and it is advantageous to do so in harsh and time-varying channel conditions. In a broadband system, the channel is frequency selective over its large system bandwidth, meaning the signal fading on each subcarrier is independent. The interference level on each subcarrier can also be different and vary uniquely with time. This results in a different signal-to-impairment level on each of the subcarriers. Hence, having an appropriate modulation format on these subcarriers helps to optimize the overall system throughput. The OFDM system inherits an adaptation of modulation formats to each of the subcarriers depending on channel conditions, and this is called channel-dependent scheduling. The link adaptation will be discussed later in this section.

A CP block copies a portion of the samples at the end of the time-domain sample block (at the IFFT output) to the beginning. Since the DFT/FFT outputs are periodic in theory, copying the samples to the beginning will make the signal continuous. The length of the CP depends on the channel delay spread, and is preferably longer than the length of the channel response. At the receiver, the prefix part of the symbol is thrown away, because it may contain ISI from its previous symbol. Hence, any effect of ISI caused by multipath signal propagation is removed. However,

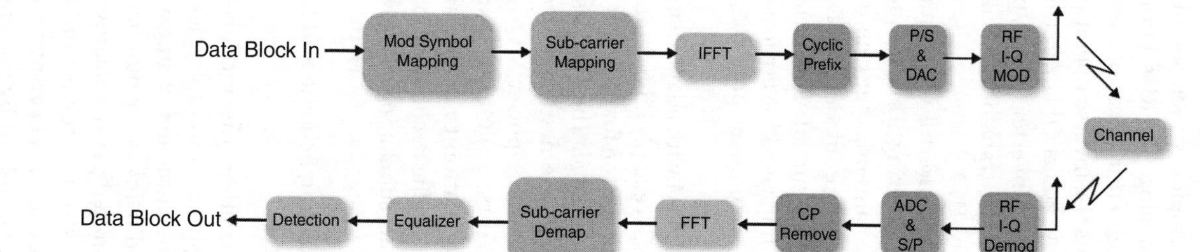

Figure 2.4 OFDM transmitter and receiver.

the prefix is an overhead in an OFDM system, as it does not carry any useful information.

The block of complex samples is then serialized in the time domain and converted to analog. The RF section modulates the I-Q samples to the final radio transmission frequency. A corresponding receiver performs the inverse operations of the transmitter (in reverse order). A typical OFDM receiver includes an RF section, analog-to-digital converter (ADC), parallel-to-serial converter, CP remover, Fourier transformer, subcarrier demapper, equalizer and detector [41, 42].

In summary, OFDM is a powerful technique when used in communications systems suffering from frequency selectivity. Combined with multiple antennas at the transmitter and receiver, as well as adaptive modulation, OFDM proves to be robust against channel delay spread. Furthermore, it leads to significant data rates and improved bit error performance over links having only a single antenna at both the transmitter and receiver.

Next, we will describe OFDM with adaptive modulation when applied to MIMO systems. We apply an optimization algorithm to obtain a bit and power allocation for each subcarrier assuming instantaneous channel knowledge. The analysis and simulation is considered in two stages. The first stage involves the application of a variable-rate, variable-power MQAM technique for a single-input single-output (SISO) OFDM system. The second stage applies adaptive modulation to a general MIMO system by making use of a singular value decomposition to separate the MIMO channel into parallel subchannels [54].

2.1.1 Modulation and Coding Scheme and Link Adaptation

In this section, we will outline the modulation and coding scheme for LTE FDD. Figure 2.5 shows a typical radio transmitter and receiver with modulation and coding at Tx and Rx [53]. The throughput over a radio link is the number of data bits that can be successfully transmitted per modulation symbol. Coding (more specifically, forward error correction) adds redundant bits to the data bits. These can correct errors in the received bits. The degree of coding is determined by its *rate*, the proportion

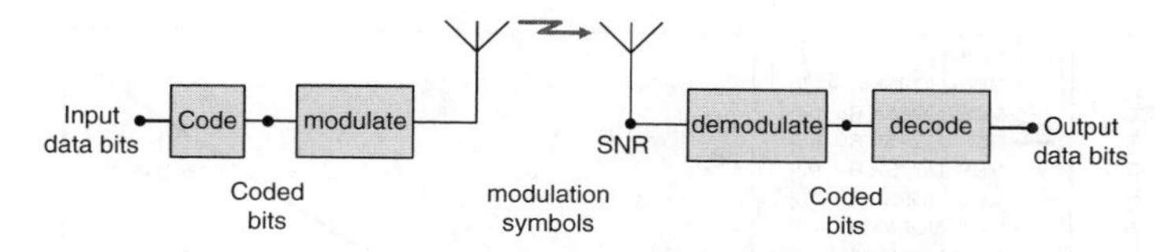

Figure 2.5 Coding and modulation for transmission of data over a radio link.

of data bits to coded bits. This typically varies from 1:8 to 4:5. Coded bits are then converted into modulation symbols. The order of the modulation determines the number of coded bits that can be transmitted per modulation symbol. Typical examples are QPSK, 16QAM, 64QAM, and 256QAM, which have 2, 4, 6, and 8 bits per modulation symbol, respectively.

The efficiency of a given modulation and coding scheme (MCS) is the product of the rate and the number of bits per modulation symbol. Throughput has units of data bits per modulation symbol. This is commonly normalized to a channel of unity bandwidth, which carries one symbol per second. The units of efficiency then become bits per second per hertz. A given MCS requires a certain signal-to-interference-plus-noise ratio (SINR) at the receiver antenna to operate with an acceptably low bit error rate (BER) in the output data. An MCS with a higher throughput needs a higher SINR to operate. Adaptive modulation and coding (AMC) works by measuring and feeding back the channel SINR to the transmitter, which then chooses a suitable MCS from a 'codeset' to maximize throughput (efficiency) at that SINR and to maintain a target BER. A codeset contains many MCSs so that it can cover a range of SINRs. An example of a codeset is shown in Figure 2.6. Each MCS in the codeset has the highest throughput for a 1–2 dB range of SNIR.

The Shannon bound represents the maximum theoretical throughput than can be achieved over an AWGN channel for a given SINR, as shown in Figure 2.7. An AMC system achieves around 0.75 times the throughput of the Shannon bound, over the range of SINR in which it operates. Figure 2.7 shows the baseline E-UTRA DL and UL spectral efficiency (bps/Hz) versus SINR based on the parameters in 3GPP along with the Shannon bound. Figure 2.6 indicates that 4G systems do not

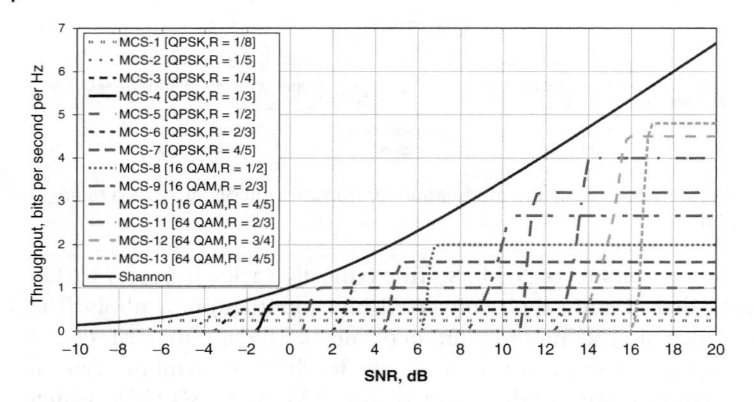

Figure 2.6 Throughput of a set of coding and modulation combinations, AWGN channels assumed.

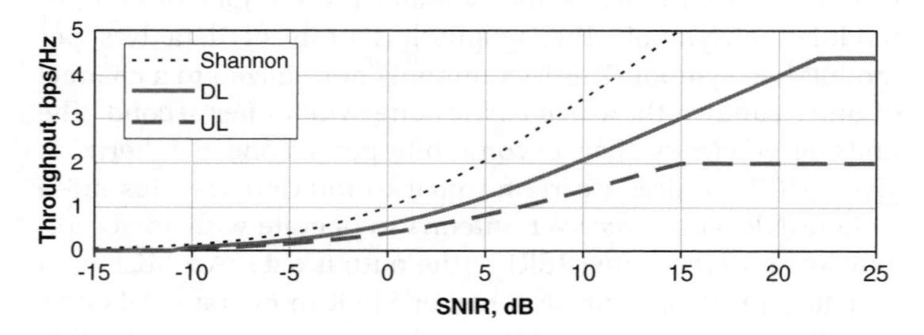

Figure 2.7 Spectral efficiency versus SNIR for baseline E-UTRA.

achieve the Shannon theory capacity and there is still room for improvement. This is one of the main motivations for 5G systems, which will adopt several enhancements to improve the system capacity with the aim of hitting the Shannon limits, especially at high SINRs. These techniques will be presented later in this chapter.

2.1.2 Link Adaptation

Link adaptation is mandatory to select the optimum MCS for the channel conditions and user equipment (UE) capability. Link adaptation is mandatory to guarantee performance metrics such as data rates, packet error rates, and latency. The AMC is the best-known link adaption technique, with various modulation schemes and channel coding rates available depending on the channel conditions. Users near the eNB, with high SINR values, will be assigned a higher MCS index value. As the user moves away from the eNB, the SINR is degraded and the interference is increased, and the assigned MCS will be decreased. The same MCS should be applied to all groups of RBs belong to the same Layer 2 protocol data unit scheduled to one user within one transmission time interval (TTI) by a single antenna stream.

2.2 DS-CDMA Basic Formulation

A baseband asynchronous DS-CDMA system is modeled as shown in Figure 2.8 [1–6]. In this model, we have K mobile users transmitting simultaneously to the base station. Each user symbols is assumed for simplicity and without loss of generality to be binary phase shift keying (BPSK) with arbitrary power and timing. In practice, the adopted modulations are QPSK, 16QAM, 64QAM and 256QAM. In UMTS with HSPA+, the first three modulations are adopted in the DL and only QPSK and 16QAM in the UL.; 256QAM is not practically deployed yet. In 4G/LTE, the DL and UL adopt the three modulations (QPSK, 16QAM, and 64QAM). The 256QAM modulation was recently introduced in the DL with Category 11 and Category 12 terminals.

Each user's symbol is spread by a unique spreading waveform $c_j = \left[c_j(0) \cdots c_j(L-1)\right]^T$ of length L, which is common to all users. It is assumed that the spreading codes are periodic, with a period L (short codes). Although most of spreading process in WCDMA depends on long codes, extension from short codes to long codes can be easily done. In 3GPP specifications for UMTS terrestrial radio access network (UTRAN), the air interface consists of two separate operations: channelization and scrambling. Channelization uses orthogonal codes, while scrambling uses pseudo-noise (PN) codes [7] to improve the cross-correlation properties of channelization codes. The chip period $T_c = T/L$, where T is the symbol period. The continuous

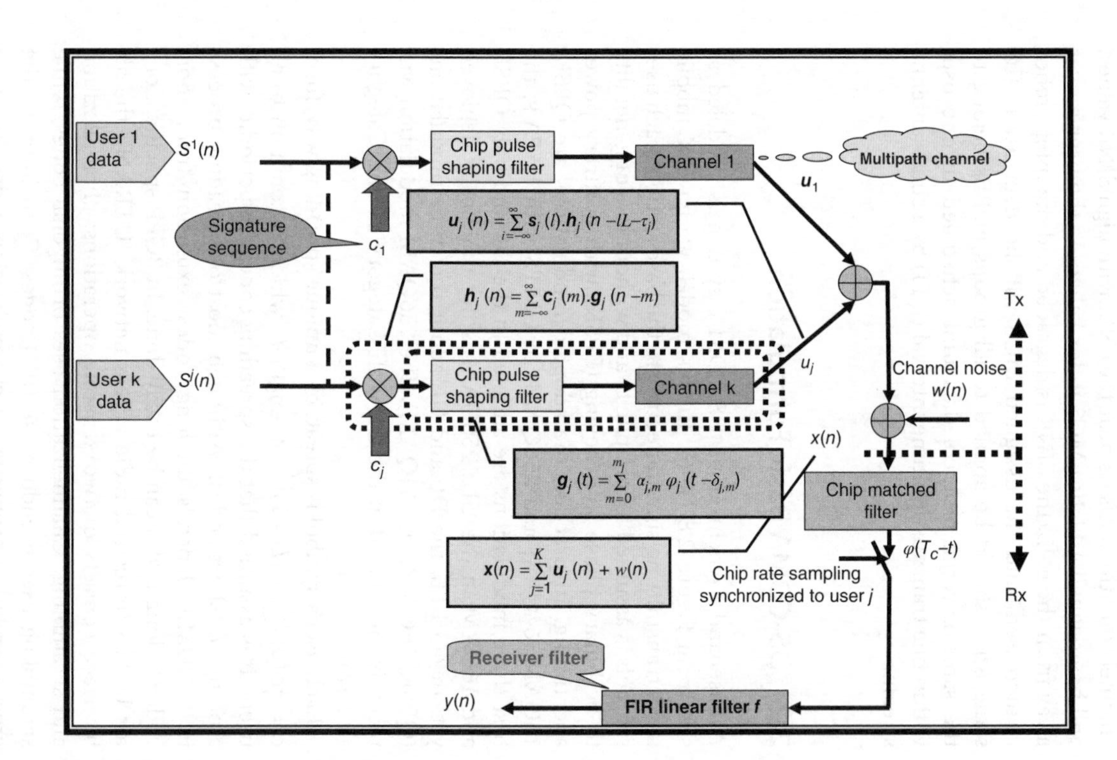

Figure 2.8 Discrete time DS-CDMA System model.

time-spreading waveform of user j can be expressed as:

$$c_j(t) = \sum_{i=0}^{L-1} c_j(i)\varphi(t - iT_c) \tag{2.1}$$

where $\varphi(t)$ is the chip pulse-shaping filter.

The spreading sequence may be complex, since the elements of the symbol or chip sequences may be chosen from a two-dimensional constellation. For example, in the IS-95 downlink, the spreading sequence is based on a complex quadrature phase shift keying (QPSK) constellation, but the symbol sequence is from BPSK [8, 9]. Another example is the UMTS system, where there are two dedicated channels and one common channel on the uplink. User data is transmitted on the dedicated physical-data channel (DPDCH), and control information is transmitted on the dedicated physical-control channel (DPCCH). The DPCCH is needed to transmit pilot symbols for coherent detection, power-control signaling bits, and rate information for rate detection. Combined time multiplexing (IQ multiplexing) and code multiplexing (dual-channel QPSK) is used in the WCDMA UL to avoid electromagnetic compatibility problems with discontinuous transmission/reception (DTX/DRX). DTX and DRX are power-saving mechanisms that are adopted in UMTS and OFDM systems to extend battery lifetimes for smartphones and portable devices [10, 53]. For more details on DRX practical deployment refer to the literature [53, 55].

I/Q code multiplexing leads to parallel transmission of two channels using a complex spreading circuit, as shown in Figure 2.9 [1, 7, 11, 12]. There is an intrinsic delay between transmission and reception of at least two TTI.

The I/Q modulation used instead of simple BPSK allows two data streams to be sent on one RF carrier, at the same BER vs Eb/No performance as BPSK (note that I and Q are orthogonal, so QPSK corresponds to two BPSK transmissions). A raised-cosine filter is used prior to the I/Q modulator. This provides spectral attenuation with controlled ringing. The ringing period is carefully chosen so that ringing will equal zero at all data sampling instants in the future. This minimizes ISI.

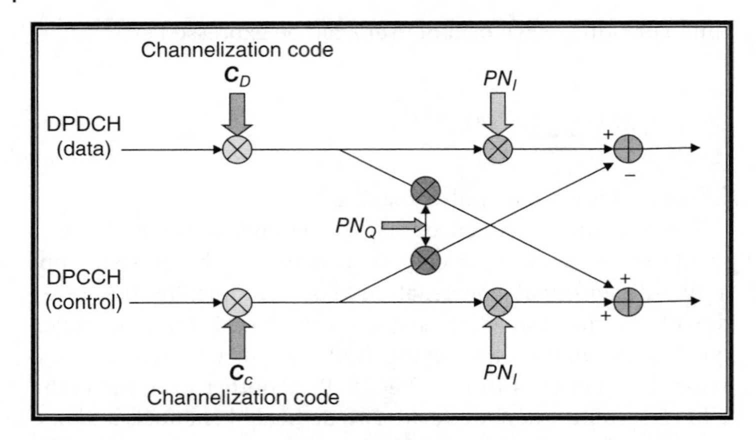

Figure 2.9 I/Q code multiplexing with complex spreading circuit.

2.2.1 Pulse-shaping Filter

All users are assumed to use the same chip pulse-shaping filter, denoted by $\varphi(t)$, which is assumed to be time limited to $[0, T_c)$. Pulse-shaping filtering is applied to each chip at the transmitter. The impulse response of this chip pulse is a root-raised cosine. The corresponding raised cosine impulse is given by:

$$\varphi(t) = \frac{\sin\left(\pi \frac{t}{T_c}\right)}{\pi \frac{t}{T_c}} \cdot \frac{\cos\left(\alpha\pi \frac{t}{T_c}\right)}{1 - 4\alpha^2 \frac{t^2}{T_c^2}} \tag{2.2}$$

where α is the roll-off factor. Figure 2.10 shows the root-raised cosine chip pulse with roll-off factor $\alpha = 0.22$ as in the UMTS system [1, 12].

2.2.2 Discrete Time Model

The received signal is the superposition of all users' contribution signals plus noise and is sampled at the chip rate. The aggregate channel noise is assumed to be zero-mean Gaussian with variance σ_w^2 and independent from source symbols. It is denoted by $\boldsymbol{w}(n)$. Therefore, the sampled received signal can be written as follows:

$$\boldsymbol{x}(n) = \sum_{j=1}^{K} \boldsymbol{u}_j(n) + \boldsymbol{w}(n) \tag{2.3}$$

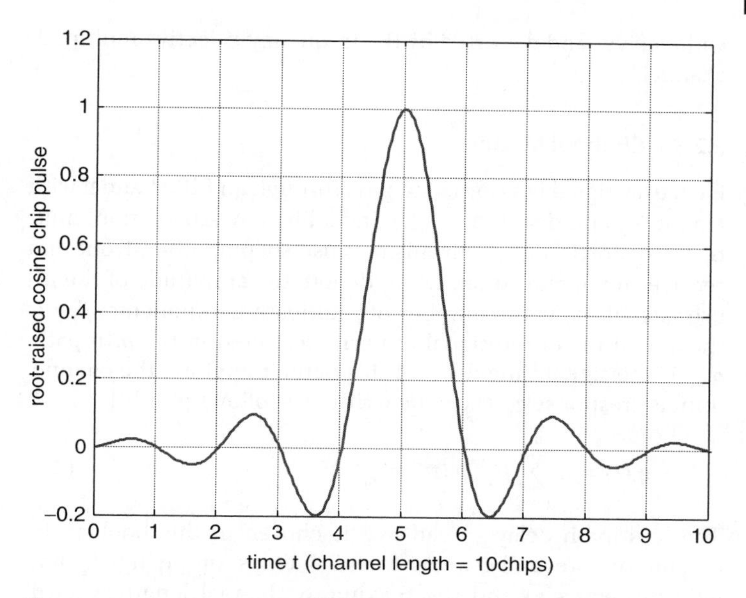

Figure 2.10 Impulse response of chip pulse shaping filter with $\alpha = 0.22$.

The $\boldsymbol{u}_j(n)$ is the jth user's contribution to the received vector and it can be expressed as:

$$\boldsymbol{u}_j(n) = \sum_{l=-\infty}^{\infty} \boldsymbol{s}_j(l).\boldsymbol{h}_j(n - lL - \tau_j) \tag{2.4}$$

τ_j and $\boldsymbol{s}_j(n)$ represent, respectively, the user delay and transmitted data bits of user j. Each user-information-bearing sequence $\boldsymbol{s}_j(n)$ is zero mean, i.i.d, and independent of other users. The effective signature waveform $\boldsymbol{h}_j(n)$ of user j represents its spreading sequence transmitted through a channel characterized by impulse response $\boldsymbol{g}_j(t)$. Therefore, the effective signature waveform of user j is the convolution of its code with the multipath channel and can be modeled as follows:

$$\boldsymbol{h}_j(n) = \sum_{m=-\infty}^{\infty} \boldsymbol{c}_j(m).\boldsymbol{g}_j(n - m) \tag{2.5}$$

The FIR channel (the channel impulse response) $\boldsymbol{g}_j(t)$ of user j is the chip waveform that has been filtered at the transmitter and receiver and distorted by the frequency-selective multipath channel.

2.2.3 Channel Model

Each user signal is assumed to pass through an FIR channel with length N_g and denoted by $g(t)$, including any attenuation, multipath distortion, asynchronism, pulse shaping filter, front end receiver filter, and so on. Let a_j denote the amplitude of the jth user signal, m_j is the number of multipath components, $\beta_{j,m}$ is the amplitude of the signal of user j scattered in the mth path. $\sigma_{j,m}$ is the signal time delay of this path. Therefore, the channel impulse response $\mathbf{g}_k(t)$ can be written as follows [13, 14]:

$$\mathbf{g}_j(t) = a_j \sum_{m=0}^{m_j} \beta_{j,m} \varphi_j(t - \sigma_{j,m}) \tag{2.6}$$

The multipath delay spread $\sigma_{j,m}$ is chosen in this book to be a random integer between 1 and 8 chips (implying T_c and $8T_c$, respectively) and the maximum channel length N_g will be limited to $10T_c$ as a result of the multipath propagation. The channel length N_g identifies the maximum number of multipath fingers in the practical CDMA rake receiver. This represents a practical CDMA multipath delay spread, with an elliptical channel model to describe the multipath propagation condition. For example, the channel delay spread will be ≈ 8 μs for a CDMA2000 system with $T_c = 0.813$ μs (that is, chip rate = 1.2288 Mcps) and the channel delay spread will be ≈ 3 μs for UMTS/WCDMA with $T_c = 0.260$ μs (chip rate = 3.84 Mcps) [10, 13, 56]. The conducted simulations in this book are based on 31chips Gold code length. Therefore, the maximum delay spread adopted in this book is about one third of the code length. As a result, for long codes such as the 256 chip-length code in the UMTS system, a delay spread of order 20 μs can be considered. It is noteworthy that 3GPP specifies a channel length of up to 20 μs. For the TD-SCDMA system, with a maximum 16 chip code length, and according to the 31 chips Gold code used in this book, the maximum multipath delay spread is 5 chips, which means 3.125 μs where $T_c = 0.625$ μs.

Code allocation in commercial CDMA systems is an important topic. The code allows the UE to distinguish between BSs in the scrambling code; a 38,400-chip sequence of a 2^{18} length Gold code. Each cell is assigned one of 512 primary codes. The code that allows the UE to distinguish between data channels from one BS is the channelization code, which is derived from the Hadamard code tree. The code that allows the BS to distinguish between UEs is the scrambling code; a 38,400-chip sequence of a 2^{25} long Gold code. Each UE is assigned one of 2^{25} possible codes by the UTRAN system. The code used to allow the BS to distinguish between data channels from one UE is the channelization code, also derived from the Hadamard code tree. Some channels are sent on the I-channel, others on the Q-channel.

The channel model is characterized by delay profiles that are selected to be representative of low-, medium- and high-delay spread environments. The delay profiles of the channel models are provided in Table 2.3 and the tapped delay line models for typical propagation channel models used for LTE are defined in Table 2.4 [53]. UE speeds and multipath profiles depend on the type of environment (say dense urban, urban, or rural). In the link-budget calculation, the EPA model at 3 km/h is used for dense urban and urban morphologies. The ETU model at 120 km/h and Extended vehicular A model (EVA) at 120 km/h are used for suburban and rural environments, respectively. Selection of a multipath channel model for a certain morphology is done for simulation purposes only. In practice, various multipath channel conditions occur concurrently within the same cell environment. Therefore, a mix of multipath channel

Table 2.3 Delay profiles for E-UTRA channel models.

Model	Number of channel taps	Delay spread (rms) (ns)	Maximum excess tap delay (span) (ns)
Extended pedestrian A (EPA)	7	45	410
Extended vehicular A model (EVA)	9	357	2510
Extended typical urban model (ETU)	9	991	5000

Table 2.4 Typical propagation channel models used for LTE.

Channel model	Extended pedestrian A model (EPA)		Extended vehicular A model (EVA)		Extended typical urban model	
Number of taps	7		9		9	
Excess tap delay [ns] / Relative power [dB]	0	0	0	0	0	−1
	30	−1	30	−1.5	50	−1
	70	−2	150	−1.4	120	−1
	90	−3	310	−3.6	200	0
	110	−8	370	−0.6	230	0
	190	−17.2	710	−9.1	500	0
	410	−20.8	1090	−7	1600	−3
			1730	−12	2300	−5
			2510	−16.9	5000	−7

models represents the real channel characteristics. However, for link budget purposes, the worst-case channel model is considered and therefore the worst-case cell radius is estimated for each cluster.

In wireless systems, the delay spread is a measure of the multipath richness of a communications channel. It can be interpreted as the difference between the time of arrival of the earliest significant multipath component (typically the line-of-sight component) and the time of arrival of the latest multipath components. The delay spread is mostly used in the characterization of wireless channels, but it also applies to any other multipath channel (for example in optical fibers).

Delay spread can be quantified through different metrics, although the most common one is the root mean square (rms) delay spread. Denoting the power delay profile of the channel by $A_c(\tau)$, the mean delay of the channel is:

$$\overline{\tau} = \frac{\int_0^\infty \tau A_c(\tau)d\tau}{\int_0^\infty A_c(\tau)d\tau} \tag{2.7}$$

and the rms delay spread is given by:

$$\tau_{\text{rms}} = \sqrt{\frac{\int_0^\infty (\tau - \overline{\tau})^2 A_c(\tau)d\tau}{\int_0^\infty A_c(\tau)d\tau}} \tag{2.8}$$

The importance of delay spread is in the way it affects the ISI [57]. If the symbol duration is long enough compared to the delay spread (typically ten times as big would be good enough), one can expect an equivalent ISI-free channel. The correspondence with the frequency domain is the notion of coherence bandwidth, which is the bandwidth over which the channel can be assumed flat. Coherence bandwidth is related to the inverse of the delay spread. The shorter the delay spread, the larger is the coherence bandwidth. Delay spread has a significant impact on ISI.

Because of multipath reflections, the channel impulse response of a wireless channel looks like a series of pulses, as

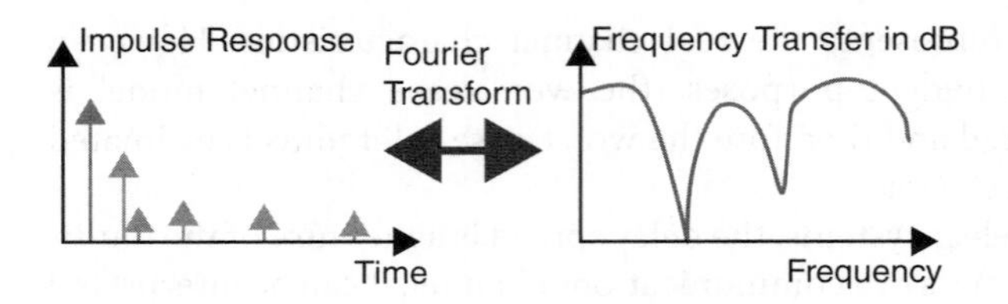

Figure 2.11 Example of impulse response and frequency transfer function of a multipath channel.

shown in Figure 2.11. In practice, the number of pulses that can be distinguished is very large, and depends on the time resolution of the communication or measurement system [58]. In system evaluations, we typically prefer to address a class of channels with properties that are likely to be encountered, rather than one specific impulse response. Therefore, we define the (local-mean) average power that is received with an excess delay that falls within the interval (T, T + dt). Such characterization for all T gives the "delay profile" of the channel. The delay profile determines the frequency dispersion: the extent to which the channel fading at two different frequencies f1 and f2 is correlated.

The maximum delay time spread is the total time interval over which reflections with significant energy arrive. The rms delay spread T_{rms} is the standard deviation (or root-mean-square) value of the delay of reflections, weighted proportionally to the energy in the reflected waves.

In macro-cellular mobile radio, delay spreads are mostly in the range from T_{rms} from 100 ns–10 μs. A typical delay spread of 0.25 μs corresponds to a coherence bandwidth of about 640 kHz. Measurements made in the US indicate that delay spreads are usually less than 0.2 μs in open areas, about 0.5 μs in suburban areas, and about 3 μs in urban areas. Measurements in the Netherlands showed that delay spreads are relatively large in European-style suburban areas, but rarely exceed 2 μs. However, large distant buildings such as blocks of flats occasionally cause reflections with excess delays of the order of 25 μs. In indoor and microcellular channels, the delay spread is usually smaller, and rarely exceeds a few hundred nanoseconds. Seidel and Rappaport reported delay spreads in four European cities of less than 8 μs in macro-cellular channels, less than

2 μs in microcellular channels, and between 50 and 300 ns in picocellular channels [58].

The delay profile is the expected power per unit of time received with a certain excess delay. It is obtained by averaging a large set of impulse responses. In an indoor environment, early reflections often arrive with almost identical power. This gives a fairly flat profile up to some point, and a tail of weaker reflections with larger excess delays. Besides the normal reflections from nearby obstacles (which cause reflections with short excess delay), remote high-rise buildings cause strong reflections with large excess delays. The combined effects often result in multiple clusters of reflections.

A wideband signal with symbol duration T_c (such as (DS)-CDMA signals with chip time T_c), can “resolve” the time dispersion of the channel with an accuracy of about T_c. For DS-CDMA, the number of resolvable paths is [58]:

$$N = \left\lceil \frac{T_{Delay}}{T_{Chip}} \right\rceil + 1 \tag{2.9}$$

where $\lceil x \rceil$ is the largest integer value smaller than x and T_{Delay} is total length of the delay profile. A DS-CDMA rake receiver can exploit N-fold path diversity. Table 2.5 illustrates how different systems handle the channel delay spread [58].

In LTE systems, the CP is selected to mitigate the delay spread (ISI) impact as follows:

Urban environments (maximum delay spread of 15 μs**):** Among the three possible configurations listed in the LTE standard, in this kind of environment the chosen parameters would be Δf = 15 KHz and CP = 17 μs (extended CP mode). This value of Δf is the most common and is used to serve users moving at velocities of up to 500 km/h. The chosen length of the CP is enough to avoid ISI.

Indoor environments (maximum delay spread of 1μs**):** In this case, the chosen parameters would be Δf = 15 KHz and CP = 5 μs (normal parametrization). Again, the chosen value of Δf would be enough to cover speeds of up to 500 km/h and a shorter CP is enough to avoid ISI with a delay spread of 1 μs.

Table 2.5 How systems handle delay spreads.

System	Delay spread mitigation.
Analog	• Narrowband transmission
GSM	• Adaptive channel equalization • Channel estimation training sequence
DECT	• Use the handset only in small cells with small delay spreads • Diversity and channel selection can help a little (pick channel where late reflections are fading)
IS95 Cellular CDMA	• Rake receiver separately recovers signals over paths with excessive delays • CDMA array processing can further improve performance, because it also exploits angle spreads
LTE	• OFDM multicarrier modulation: radio channel is split into many narrowband (ISI-free) subchannels

2.2.4 Matrix Formulation for DS/CDMA System Model

The user contribution in Figure 2.8 can be represented as a matrix-vector product as follows:

$$\boldsymbol{u}_j(n) = \boldsymbol{G}^{(j)}\boldsymbol{C}^{(j)}\boldsymbol{S}^{(j)}(n) \tag{2.10}$$

where $\boldsymbol{G}^{(j)}$ is a $N_f \times [N_f + N_g - 1]$ Toeplitz matrix, representing the baud-spaced channel convolution matrices of user j, which incorporates all distortion types. That is,

$$\boldsymbol{G}^{(j)} = \begin{bmatrix} g_0^j & g_1^j & \cdots & g_{N_g-1}^j & 0 & \cdots & 0 \\ & g_0^j & g_1^j & \cdots & g_{N_g-1}^j & \cdots & 0 \\ & & \ddots & & & & 0 \\ 0 & 0 & \cdots & g_0^j & g_1^j & \cdots & g_{N_g-1}^j \end{bmatrix}_{N_f \times N_f + N_g - 1} \tag{2.11}$$

where $\boldsymbol{S}^{(j)}$ contains the contribution symbols for user j, and the matrix $\boldsymbol{C}^{(j)}$ is $N_f + N_g - 1 \times N_s$ representing the spreading operation and can be structured as follows [1, 5]:

$$\mathbf{C}^{(j)} = \begin{bmatrix} \begin{pmatrix} c^j_{a_k} \\ \vdots \\ c^j_{L-1} \end{pmatrix} & & & 0 \\ & \begin{pmatrix} c^j_0 \\ \vdots \\ c^j_{L-1} \end{pmatrix} & & \\ & & \ddots & \\ 0 & & & \begin{pmatrix} c^j_0 \\ \vdots \\ c^j_{b_k} \end{pmatrix} \end{bmatrix}_{N_f + N_g - 1 \times N_s} \tag{2.12}$$

where

$$a_j = \langle 1 - N_g - \tau_j \rangle_L \tag{2.13}$$

$$b_j = \langle N_f - 1 - \tau_j \rangle_L \tag{2.14}$$

N_f is the detector length and N_s represents the number of interfering bits (channel ISI span). The value of N_s depends on the channel impulse response length N_g and is related to the length of the chip sequence L and the detector length N_f and can be given by [1, 5]

$$N_s = \left\lceil \frac{N_f + N_g - 1}{L} \right\rceil \tag{2.15}$$

The expression for $\boldsymbol{x}(n)$ may be further clarified through the definition of the global channel matrix $\boldsymbol{H}$ and multiuser source code $\boldsymbol{S}(n)$. This yields:

$$\boldsymbol{x}(n) = \boldsymbol{H}.\boldsymbol{S}(n) + \boldsymbol{w}(n) \tag{2.16}$$

and hence

$$\boldsymbol{x}(n) = \sum_{j=1}^{K} \mathbf{G}^{(j)}.\mathbf{C}^{(j)}.\boldsymbol{S}^{(j)}(n) + \boldsymbol{w}(n) \tag{2.17}$$

where $\boldsymbol{H}$ represents the multiuser code-channel matrix and $\boldsymbol{S}(n)$ represents multiuser source vector, defined respectively by:

$$\boldsymbol{H} = \left(\boldsymbol{G}^{(1)}\boldsymbol{C}^{(1)},\ \boldsymbol{G}^{(2)}\boldsymbol{C}^{(2)}\ \cdots\ \boldsymbol{G}^{(K)}\boldsymbol{C}^{(K)}\right)_{N_f \times KN_S} \tag{2.18}$$

$$\boldsymbol{S}(n) = \left(\boldsymbol{S}_1^T(n)\ \boldsymbol{S}_2^T(n)\ \cdots\ \boldsymbol{S}_k^T(n)\right)^T_{KN_s \times 1} \tag{2.19}$$

A chip-rate linear receiver can be designed by collecting N_f samples from the received signal vector $\boldsymbol{x}(n)$, which is at least as long as the signature waveform plus delay spread N_g. Assuming the user delay τ_j is known, the contribution from this user to the received signal can be expressed as follows:

$$\boldsymbol{u}_j(n) = \boldsymbol{h}^j_\delta s^j(n-\delta) + \sum_{\substack{m=1 \\ m \neq \delta}}^{N_s} \boldsymbol{h}^j_m s^j(n-m) \tag{2.20}$$

where $\boldsymbol{h}^j_\delta$ is the effective signature waveform of the jth user at delay δ. We should distinguish δ from τ_j where the first delay is an integer $0 \leq \delta \leq N_s - 1$ and represents number of delayed bits, and the second delay τ_j is the synchronization delay of user j in chips. Without loss of generality, we will assume in this book that $\tau_j \leq L$. In the rest of this book, we will assume that the delay τ_j of the particular user is known and hence the system is synchronized to it while the multipath components delays $\sigma_{j,m}$ are unknown. The software provided with the book can simulate the asynchronous DS-CDMA system as well.

The received vector can be reformulated as follows:

$$\boldsymbol{x}(n) = \boldsymbol{h}^j_\delta s^j(n-\delta) + \boldsymbol{H}_i.\boldsymbol{S}_i(n) + \boldsymbol{w}(n) \tag{2.21}$$

where $\boldsymbol{S}^j(n-\delta)$ are the data bits of the user of interest and $\boldsymbol{S}_i(n)$ is the user's interference matrix with regard to $\boldsymbol{S}^j(n-\delta)$, including ISI from the same user and multiple access interference (MAI) from other users. $\boldsymbol{H}_i$ is the global channel matrix $\boldsymbol{H}$ without the user of interest contribution at delay δ. Therefore, $\boldsymbol{H}_i$ is the effective signature matrix of both MAI and ISI, with columns corresponding to symbols in $\boldsymbol{S}_\mathrm{i}(n)$.

We can reformulate $\boldsymbol{h}^j_m$ in another form, which will be useful in developing code-constrained receivers, as follows:

$$\boldsymbol{h}^j_m = \boldsymbol{C}^j_m.\boldsymbol{g}_j \tag{2.22}$$

where $\boldsymbol{h}^j_m$ is the mth column of the matrix $\boldsymbol{H}^{(j)} = \boldsymbol{G}^{(j)}\boldsymbol{C}^{(j)}$ and can be estimated by constructing the convolution matrix of the mth column in matrix $\boldsymbol{C}^{(j)}$ as follows:

$$\overline{\boldsymbol{C}}^j_m = \begin{bmatrix} & 0_{((m-1)L+(L-1-a_k))\times N_g} & \\ c^j_{L-1} & & \\ & \ddots & \\ & & c^j_{L|b_k} \\ c^j_{0|a_k} & & \\ & & \vdots \\ & \ddots & \\ & & c^j_0 \\ & 0_{((m-1)L+b_k)\times N_g} & \end{bmatrix}_{(N_f+2N_g-1)\times N_g} \tag{2.23}$$

Therefore,

$$\boldsymbol{C}^j_m = \overline{\boldsymbol{C}}^j_m(N_g : N_f + N_g - 1) \tag{2.24}$$

$$\boldsymbol{g}_1 = \begin{bmatrix} g^1_{N_g-1} \\ \vdots \\ g^1_0 \end{bmatrix}_{N_g\times 1} \tag{2.25}$$

The vector $\boldsymbol{g}_1$ is the unknown channel vector that represents multipath propagation and other distortions.

2.2.5 Synchronous DS/CDMA System

In this system, the decision is taken based on one bit period, the detector length $N_f = L$ and hence $\delta = 0$ for all users. This means that the first element of the detector lines up with the first chip in the spreading sequence. The spreading operation matrix $\boldsymbol{C}^{(j)}$

is $N_f + N_g - 1 \times N_s$ and can be represented as follows:

$$C^{(j)} = \begin{bmatrix} \begin{pmatrix} c_0^j \\ \vdots \\ c_{L-1}^j \end{pmatrix} & & & 0 \\ & \begin{pmatrix} c_0^j \\ \vdots \\ c_{L-1}^j \end{pmatrix} & & \\ & & \ddots & \\ 0 & & & \begin{pmatrix} c_0^j \\ \vdots \\ c_Q^j \end{pmatrix} \end{bmatrix}_{N_f+N_g-1 \times N_s} \tag{2.26}$$

where

$$Q = \langle N_f + N_g - 1 \rangle_L \tag{2.27}$$

Without loss of generality, we will assume that user number one is the desired user and is used as the timing reference. In addition, we assume for simplicity $N_s = 1$. This means the ISI spans one bit only. The received data vector for each symbol can be reformulated as follows [14]:

$$\boldsymbol{x}(n) = \boldsymbol{h}_1 \boldsymbol{s}_1(n) + \overline{\boldsymbol{H}}_i \boldsymbol{s}_i(n) + \boldsymbol{w}(n) \tag{2.28}$$

We can write the effective signature waveform of the required user $\boldsymbol{h}_1$ in matrix form as follows:

$$\boldsymbol{h}_1 = \boldsymbol{C}_1 . \boldsymbol{g}_1 \tag{2.29}$$

where

$$\boldsymbol{C}_1 = \begin{bmatrix} \boldsymbol{c}_1(L-Ng) & \cdots & \boldsymbol{c}_1(L-1) \\ & \ddots & \\ \boldsymbol{c}_1(0) & & \\ & \ddots & \boldsymbol{c}_1(0) \\ 0 & & 0 \end{bmatrix}_{N_f \times N_g}, \quad \boldsymbol{g}_1 = \begin{bmatrix} \boldsymbol{g}_1(N_g-1) \\ \vdots \\ \boldsymbol{g}_1(0) \end{bmatrix}_{N_g \times 1} \tag{2.30}$$

The formulation above will act as the basis for developing the code-constraint detectors that will be described in Chapters 3–6. The 3GPP standard did not define specific receiver algorithms, although the specification has been defined in such a way that a RAKE receiver will satisfy most cases. Advanced receivers will improve DS/CDMA performance, increase the capacity, and minimize interference.

2.3 Performance Evaluation

The algorithms developed for DS/CDMA systems are based on minimizing the mean square error (MSE) or the BER. Two performance measures are adopted for these type of the algorithms, the SINR and BER. MSE can also be used as indicator of the performance. However, the SINR and BER are the main performance indicators, especially for BER-based detectors that target the BER performance of the system rather the MSE cost function.

We focus our consideration in this book on linear detection, where the linear detector output is a linear combination of the received chip sampled signals. That is:

$$y(n) = \boldsymbol{f}^H(n)\boldsymbol{x}(n) \tag{2.31}$$

The output of the detector is $y(n)$, and $\boldsymbol{f}^H(n)$ is a vector consisting of the weights.

In BPSK, the bit decision is made according to:

$$\hat{s}_1(n) = \text{sgn}(\text{Re}\{y_1(n)\}) \tag{2.32}$$

where sgn(.) denotes the sign function, and $\text{Re}\{y_1(n)\}$ is the real part of the output $y_1(n)$.

2.3.1 Signal to Interference plus Noise Ratio

We will develop an expression for the output SINR based on the proposed generalized system model. The SINR is defined as the energy of the desired user divided by the energy of interference users (including MAI and ISI), the ISI of the interested user, plus an AWGN term [13, 14]. This yields:

$$SINR = \frac{E\left\{|\boldsymbol{f}^T\boldsymbol{s}^1(n-\delta)\boldsymbol{h}_\delta^1|^2\right\}}{E\left\{\left\|\boldsymbol{f}^T\left(\underbrace{\sum_{k=2}^{K}\sum_{m=0}^{N_s-1}\boldsymbol{s}_k(n-m)\boldsymbol{h}_m^k}_{\substack{MAI\,+\,ISI \\ \text{of interfering users}}}+\underbrace{\sum_{\substack{m=0\\m\neq\delta}}^{N_s-1}\boldsymbol{h}_m^1 s_1(n-m)}_{\substack{ISI \\ \text{of interested user}}}+\underbrace{\boldsymbol{w}(n)}_{AWGN}\right)\right\|^2\right\}} \tag{2.33}$$

Assuming the transmitted data from each user is independent from other users and also from the noise, the SINR can be estimated as follows:

$$SINR = \frac{|\boldsymbol{f}^H\boldsymbol{h}_\delta^1|^2}{\sum_{k=2}^{K}\sum_{m=0}^{N_s-1}\boldsymbol{f}^H\boldsymbol{h}_m^k(\boldsymbol{h}_m^k)^H\boldsymbol{f}+\sum_{\substack{m=0\\m\neq\delta}}^{N_s-1}\boldsymbol{f}^H\boldsymbol{h}_m^1(\boldsymbol{h}_m^1)^H\boldsymbol{f}+\sigma_w^2\boldsymbol{f}^H\boldsymbol{f}} \tag{2.34}$$

For the synchronous system addressed in Section 2.2.5, the SINR expression can be reduced to the following simple form [13]:

$$SINR = \frac{|\boldsymbol{f}^H\boldsymbol{h}_1|^2}{\sum_{k=2}^{K}\boldsymbol{f}^H\boldsymbol{h}_k\boldsymbol{h}_k{}^H\boldsymbol{f}+\sigma_w^2\boldsymbol{f}^T\boldsymbol{f}} \tag{2.35}$$

2.3.2 Bit Error Rate

The detector output $\mathbf{y}_1(n)$ can be expressed as:

$$y_1(n) = \boldsymbol{f}^H(\overline{\boldsymbol{x}}(n)+\boldsymbol{w}(n)) = \overline{y}(n)+e(n) \tag{2.36}$$

where $e(n) = \boldsymbol{f}^H\boldsymbol{w}(n)$ is Gaussian, with zero mean and variance $E[|e(n)|^2] = 2\sigma_w^2\boldsymbol{f}^H\boldsymbol{f}$.

To derive the BER expression for a linear detector with weight vector $\boldsymbol{f}$, let us define the signed variable:

$$y_s(n) = \text{sgn}(s_1(n))\text{Re}\{\boldsymbol{y}_1(n)\} \tag{2.37}$$

The probability density function (PDF) of $y_s(n)$ is a mixed sum of Gaussian distributions [15–18]:

$$p(y_s) = \frac{1}{N_b\sqrt{2\pi\sigma_w^2 \boldsymbol{f}^H\boldsymbol{f}}}\sum_{j=1}^{N_b} e^{\left(-\frac{\left(y_s - \mathrm{sgn}\left(s_1^{(j)}\right)\mathrm{Re}\{y_j\}\right)^2}{2\sigma_w^2 \boldsymbol{f}^H\boldsymbol{f}}\right)} \tag{2.38}$$

where $N_b = 2^K$ denotes the possible combinations of transmitted user symbols and K is the number of users. $s_1^{(j)}$ is the first user of $\boldsymbol{s}^{(j)}(k)$, where

$$\boldsymbol{s}^{(j)}(k) = \begin{bmatrix} \boldsymbol{s}^{(j)}(k) \\ \boldsymbol{s}^{(j)}(k-1) \\ \\ \boldsymbol{s}^{(j)}(k-N_s+1) \end{bmatrix}, \quad 1 \le j \le N_b \tag{2.39}$$

and $\boldsymbol{y}_j$ is the corresponding detector output, given by:

$$\boldsymbol{y}_j = \boldsymbol{f}^H\boldsymbol{r}_j, \quad 1 \le j \le N_b \tag{2.40}$$

where $\boldsymbol{r}_j$, $1 \le j \le N_b$, are equiprobable noise-free received signal samples. Therefore, the error probability of the linear detector is given by:

$$P_E(\boldsymbol{f}) = \frac{1}{N_b\sqrt{2\pi}}\sum_{j=1}^{N_b}\int_{q_j(\boldsymbol{f})}^{\infty} \exp\left(-\frac{x_j^2}{2}\right)dx_j = \frac{1}{N_b}\sum_{j=1}^{N_b} Q(q_j(\boldsymbol{f})) \tag{2.41}$$

where

$$Q(x) = \frac{1}{\sqrt{2\pi}}\int_x^{\infty} \exp\left(-\frac{y^2}{2}\right)dy \tag{2.42}$$

and

$$q_j(\boldsymbol{f}) = \frac{\mathrm{sgn}\left(s_1^{(j)}(n)\right)\boldsymbol{y}_j(n)}{\sigma_w\sqrt{\boldsymbol{f}^H\boldsymbol{f}}} = \frac{\mathrm{sgn}\left(s_1^{(j)}(n)\right)\boldsymbol{f}^H\boldsymbol{r}_j}{\sigma_w\sqrt{\boldsymbol{f}^H\boldsymbol{f}}} \tag{2.43}$$

In practice, the set $\boldsymbol{r}_j$ is not available. Kernel density estimation is known to produce reliable PDF estimates with short data records. Given a block of N training samples $\{\mathbf{r}(n), s(n)\}$, a kernel density estimate of the PDF is:

$$\hat{p}(y_s) = \frac{1}{N\sqrt{2\pi}\rho}\sum_{n=1}^{N} e^{\left(-\frac{(y_s - \mathrm{sgn}(s_1(n))y(n))^2}{2\rho^2}\right)} \tag{2.44}$$

where the radius parameter ρ is related to the noise standard deviation σ_w [17]. Arasaratnam *et al.* suggested a lower bound ρ as follows [19]:

$$\rho = \left(\frac{4}{3P}\right)^{1/5} \sigma_\eta \tag{2.45}$$

An alternative BER expression is determined by the receiver output SINR and the Euclidean distance between the receiver output and the decision boundary; that is,

$$P_e = E_{\{s_k(n)\}}\left\{P_e(s_k(n))\right\} \tag{2.46}$$

$$P_e(s_k(n)) = Q\left(\frac{\sum_{k=2}^{K}\sum_{m=0}^{N_s-1} \boldsymbol{f}^H \boldsymbol{h}_m^k s_k(n-m) + \sum_{\substack{m=0\\ m\neq\delta}}^{N_s-1} \boldsymbol{f}^H \boldsymbol{h}_m^1 s_1(n-m) + \boldsymbol{f}^H \boldsymbol{h}_\delta^1}{\sigma_w(\boldsymbol{f}^H\boldsymbol{f})^{1/2}}\right) \tag{2.47}$$

For the synchronous system addressed in Section 2.2.5, the BER can be expressed as follows [13]:

$$P_e(s_k(n)) = Q\left(\frac{\boldsymbol{f}^H \boldsymbol{h}_1 + \sum_{k=2}^{K} \boldsymbol{f}^H \boldsymbol{h}_k s_k(n)}{\sigma_w\left(\boldsymbol{f}^H\boldsymbol{f}\right)^{1/2}}\right) \tag{2.48}$$

2.4 MIMO/OFDM System Model

MIMO-OFDM is a key technology for next-generation cellular communications (3GPP LTE and 5G) as well as wireless LAN (IEEE 802.11ac, IEEE 802.11ad), wireless PAN (MB OFDM), and broadcasting (DAB, DVB, DMB). MIMO systems are point-to-point communication links with multiple antennas at both the transmitter and receiver. The use of multiple antennas at both transmitter and receiver provides enhanced performance over 'diversity' systems: those where either the transmitter (Tx diversity) or receiver (Rx diversity), but not both, have multiple antennas. MIMO can significantly increase the data rates of wireless systems without increasing transmit power or bandwidth. The cost of this increased rate is the added cost of deploying multiple antennas, the space requirement to house them, and the added complexity of multidimensional signal processing. MIMO 2×2 was adopted with the first release of LTE and MIMO 4×4 is now being deployed and has been tested in several networks. MIMO techniques will be evolved in 5G to include massive MIMO and 3D beamforming [43].

Frequency division multiplexing (FDM) has been widely used for signal transmission in frequency-selective channels. FDM divides the channel bandwidth into subchannels and transmits multiple, relatively low-rate signals by carrying each signal on a separate carrier frequency. To facilitate separation of the signals at the receiver, the carrier frequencies are spaced sufficiently far apart to ensure that signal spectra do not overlap. Moreover, in order to separate the signals with readily sizeable filters, empty spectral regions are placed between the signals (so called 'guard bands'). As such, the resulting spectral efficiency of the system is quite low. Even the LTE system, which relies on OFDM, introduces guardbands at both side of the LTE channel (20 MHz) to avoid interference from adjacent channels, as shown in Figure 2.12. The guardband at one edge is the channel bandwidth minus the transmission bandwidth divided by two.

In order to solve the bandwidth efficiency problem, orthogonal frequency division multiplexing was proposed. This technique employs orthogonal tones to modulate the signals [20]. The tones are spaced at frequency intervals equal to the symbol rate and are capable of separation at the receiver. This carrier spacing provides optimum spectral efficiency. Although OFDM was proposed in the 1960s, it was not widely employed until the 1990s, largely because of significant circuit design issues, such as spurious frequency components and linearity of amplifiers. Today, OFDM is a major contender for 4G and 5G wireless applications, offering significant potential performance enhancements over existing wireless technologies. OFDM will be evolved further in 5G, via several enhancements that are discussed later in this chapter.

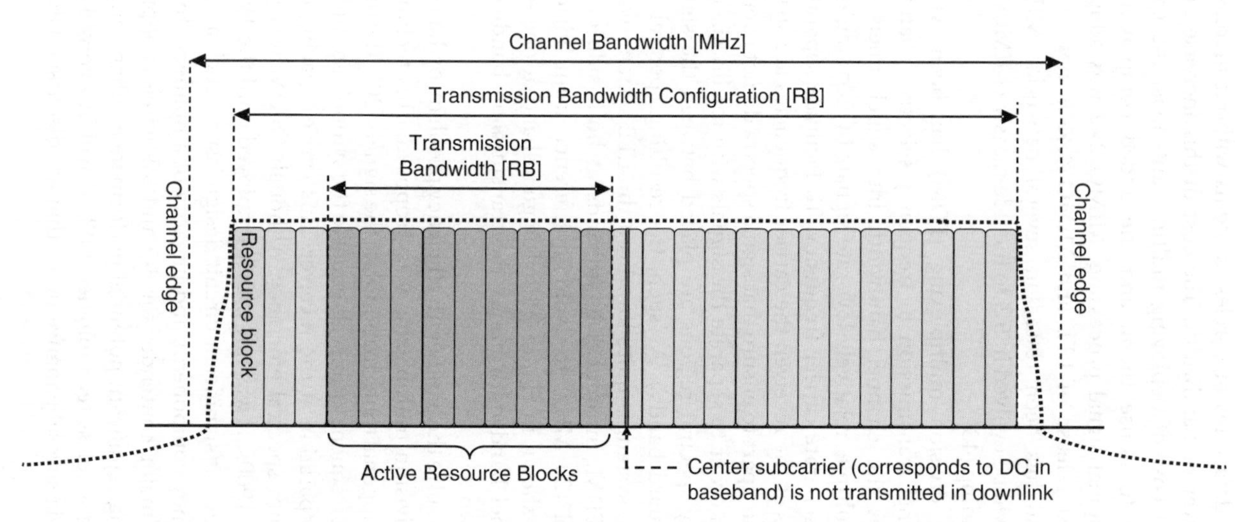

Figure 2.12 Channel bandwidth and transmission bandwidth configuration for one LTE carrier.

2.4.1 FFT and IFFT

The simplified OFDM system adopted in this book is illustrated in Figure 2.13 [21, 22, 54]. The system assembles the input bits and maps them onto complex numbers (in the modulator blocks). This determines the constellation points of each subcarrier. The number of bits assigned to each subcarrier is variable, based on the variability of the SNR across the frequency range. Optimization of this bit assignment is based on the technique described by Bansal and Brzeinski [54]. The number of subcarriers N used in an OFDM system is a trade-off between the frequency offset of adjacent carriers and the adjacent channel interference. A greater number of subcarriers implies less adjacent channel interference, but increased susceptibility to frequency offset, and vice-versa.

The key components of an OFDM system are the IFFT at the transmitter and the FFT at the receiver. These operations perform reversible linear mappings between N complex data symbols and N complex OFDM symbols. An N-point FFT requires only of the order of $N \log N$ multiplications rather than N^2, as in a straightforward computation. Because of this, an OFDM system typically requires fewer computations per unit time than an equivalent system with equalization. Transmission of data in the frequency domain using an FFT, as a computationally efficient orthogonal linear transformation, results in robustness against ISI in the time domain.

A downlink OFDMA system with N subcarriers is shown in Figure 2.14 [60]. It is assumed that there are K users served by a BS, and that the BER_k and minimum data rate R_k bits per OFDM symbol are the QoS parameters required for the user k_{th}. In the transmitter, the data from K users are fed into a subcarrier-bit-and-power allocation block. Adaptive subcarrier bit and power allocation can be applied if the instantaneous channel conditions for all users are known at the transmitter. The available modulation modes include BPSK, QPSK, 16QAM and 64QAM. After IFFT, the OFDM signals are transmitted via a downlink degraded channel. When the length of the CP is longer than maximum time delay, the channel gain of every subcarrier can be seen as a constant: that is, a flat fading

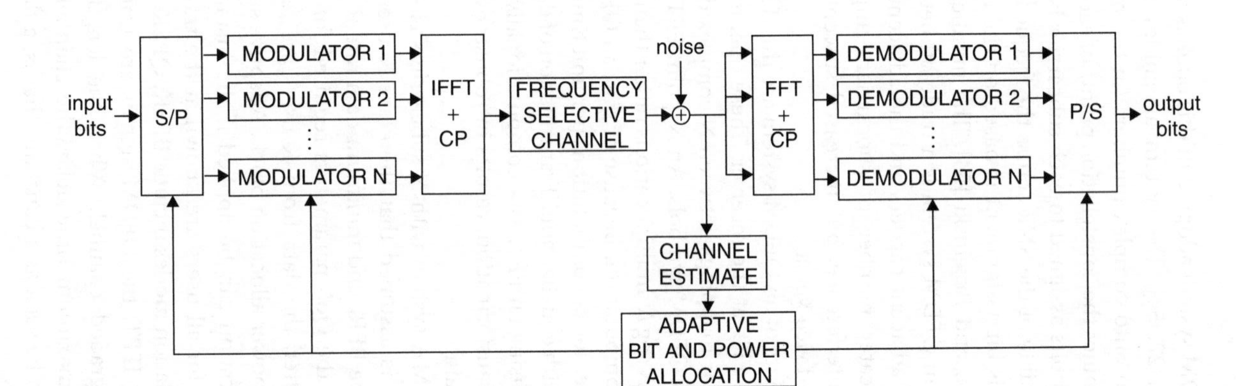

Figure 2.13 OFDM system block diagram.

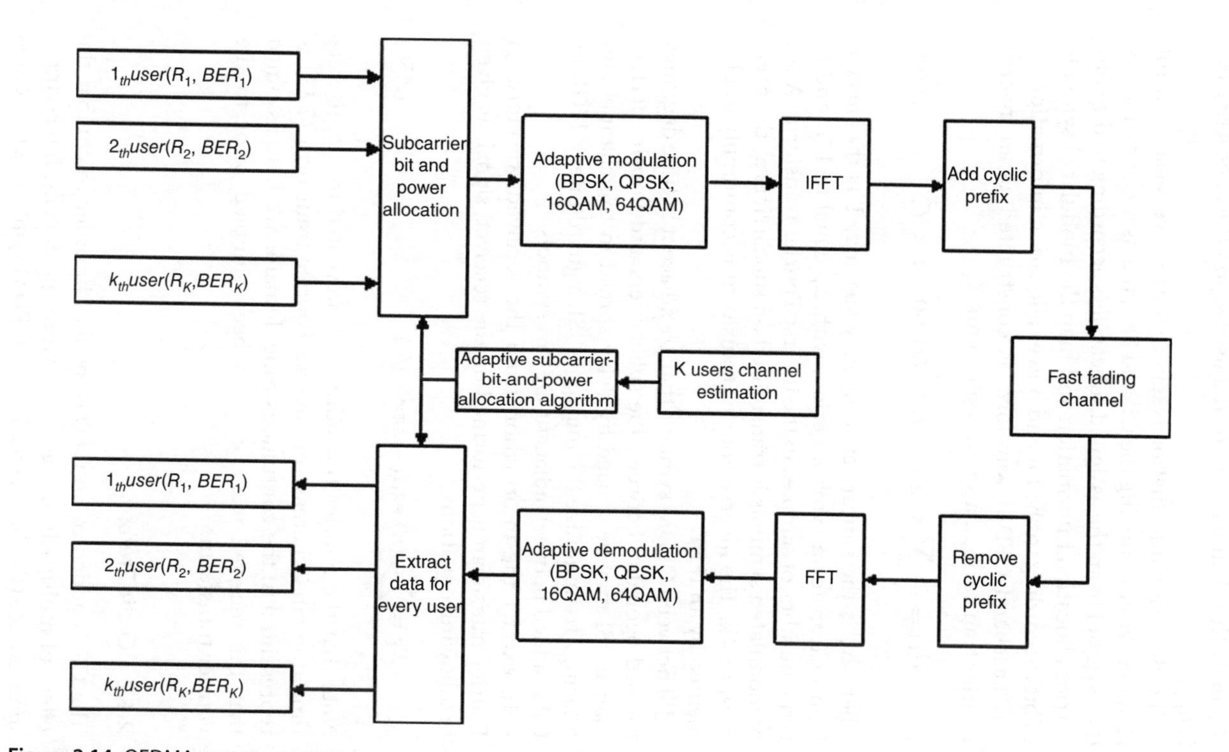

Figure 2.14 OFDMA system structure.

channel. The channel gain for the user k_{th} on the subcarrier n_{th} is $h_{k,n}$.

The allocation information is sent to the receivers via a control channel. After removing the CP and FFT, the user can extract its data symbol from the assigned subcarriers according to the subcarrier allocation information and map the modulated symbols to bits according to the bit and power allocation information.

A baseband OFDM waveform is constructed as an inverse Fourier transform of a set of coefficients X_k,

$$r(t) = \frac{1}{N}\sum_{k} X_k \exp\ (\mathrm{j}2\pi k\Delta Ft) \quad 0 \le t < T \tag{2.49}$$

where Δ_F is the subcarrier frequency spacing, T is the inverse Fourier transform symbol period, with Δ_F equal to 1/T, and N is the number of samples in the inverse Fourier transform. A set of modulated symbols is transmitted on subcarriers as the coefficients Xk. The inverse Fourier transform is commonly implemented by an IFFT.

ISI between OFDM symbols that are adjacent in time degrades the orthogonality between the subcarriers and impairs performance. ISI may be caused by delay spread in the channel and filtering. To minimize the impact of ISI, a guard interval (GI) or CP is added between adjacent OFDM symbols.

To extract the information from the received waveform, a Fourier transform is performed on the received signal, as given in following equation:

$$X_k = \sum_{n} r(n) \exp(-\mathrm{j}2\pi\kappa \cdot {}^{n}/_{N}) \tag{2.50}$$

Note that the Fourier transform is represented in the discrete time domain, whereas the inverse Fourier transform (2.13) is represented in the continuous time domain. We have assumed that the received waveform has been sampled prior to the Fourier transform.

2.4.2 Cyclic Prefix

The CP is added to an OFDM symbol in order to combat the effect of multipath propagation. More specifically, intersymbol interference (ISI) between adjacent OFDM symbols is prevented by introducing a guard period (interval) in which the multipath components of the desired signal are allowed to die out, after which the next OFDM symbol is transmitted.

A useful technique to help reduce the complexity of the receiver is to introduce a guard symbol during the guard period. This guard symbol is chosen to be a prefix extension to each block. The reason for this is to convert the linear convolution of the signal and channel to a circular convolution, thereby causing the FFT of the circularly convolved signal and channel to simply be the product of their respective FFTs. However, in order for this technique to work, the guard interval should be greater than the channel delay spread. Thus we see that the relative length of the CP depends on the ratio of the channel delay spread to the OFDM symbol duration. As indicated earlier, the LTE system introduces normal or extended CP depending on the deployment scenario.

2.4.3 Single-user MIMO/OFDM

A great deal of research work has been devoted to combining MIMO, as a spatial scheme, with OFDM systems. Such a system combines the advantages of both techniques, simultaneously providing increased data rates and elimination of the effects of delay spread [44–51].

Power control for subchannels on a MIMO/OFDM system can be crucial in enhancing the spectral and power efficiency. Without any interference, the best power-control approach to optimizing the transmission is the water-filling solution. However, this is not practical and adaptive loading algorithms may be deployed to characterize the practical performance of a MIMO-OFDM system with a single-antenna OFDM system [22, 44].

Let us start with a basic communication system as follows:

$$y = \sqrt{\rho} \cdot h \cdot x + z \tag{2.51}$$

where x is the transmitted data with unity mean expected power, h is the channel fading coefficient, z is independent, complex AWGN with zero mean and unit variance, ρ is the average SNR, and y is the received signal. Typically, h is modeled as Rayleigh fading and is defined as complex, zero mean, unit variance

Gaussian: Normal $(0,1/\sqrt{2}) + \sqrt{-1}$ Normal $(0,1/\sqrt{2})$. The notation Normal (x, y) defines a Gaussian distributed random variable with a mean of x and a variance of y.

In this model, both the transmitter and receiver are configured with one antenna. This is termed SISO, with reference to the single input to the channel and the single output from the channel. The capacity for a general SISO system is given by the Shannon capacity formula:

$$C(\text{bps/Hz}) = \log_2 (1 + \rho \cdot |h|^2) \tag{2.52}$$

In the frequency domain of an OFDM/SISO system (after the FFT), each subcarrier may be described by the SISO model (2.51). Subsequently, the receiver extracts the required information by equalizing the received signal, as follows:

$$\begin{aligned} \hat{x} &= (\sqrt{\rho} \cdot h)^{-1} \cdot y \\ &= x + (\sqrt{\rho} \cdot h)^{-1} \cdot z \end{aligned} \tag{2.53}$$

where $\hat{x}$ is the noisy estimate of the transmitted signal x.

MIMO describes a system with a transmitter with multiple antennas transmitting through the propagation environment to a receiver with multiple receive antennas, as illustrated in Figure 2.15 for MIMO 2×2. If the transmitter transmits two different streams from two different antennas, the system deploys spatial division multiplexing (SDM) and is termed a MIMO/SDM system. In this scenario, the system exploits space diversity to increase the system capacity. If the system transmits the same data on the same Tx antennas (x1 = x2), then the system deploys transmit diversity to extend the coverage and reliability of the system. If multiple antennas are adopted at the transmitter then transmit beamforming may be used. This is usually applied as a mode of MIMO on the cell edge in order to enhance the coverage rather than increase the throughput. In an LTE system, there are different MIMO modes, as illustrated in Figure 2.16. The system may deploy maximum ratio combining (MRC) or other enhanced detectors at the receiver side to exploit receive diversity and enhance the receiver sensitivity: by almost 3 dB in the case of two receive antennas.

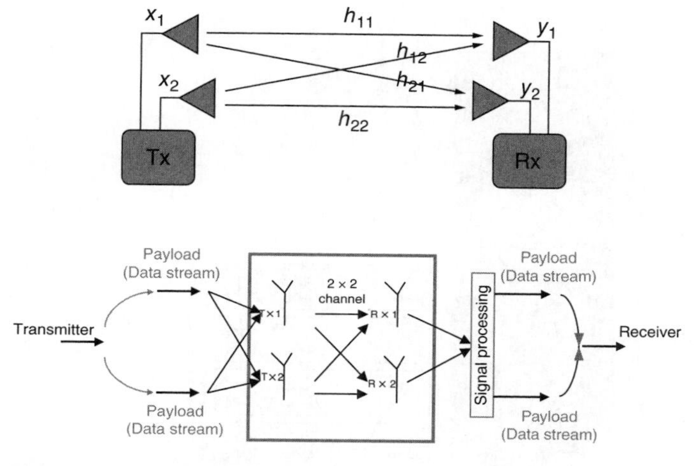

Figure 2.15 MIMO system.

2.4.3.1 3GPP LTE MIMO

In an LTE system with a MIMO 2×2 configuration, Rank-2 multiplexing gains are achieved with good channel conditions; there is little interference in a rich multipath environment. With good channel conditions, the data streams are transmitted simultaneously, with the total transmission power shared among multiple data streams. Therefore, the total SNR is also shared among multiple data streams, resulting in lower SNR for each individual data stream. If the total SNR is low, then SNR on each individual stream will be small and the throughput on each data stream will suffer. This indicates that the spatial multiplexing gain for MIMO is mostly achieved in the high SNR region, where good throughputs can be achieved on each of the independent data streams.

In open-loop spatial multiplexing operations, the network receives minimal information from the UE: only a rank indicator (RI) and a channel quality indicator (CQI). The RI indicates the number of streams (the term "stream" is used for now, but in later subsections "layers" will be used) that can be supported under the current channel conditions and modulation scheme. CQI indicates the channel conditions under the current transmission scheme, roughly indicative of the corresponding SNR. Therefore, only one CQI is reported by the UE, which is the

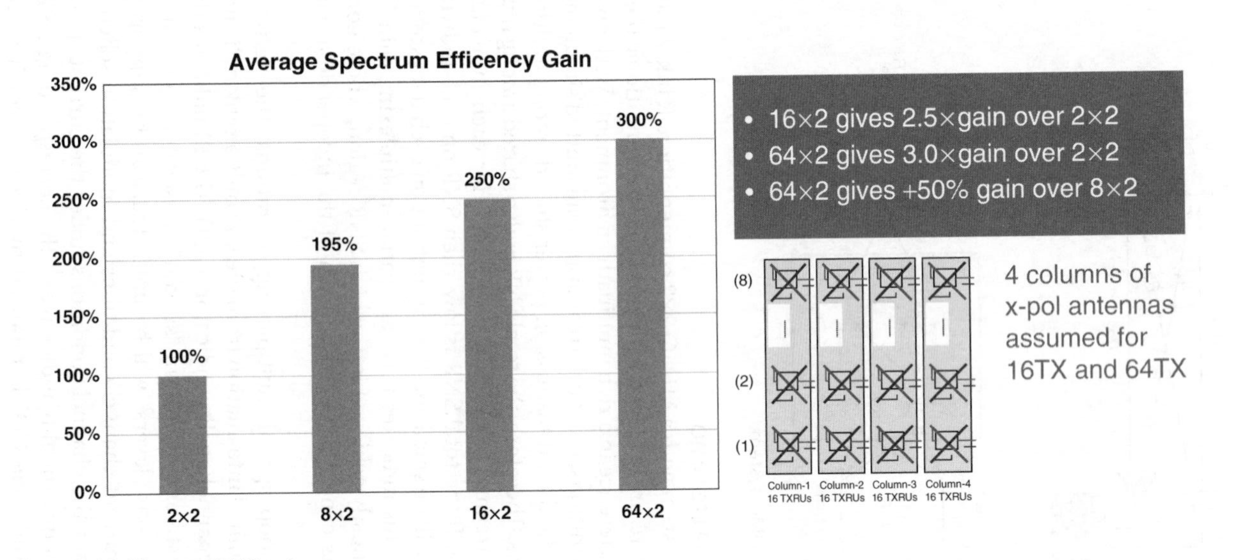

Figure 2.16 Massive MIMO gain.

Table 2.6 MIMO DL TxM mode.

Downlink transmission mode	PDSCH transmission on	UE feedback	3GPP release
Mode 1	Single antenna port (SISO, SIMO)	CQI	Rel-8
Mode 2	Transmit diversity	CQI	
Mode 3	Open-loop spatial multiplexing	CQI, RI	
Mode 4	Closed-loop spatial multiplexing	CQI, RI, PMI	
Mode 5	MU-MIMO (Rank 1 to the UE)	CQI, PMI	
Mode 6	Closed-loop, with Rank 1 spatial multiplexing (precoding)	CQI, PMI	
Mode 7	Beamforming with single antenna port (non-codebook based)	CQI	
Mode 8	Dual-layer beamforming	CQI, RI, PMI	Rel-9
Mode 9	Non-codebook precoding with seamless switching between SU-MIMO and MU-MIMO up to Rank 8	CQI, RI, PMI	Rel-10

RI, rank indicator; PMI, precoding matrix indicator; CQI, channel quality indicator.

spatial average of all the streams. The network scheduler then uses the CQI to select the corresponding MCS for the channel conditions. The network adjusts its transmission scheme and resources so that the UE matches the reported CQI and RI with an acceptable block error rate.

On the other hand, at cell edges or in other low-SNR or poor multipath conditions, instead of increasing data rates or capacity, MIMO is used to exploit diversity and increase the robustness of data transmission. In transmit diversity mode, MIMO functions much like a MISO system. Each antenna transmits essentially the same stream of data, so the receiver gets replicas of the same signal. This increases the SNR at the receiver side and thus the robustness of data transmission, especially in fading scenarios. Typically, an additional antenna-specific coding is applied to the signals before transmission to increase the diversity effect and to minimize co-channel interference. The UE receives the signals from both Tx at both Rx and reconstructs a single data stream from all multipath signals.

The most popular open-loop transmit diversity scheme is space/time coding, where a code known to the receiver is applied at the transmitter. Of the many types of space/time codes, the most popular is orthogonal space/time block codes (OSTBCs), or Alamouti code. Its popularity is a function of its ease of implementation and the linearity at both the transmitter and the receiver.

Dual-layer beamforming combines beamforming with 2×2 MIMO spatial multiplexing. It has been adopted in 3GPP versions for both MU-MIMO and SU-MIMO. These beamforming techniques require deployment of beamforming antenna arrays as well as special configurations of networks and UEs. The four types of MIMO in LTE are summarized in Table 2.7.

A codeword is the output from a channel coder. With multiple-layer transmissions, data arrives from higher-level processes in one or more codewords. In Release 8/9, one codeword is used for Rank-1 transmission, and two codewords for Rank-2/3/4 transmissions. Each codeword is then mapped onto one or more layers. The number of layers depends on the number of transmit antennas ports and channel rank report by UE (RI). There is a fixed mapping scheme of codewords to layers depending on the transmission mode used. Each layer is then mapped onto one or more antennas using a precoding matrix. In Release 8/9, there are a maximum of four antenna ports, which potentially form up to four layers. Precoding is used to support spatial multiplexing. When the UE detects a similar SNR from both Tx, the precoding matrix will map each layer onto a single antenna. However, when one Tx has a high SNR and another has a low SNR, the precoding matrix will divide the layers between the Tx antennas, in an effort to equalize SNR between the layers. In Release 10, a non-codebook precoding

Table 2.7 MIMO types in LTE.

Open-loop MIMO	Closed-loop MIMO	Multi-user MIMO	Beamforming
• Supports transmit diversity and open-loop spatial multiplexing • Use of one of these two mechanisms depends on the channel conditions and rank • In Rank-2 channel, a single codeword is sent and TxD gains are achieved • In Rank-1 channel, two codewords are sent and spatial multiplexing gains are achieved with higher throughput • No PMI reported	• Similar to open loop, closed loop also supports transmit diversity and closed-loop spatial multiplexing • UE provides an RI as well as a PMI, which determines the optimum precoding for the current channel conditions • PMI can cause higher overhead on UL, but better channel estimation generally	• While SU-MIMO increases the data rate of one user, MU-MIMO allows increase in overall system capacity	• Similar to Rank-1 spatial multiplexing (Closed-loop MIMO), beamforming gain is realized because both antenna paths carry the same information in low SNR or in less multipath conditions • Beamforming and spatial multiplexing have conflicting antenna configuration requirements • Beamforming requires antenna to be correlated with same polarization. Spatial multiplexing requires transmit antennas to be decorrelated, with cross-polarization

PMI, precoding matrix indicator; RI, rank indicator.

with seamless switching between SU-MIMO and MU-MIMO of up to Rank 8 is defined. The MIMO transmission modes are shown in Table 2.6.

MIMO 2×2 is commonly used with:

- TM3: open-loop (OL) spatial multiplexing
- TM4: closed-loop (CL) spatial multiplexing.

CL-MIMO is better suited to low-speed scenarios when the PMI feedback is accurate, while OL-MIMO provides robustness in high-speed scenarios when the feedback may be less accurate. The advantage of CL-MIMO over OL-MIMO is limited due to the small number of PMI choices for 2×2 configurations in the current LTE standard. The adaptive mode selection between modes 2, 3, 4, and 6 requires the eNB to reconfigure the mode through RRC messages, which can increase signaling load, and additional delay in adapting to the best RF conditions suitable for the selected mode.

Thus, with MIMO/SDM [21], the maximum data rate of the system increases as a function of the number of independent data streams. The system must contain at least the same number of Tx antennas as data streams. With a linear receiver, the system must contain at least the same number of Rx antennas as data streams. In other words, the data rate of the system increases by min(Tx antennas, Rx antennas, data streams). An M × N MIMO/SDM system is represented more generally by (2.54) and (2.55). Here it has been assumed that the total transmit power is equally divided over the M transmit antennas:

$$\begin{bmatrix} y_1 \\ y_2 \\ \vdots \\ y_N \end{bmatrix} = \sqrt{\rho/_M} \begin{bmatrix} h_{11} & h_{12} & \cdots & h_{1M} \\ h_{21} & h_{22} & \cdots & h_{2M} \\ \vdots & \vdots & \ddots & \vdots \\ h_{N1} & h_{N2} & \cdots & h_{NM} \end{bmatrix} \begin{bmatrix} x_1 \\ x_2 \\ \vdots \\ x_M \end{bmatrix} + \begin{bmatrix} z_1 \\ z_2 \\ \vdots \\ z_N \end{bmatrix} \tag{2.54}$$

$$Y_N = \sqrt{\rho/M}.\ H_{N\times M} X_M + Z_N \tag{2.55}$$

In case of MIMO/OFDM, each subcarrier is described by the above equation and hence a generalization of the Shannon capacity formula for M transmit antennas and N receive antennas is given by:

$$C(\text{bps/Hz}) = \log_2 \left[\det \left(I_N + \rho/_M \cdot H\ H^*\right)\right] \tag{2.56}$$

It seems very interesting with such simplicity that we may increase the data rate of a system by merely adding additional transmit and receive antennas. However, there are bounds on this increase, linked to the channel matrix. If we assume that each Tx port is connected to each Rx port and therefore the channel matrix is singular, MIMO/SDM cannot be realized. As a result, a rich multipath environment allows MIMO to exploit SDM to the maximum extent. Therefore, MIMO gain depends on several factors, including the multipath environment, the distance between Tx and Rx, and other factors. When a user is near the base station in a rich multipath environment, a full realization of MIMO is achieved and the user throughput will be doubled. As the user moves away from the base station, the MIMO gain will be reduced in order to maintain a robust link; it will be converted to TX diversity, extending the coverage by almost 3 dB for two transmit antenna.

Consider an uplink SU-MIMO/OFDM Rayleigh block fading channel, corresponding to a rich scattering environment with time variation characterized by the fade time. The user and the BS are equipped with M_t transmit and M_r receive antennas, respectively. The frequency band is divided into M subcarriers. Assuming the length of the CP is longer than the maximum time dispersion of the channel response time, which leads to flat fading channel in each subcarrier. The $M_r \times 1$ received vector at subcarrier m, can be modeled as:

$$\boldsymbol{r}_m(n) = \sum_{z=-\infty}^{\infty} \boldsymbol{H}_m(n-z)\boldsymbol{s}_m(n) + \boldsymbol{\eta}_m(n) \tag{2.57}$$

where $\boldsymbol{\eta}_m(n)$ is an $M_r \times 1$ vector corresponding to the AWGN with variance $E(\eta(n)\eta^H(n)) = \sigma_\eta^2 I_{M_r}$, where I_{M_r} is the identity matrix with size M_r, $\boldsymbol{s}_m(n)$ is the $M_t \times 1$ complex modulated transmitted vector, and $\boldsymbol{H}_m(n)$ is the $M_r \times M_t$ MIMO channel transfer function, and is given by:

$$\boldsymbol{H}_m(n) = \begin{bmatrix} h_{11}(n) & h_{12}(n) & \cdots & h_{1M_t}(n) \\ h_{21}(n) & \ddots & \ddots & \vdots \\ \vdots & \ddots & \ddots & \vdots \\ h_{M_r1}(n) & \cdots & \cdots & h_{M_rM_t}(n) \end{bmatrix} \tag{2.58}$$

Here, h_{ij} is the channel response from the jth transmit antenna and the ith receive antenna and is given by:

$$h_{ij}(n) = \sqrt{\Gamma}\sum_{l=1}^{L} \gamma_l \beta_l(n) a_i^r(\theta_l^r) a_j^t(\theta_l^t) \tag{2.59}$$

where L is the number of paths, Γ is the product of path loss, γ_1 is path amplitude, and $\beta_l(n)$ is the Rayleigh fading factor. $a_i^r(\theta_l^r)$

and $a_j^t(\theta_l^t)$ are the array response at arrival angle θ_l^r and departure angle θ_l^t, respectively, for the lth path. For the ith antenna, d is the spacing between antennas and λ is the wavelength. The array response is given by:

$$a_i(\theta_l) = \exp\left(-j\,2\pi\frac{(i-1)d}{\lambda}\sin(\theta_l)\right) \tag{2.60}$$

With FFT and by dropping the subscript m for notational simplicity, (2.57) can be written as:

$$\boldsymbol{r}(f) = \boldsymbol{H}(f)\boldsymbol{s}(f) + \boldsymbol{\eta}(f) \tag{2.61}$$

If we consider the case of perfect channel state information at the transmitter and receiver, we can decompose the MIMO channel on each subcarrier into parallel non-interfering SISO channels using SVD as:

$$\boldsymbol{H} = \boldsymbol{U}\boldsymbol{\Lambda}\boldsymbol{V}^H \tag{2.62}$$

where the superscript $(.)^H$ is defined as the Hermitian, and $\boldsymbol{U}$ and $\boldsymbol{V}$ are the left and right singular matrices. $\boldsymbol{\Lambda}$ is a diagonal matrix, which contains the eigenvalues in descending order. SVD is used to coordinate the space domain to collect the signal power in order to maximize the SINR; in this sense, the SVD filter is a spatial maximal-ratio combining (MRC) filter [21].

The received signal at each subcarrier can be expressed as follows:

$$\begin{aligned}
\boldsymbol{r}(f) &= \boldsymbol{H}(f)\boldsymbol{V}\boldsymbol{s}(f) + \boldsymbol{\eta}(f)\\
\boldsymbol{r}(f) &= \boldsymbol{U}\boldsymbol{\Lambda}\boldsymbol{V}^H\boldsymbol{V}\boldsymbol{s}(f) + \boldsymbol{\eta}(f)\\
\boldsymbol{r}(f) &= \boldsymbol{U}\boldsymbol{\Lambda}\boldsymbol{s}(f) + \boldsymbol{\eta}(f)
\end{aligned} \tag{2.63}$$

From the above formulation, the singular values on the diagonal matrix $\boldsymbol{\Lambda}$ can be represented as the received signal strength for each element $\boldsymbol{s}(f)$ at the receiver.

For a standard single antenna transmission system, the channel capacity can be estimated using Shannon theory as follows:

$$C = B \cdot log_2(1+\rho)$$

where C is the channel capacity (bits/second)

B is the occupied bandwidth

ρ *is the SNR (in linear ratio)* (2.64)

As we can see, the channel capacity is proportional to the occupied bandwidth and SNR, which means if we would like to increase the channel capacity, we can either increase the bandwidth or SNR. But the use of spectrum is normally assigned and fixed by regulatory authorities, which means that altering the bandwidth is difficult. It requires significant investment to acquire additional spectrum. Increasing SNR can be done by using a more complex modulation scheme, for example 256QAM, but this requires extra processing power, increasing the manufacturing cost. SNR only increases logarithmically.

Meanwhile, the MIMO capacity can be expressed as:

$$C = \sum_{i=1\ to\ N} B \cdot log_2 \left[1 + \frac{\rho}{N} \sigma_i^{\ 2}(H) \right] \tag{2.65}$$

where,

N is the number of pairs of transmit-receive antenna
σ_i^2 is the singularvalues of the radio channel matrix, H

Equation (2.65) is derived from (2.64) for the single antenna condition, which shows that the channel capacity is proportional to the number of pairs of transmit–receive antennas, which increase linearly. Therefore, the number of effective antennas can be raised to increase the channel capacity and transmit data rate without changing the channel bandwidth or using a complex modulation. This is a more practical and economic way to improve wireless transmissions.

In the Matlab simulation provided with this book, the MIMO channel is modelled as follows:

$$\boldsymbol{y}_n = \sum_{l=0}^{L-1} \boldsymbol{H}_l \boldsymbol{x}_{n-1} + \boldsymbol{n}_n \tag{2.66}$$

where,

$\boldsymbol{y}_n$, $\boldsymbol{x}_n$, $\boldsymbol{n}_n$, are output, input, and additive noise respectively and they can be vectors
$\boldsymbol{H}_l$ is the channel gain, it can be a matrix
L is the symbol periods of the delay spread

The MIMO gain depends on the channel gain matrix, which is linked to the multipath environment. In practical deployments, LTE and 5G will evolve to MIMO 64x64 and possibly beyond.

However, field testing indicates that gains beyond MIMO 8x8 are marginal and limited to certain scenarios. Combining MIMO with beamforming is another domain that may bring improvement. Figure 2.16 provides the expected gain from different MIMO deployment up to massive MIMO with 64 antenna elements at base station.

2.4.4 Adaptive Resource Management

The advantage of OFDM is that each subchannel is relatively narrowband and is assumed to have flat fading. However, it is entirely possible that a given subchannel has a low gain, resulting in a large BER. Therefore it is desirable to take advantage of subchannels having relatively good performance; this is the motivation for adaptive loading. In the context of time-varying channels, there is a decorrelation time associated with each frequency-selective channel instance, so a new adaptation must be implemented each time the channel decorrelates [22, 54].

The optimal adaptive transmission scheme, which achieves the Shannon capacity for a fixed transmit power, is the water-filling distribution of power over the frequency-selective channel. However, while the water-filling distribution will indeed yield the optimal solution, it is difficult to compute, and it tacitly assumes infinite granularity in the constellation size, which is not practically realizable. An efficient adaptive loading technique to achieve power and rate optimization based on knowledge of the subchannel gains can be found in the literature [22, 23].

Water filling may be a good technique if the transmission channel is perfect and free of interference, but in real-world conditions this is not true: practical wireless transmission channels may contain a lot of noise that causes interference to different carriers. In an OFDM system, data bits are modulated into different subcarriers for transmission. Since OFDM divides the available bandwidth into smaller bands, its advantage of being narrowband and flat-fading can be assumed. Interference and some channel conditions may affect some subcarriers, giving them lower gain and resulting in a bad BER if data are carried in these subcarriers for transmission. The other, unaffected subcarrier will have better performance than the degraded subcarriers. Therefore, water filling is not an optimal solution here; filling every data bit onto every available subcarrier causes the data reaching the destination to be corrupted, necessitating retransmission, and increasing the BER [22, 23].

Channel estimation inverts the effect of non-selective fading on each subcarrier. Usually OFDM systems provide pilot signals for channel estimation. In the case of time-varying channels, the pilot signal should be repeated frequently. The spacing between pilot signals in time and frequency domains depends on the channel coherence time and system bandwidth. The channel estimates are assumed to be perfect, and available to both the transmitter and the receiver. Given full knowledge of the channel, the transmitter and receiver can determine the frequency response of the channel, and the channel gains at each tone of the OFDM symbol. Given these gains, the adaptive algorithm can calculate the optimal bit and power allocation [54].

Provided perfect channel knowledge is available to both the transmitter and receiver, adaptive loading can be useful, taking advantage of the subcarrier with better performance, dynamically allocating different numbers of data bits and energy based on the different subcarrier performances. Subcarriers with higher gain can be allocated more data bits while using lower energy, and vice versa for carriers with lower gains. Since our channel is time-varying, different frequency-selective channels contain a different decorrelation time, the adaptation must be updated for every channel decorrelation. Adaptive loading achieves data-rate and power optimization, based on the knowledge of the gains of each subchannel. The technique implemented in the simulation for this book is based on the algorithms developed by Chow *et al.* and Campello de Souza [22, 23, 54].

The adaptive loading technique is an efficient way to achieve power and rate optimization based on knowledge of the subchannel gains [22, 23]. In the discrete bit-loading algorithm of Campello de Souza [22], we are given a set of M increasing convex functions $e_m(b)$ that represents the amount of energy necessary to transmit b bits on subchannel m at the desired probability of error using a given coding scheme. We will assume $e_m(0) = 0$.

The allocation problem, which we will be using can be formulated as and energy minimization problem:

$$\text{Minimize} \sum_{m=1}^{M} e_m(b_m)$$

$$\text{Subject to} \sum_{m=1}^{M} b_m = B$$

$$b_m \in \mathbb{Z},\ b_m \geq 0,\ m = 1,2,\ldots,M.$$

To initialize the bit allocation, the scheme of Chow *et al.* [23] is employed. The procedure is summarized as follows [54]:

Box 2.1 Bit allocation initialization

1. Compute the subchannel SNRs.
2. Compute the number of bits for the *i*th subchannel based on the formula:

   $$\hat{b}(i) = \log_2(1 + SNR(i)/GAP)$$

3. Round the value of $\hat{b}(i)$ down to $b(i)$.
4. Restrict $b(i)$ to take values 0, 1, 2, 4, 6 or 8 (corresponding to available modulation orders).
5. Compute the energy for the *i*th subchannel based on the number of bits initially assigned to it using the formula:

   $$e_i(b(i)) = (2^{b(i)} - 1)/GNR(i)$$

   where

   $$GNR(i) = SNR(i)/GAP$$

6. Form a table of energy increments for each subchannel. For the *i*th subchannel:

   $$\Delta e_i(b) = e_i(b) - e_i(b-1) = \frac{2^{b-1}}{GNR}$$

Consider the kth channel. Given the channel gain and noise PSD, the energy increment table will provide the incremental energies required for the subchannel to transition from supporting 0 bits to 1 bit, from 1 bit to 2 bits, from 2 bits to 3 bits and so on. Since we require our system to have a maximum of 8 bits, the energy increment required to go from 8 bits to 9 bits is set to a very high value. In addition, we require the subchannel to have only 0, 1, 2, 4, 6 or 8 bits. Thus odd numbers of bits are not supported. In order to take care of this, the energy increment table has to be changed using a clever averaging technique. It is best described by an example, as follows.

Suppose the energy increment required for supporting an additional bit from 2 bits in the nth subchannel is 30 units and that required for supporting an additional bit from 3 bits is 40 units. Then, we reassign the energy increment values to the same value, namely, the average of the two. In this case, that value is 35 units. This ensures that if a subchannel is allocated a single bit for going from 2 bits to 3 bits then, in the next iteration, the same minimum amount of additional energy is required to support another bit. This means that the same subchannel will be allocated to the next bit as well.

The same averaging procedure is repeated for all other possible bit transitions. The only exception that might arise is when the algorithm terminates, not having assigned the final bit to even out the total number of bits on that subchannel. In order to resolve this issue, we used an algorithm proposed by Campello de Souza [22], (the function "resolve last bit"), which will be discussed in the detail later in this section.

Note that we have introduced a new term, GAP. This parameter is in effect a tuning parameter. Different values for GAP yield different SNRs for a given desired number of bits B to transmit. This is because the GAP directly impacts the energy table value calculations. Thus, tuning the GAP allows us to characterize the BER performance of the system.

Given the initial bit allocation, the following algorithm optimizes the bit allocation [22, 54].

Box 2.2 Bit allocation optimization

Input:
  b, initial bit allocation.
  B, the total number of bits to be allocated.
Output:
  b, the optimized bit allocation.
Algorithm:
$B' \leftarrow 0$
for $m = 1$ to M
  $B' \leftarrow B' + b(m)$
while $(B' \neq B)$
  if $(B' > B)$
    $m = \arg\max_{1 \leq j \leq M} \Delta e_j(b_j)$
    $B \leftarrow B - 1$
    $b(m) \leftarrow b(m) - 1$
  else
    $m = \arg\min_{1 \leq j \leq M} \Delta e_j(b_j + 1)$
$B \leftarrow B + 1$
    $b(m) \leftarrow b(m) + 1$

Finally, in order to deal with a single violated bit constraint, we employ the following algorithm [22, 54].

Box 2.3 Bit allocation: resolve last bit

1. Check that the input bit allocation contains at most one violation of the bit constraint.
2. If there is a single violation, (say it is in subchannel ν), find the bit from the current bit allocation having the largest incremental energy that can be used to fill up subchannel ν. Let:

$$E_1 = \Delta e_\nu(b(\nu)) - \Delta e_i(b(i))$$

3. Find the bit that will cost the least to increment in the other subchannels, which have been allocated either 0 or 1 bit only. The reason we have this constraint is that all the other subchannels will have 2,4,6 or 8 bits and allocating a single bit to

them will violate the bit constraint. Let

$$E_2 = \Delta e_j(b(j)+1) - \Delta e_v(b(v))$$

4. Perform the change corresponding to the smallest of E_1 and E_2

Given these three algorithms, we have a complete characterization of the bit loading procedure for a given frequency-selective channel [54]. Figure 2.17 demonstrates instant channel gain and the bit and power allocation for this channel.

2.4.5 Multi-User MIMO/OFDM

Consider K users each equipped with M_t^k antennas. The received signal at the BS is the sum of all users' transmitted signals passing through their own MIMO channel.

$$\boldsymbol{r}(f) = \sum_{k=1}^{K} \boldsymbol{H}^k(f)\bar{\boldsymbol{s}}^k(f) + \boldsymbol{\eta}(f) \tag{2.67}$$

where $\bar{\boldsymbol{s}}^k = \boldsymbol{V}^k \boldsymbol{s}^k$ is pre-filtered inputs at the transmitter of user k. After FFT and CP removal, we assume the first user is the desired user. The spatially matched output vector with single matched filter (MF) $\boldsymbol{U}^H(f)$ at the receiver is given by:

$$\begin{aligned}\boldsymbol{x}^1(n) &= (\boldsymbol{U}^1(n))^H\boldsymbol{r}(n) = \boldsymbol{\Lambda}^1(n)\boldsymbol{s}^1(n)\\ &\quad + \sum_{k=2}^{K}(\boldsymbol{U}^1(n))^H\boldsymbol{U}^k(n)\boldsymbol{\Lambda}^k(n)\bar{\boldsymbol{s}}^k(n) + (\boldsymbol{U}^1(n))^H\boldsymbol{\eta}(n)\end{aligned} \tag{2.68}$$

Herein, the information $\mathbf{s}^1(n)$ is spread over the space by $V^1(n)$ and goes through the channel gain $\boldsymbol{\Lambda}^1(n)$, and is despread by $(\boldsymbol{U}^1(n))^H$ at the receiver. This spread/despread operation is performed in the space domain, not the time domain, so it does not have the bandwidth expansion that a typical CDMA-based technique suffers from. Therefore, these unitary vectors $\boldsymbol{U}(n)$ and $V(n)$ can be interpreted as spatial codewords, which are different from user to user as long as $\boldsymbol{H}^k$ are all different.

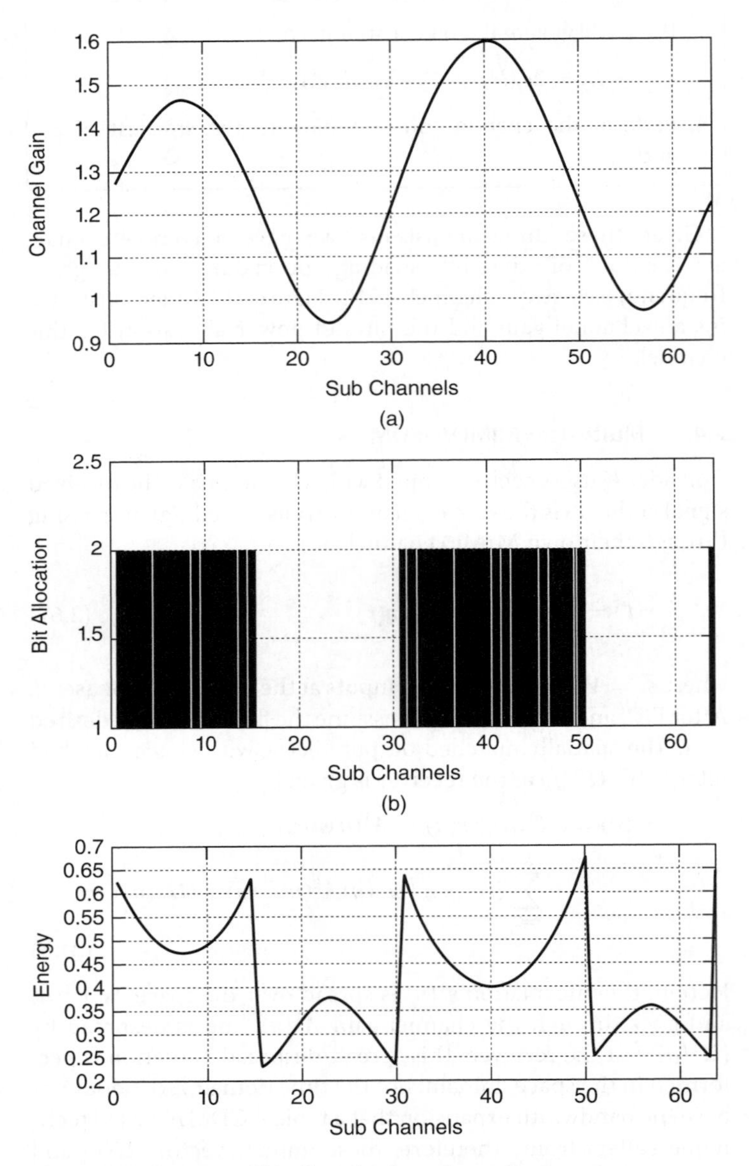

Figure 2.17 Energy and bit allocation for a channel instance.

2.4.6 Adaptive filtering in MIMO/OFDM System

In Equation (2.68), the received signals are severely corrupted by interference, due to the correlation of MFs. Linear multiuser detection (LMUD) transforms the output of the bank of MFs to obtain the decision statistic:

$$y(n) = W(n)x(n) \tag{2.69}$$

Assuming perfect interference cancellation using the MUD, adaptive resource management is conducted in the MU scenario as if we are dealing with an SU scenario. The LMUD output $\mathbf{y}(n)$ can be expressed as:

$$y(n) = W^H(\bar{x}(n) + \eta'(n)) = \bar{y}(n) + \bar{\eta}(n) \tag{2.70}$$

where, $\bar{x}(n)$ is the noiseless MF output vector, $\eta'(n) = (U^1(n))^H \eta(n)$ and $\bar{\eta}(n) = W^H \eta'(n)$ is the noise vector after being match filtered and linearly transformed.

For simplicity of the analysis and the productivity of the MBER cost function, we assume, without loss of generality, use of BPSK modulation. The bit decision is made according to the real output of the LMUD as:

$$\hat{s}(n) = \text{sgn}(\text{Re}\{y(n)\}) = \text{sgn}(y_R(n)) = \text{sgn}(\text{Re}\{W(n)x(n)\}) \tag{2.71}$$

where sgn(.) denotes the sign function.

2.4.7 Performance Evaluation of MIMO/MBER System

In a MIMO/OFDM system, we will use BER as the performance measure to develop MBER detectors rather than MSE or minimum output energy. First, we start by forming the BER cost function. Let us define the following signed variable:

$$y_{sign}(n) = \text{sgn}(s(n))y_R(n) \tag{2.72}$$

Furthermore, let $N_s = 2^{K \times M_t}$ be the number of possible transmitted vector sequences. Hence the first vector s_v, where $1 \leq v \leq N_s$, is the desired user and the real part of the LMUD output $\bar{y}_R(n)$ can only take values from the set

$Y_R(\boldsymbol{W}) \triangleq \{\bar{y}_{R,v}(\boldsymbol{W}) = \mathrm{Re}\{\bar{y}_v(\boldsymbol{W})\},\ 1 \le v \le N_s\}$. For a normalized weight matrix, the PDF of $y_{sign}(n)$ is a mixed sum of Gaussian distributions [29]:

$$P(y_{sign}) = \frac{1}{N_b\sqrt{2\pi\sigma_\eta^2}} \sum_{v=1}^{N_b} \exp\left(-\frac{(y_{sign} - \mathrm{sgn}(s(n))y_{R,v}(\boldsymbol{W}))^2}{2\sigma_\eta^2}\right) \tag{2.73}$$

The error probability of the LMUD is given by:

$$P_E(\boldsymbol{w}) = \frac{1}{N_b\sqrt{2\pi\sigma_\eta^2}} \sum_{v=1}^{N_b} \int_{q(\boldsymbol{W})}^{\infty} \exp\left(-\frac{u^2}{2}\right) du = \frac{1}{N_b} \sum_{v=1}^{N_b} Q(q(\boldsymbol{W})) \tag{2.74}$$

where $Q(.)$ is the Gaussian error function given by:

$$Q(x) = \frac{1}{\sqrt{2\pi}} \int_x^{\infty} \exp\left(-\frac{y^2}{2}\right) dy \tag{2.75}$$

and

$$q(\boldsymbol{W}) = \frac{\mathrm{sgn}(s(n))y(n)}{\sigma_\eta} = \frac{\mathrm{sgn}(s(n))\mathrm{Re}\{\boldsymbol{W}\boldsymbol{x}(n)\}}{\sigma_\eta} \tag{2.76}$$

In practice, the set Y is not available. A widely used approach for approximating the PDF is the kernel density estimate, which produces a reliable PDF estimate with short data records [59]. Given a block of Z training samples $\{\boldsymbol{x}(Z), \boldsymbol{s}(Z)\}$, the PDF can be approximated as:

$$\hat{P}(y_s) = \frac{1}{Z\sqrt{2\pi\rho^2}} \cdot \sum_{i=1}^{Z} \exp\left(-\frac{(y_s - \mathrm{sgn}(s(i))y_R(i))^2}{2\rho^2}\right) \tag{2.77}$$

where the radius parameter ρ is related to the noise standard deviation [24–26]. Therefore, the block BER can be derived from the approximated PDF as follows:

$$P_E(\boldsymbol{W}) = \frac{1}{Z} \sum_{z=1}^{Z} Q(q_z(\boldsymbol{W})) \tag{2.78}$$

Expansion of the BER cost function from BPSK modulation to QPSK modulation can be found in the literature [27].

2.5 Adaptive Antenna Array

An antenna array consists of a set of antenna elements that are spatially distributed at known locations with reference to a common fixed point [27, 28]. The antenna elements can be arranged in various geometries. Some of the popular geometrical configurations are linear, circular and planar. In a linear array, the centers of the elements of the array are aligned along a straight line. In a circular array, the centers of the elements lie on a circle. For a planar array configuration, the centers of the array lie on a single plane. Both linear and circular arrays are obviously special cases of a planar array. The radiation pattern of an array is determined by the radiation pattern of the individual elements, their orientation and relative positions in space, and the amplitude and the phase of the feeding current [27]. If each element of the array is an isotropic point source, the radiation pattern of the array will depend solely on the geometry and feeding current of the array. In that case, the radiation pattern is commonly known as the *array factor*. If each of the elements of the array is similar but nonisotropic, by the principle of pattern multiplication [61], the radiation pattern can be computed as the product of the array factor and the individual element pattern [62].

2.5.1 Uniform Linear Array

If the spacing between the elements of a linear array is equal, it is known as a uniform linear array (ULA). Figure 2.18 illustrates an N-element ULA. The spacing between the array elements is d and a plane wave arrives at the array from a direction θ off the array *broadside*. The array broadside is perpendicular to the line containing the center of the elements. The angle θ measured clockwise from the array broadside is called the direction of arrival (DOA) or the angle of arrival (AOA) of the received signal.

Let us assume that the signals originate far away from the array and the plane wave associated with the signal advances through a non-dispersive medium that only introduces a propagation delay. Under these circumstances, the signal at any other element can be represented by a time-advanced or time-delayed version of the signal at the first element. In other scenarios,

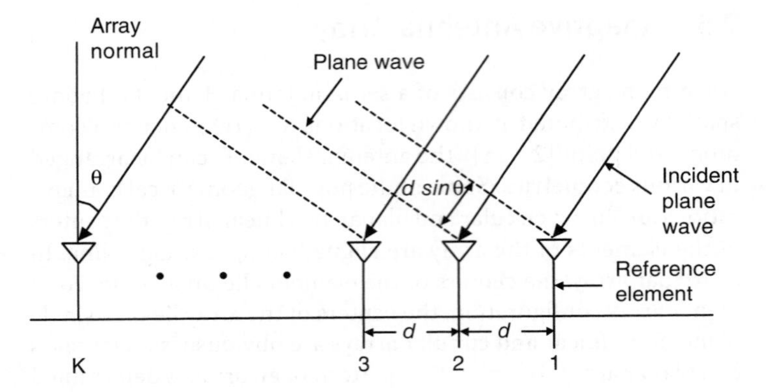

Figure 2.18 Plane wave incident on a ULA with an AOA of θ.

such as microphone arrays, when the speech source is close to the microphone array, the far-field assumptions are no longer valid and spherical wavefronts (instead of planar wavefronts) and signal attenuation have to be taken into account.

As shown in Figure 2.19, the wavefront impinging on the first element travels an additional distance $d\,sin(\theta)$ to arrive at the second element. The time delay due to this additional propagation

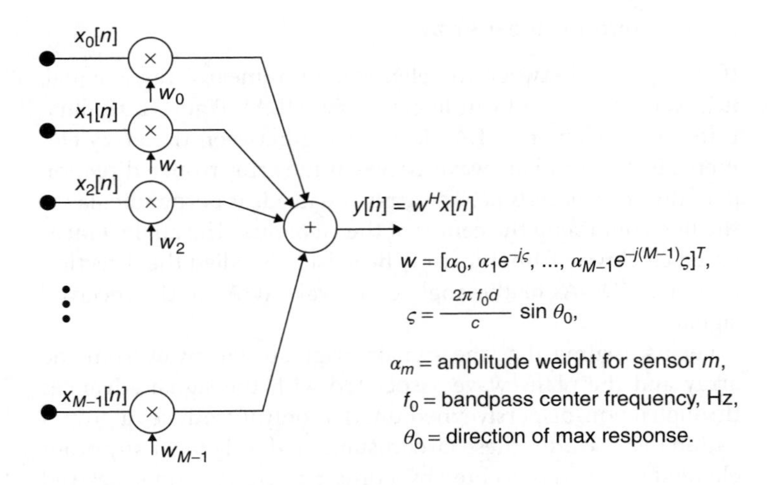

Figure 2.19 Narrowband beamformer.

distance is given by:

$$\tau = \frac{d \sin \theta}{velocity\ \ of\ \ light} \tag{2.79}$$

The time delay of the signal can now be represented by a phase shift [63]. Hence, we can define a(θ), the array response vector or the steering vector of an ULA, as:

$$a(\theta) = \left[1 \quad e^{-j\left\{2\pi \frac{d}{\lambda} \sin \theta\right\}} \quad \dots \quad e^{-j\left\{2\pi \frac{d}{\lambda} (N-1) \sin \theta\right\}} \right]^T \tag{2.80}$$

The array response vector is a function of the AOA, individual element response, the array geometry and the signal frequency. We will assume that for the range of operating carrier frequencies, the array response vector does not change. Since we have already fixed the geometry (it is a ULA) and the individual element responses (identical isotropic elements), the array response vector is a function of the AOA only.

For K users with a single path for each user, each user has its own received angle, and hence we have an array vector for each user. The set of array vectors for the K users is:

$$\boldsymbol{A} = (a_1(\theta_1) \quad a_2(\theta_2) \quad \cdots \quad a_K(\theta_K))^T \tag{2.81}$$

The sampled received signal vector can now be written as:

$$x(n) = \sum_{j=1}^{K} A_j(\theta_j) u_j(n) + \eta'(n) \tag{2.82}$$

$$x(n) = \sum_{j=1}^{K} A_j(\theta_j) \boldsymbol{H}^{(j)} . S^{(j)}(n) + \eta'(n) \tag{2.83}$$

$$x(n) = \sum_{j=1}^{K} \boldsymbol{H}'^{(j)} . S^{(j)}(n) + \eta'(n) \tag{2.84}$$

$$x(n) = \boldsymbol{H}' . S(n) + \eta'(n) \tag{2.85}$$

where $\boldsymbol{H'}$ represents the system matrix with the steering vector effect, and η' (t) is the noise vector *($N \times 1$)* in each antenna element.

$$\boldsymbol{H} = (G^{(1)} C^{(1)} \boldsymbol{A}^{(1)}(\theta_1),\ G^{(2)} C^{(2)} A^{(2)}(\theta_2) \cdots G^{(K)} C^{(K)} A^{(K)}(\theta_K))_{N_f \times KN_s} \tag{2.86}$$

For a multipath system, the array steering vector becomes more complex and we will have now a matrix steering vector for each user:

$$\boldsymbol{A}^{(j)} = (A^{(j)}(\theta_1) \quad A^{(j)}(\theta_2) \quad \cdots \quad A^{(j)}(\theta_{N_g}))^T \tag{2.87}$$

We can now expand the effect of this multipath in the system matrix as:

$$\boldsymbol{H} = (\boldsymbol{H}_1 \quad \boldsymbol{H}_2 \quad \cdots \quad \boldsymbol{H}_K) \tag{2.88}$$

where

$$\boldsymbol{H}_j = (G^{(j)}C^{(j)}A^{(j)}(\theta_1), \quad G^{(j)}C^{(j)}A^{(j)}(\theta_2) \quad \cdots \quad G^{(j)}C^{(j)}A^{(j)}(\theta_{N_g}))_{N_f \times N_g} \tag{2.89}$$

So far we have assumed that the bandwidth of the impinging signal is much smaller than the reciprocal of the propagation time across the array. This assumption, commonly known as the *narrowband assumption* [61] for the signal, made it possible to represent the propagation delay within the elements of the array by phase shifts in the signal. The narrowband model is exact for sinusoidal signals, and it is usually a good approximation for a situation where the bandwidth of the signal is very small compared to the inverse of the propagation time across the array. Any deviation from the narrowband model is detrimental to the performance of a narrowband beamformer, usually manifesting as a limit in the ability to nullify interferers [62]. In such a scenario, a wideband beamformer [27, 61] must be developed.

In this book, we will assume that the signal satisfies the narrowband assumption. The delay the wavefront experiences to propagate from the first element to the Nth element is given by:

$$\tau_{\max} = \frac{(N-1)d\sin\theta}{\textit{velocity of light}} \tag{2.90}$$

If the spacing between the elements is half the carrier wavelength, then

$$\begin{aligned} \tau_{\max} &= \frac{(N-1)\frac{\lambda}{2}}{\textit{velocity of light}}; \quad \max(\sin(\Theta)) = 1 \\ &= \frac{(N-1)\frac{\textit{velocity of light}}{2f_c}}{\textit{velocity of light}} = \frac{(N-1)}{2f_c} \end{aligned} \tag{2.91}$$

where f_c is the carrier frequency. If there are four elements and the carrier frequency is 2 GHz, $\tau_{\max} = \frac{3}{2 \times 2000 \times 10^6}$ s. So, for a WCDMA signal with a bandwidth of 5 MHz, the ratio of the reciprocal of the maximum delay and the signal bandwidth is given by: $x_{\max} = \frac{3 \times 5 \times 10^6}{2 \times 2000 \times 10^6} = 0.0037$, Therefore the narrowband assumption holds for this WCDMA signal. Also, for LTE with 20 MHz bandwidth, $\chi_{\max} = 0.0148$, so the narrowband beamformer can also be considered in such systems. With a 5G system at 3.5 GHz and a channel bandwidth of 100 MHz, $\chi_{\max} = 0.042$. In this case, the narrowband bandwidth can still be used. Finally, with 5G using mmWave technology, for example 28 GHz, a 64-antenna element (8×8) and channel bandwidth of 1 GHz, $\chi_{\max} = \frac{7x1000x10^6}{2x28000x10^6} = 0.125$ and therefore a narrowband beamformer can still be used.

For the narrowband assumption, let s(t) be a bandpass signal with $BW << c/(\text{M-1})\ d$ Hz. This means the phase difference between upper and lower band edges for propagation across the entire array is small, say less than $<\pi/100$ radians. As explained, most communications signals fit this model.

Narrowband beamforming is used when the desired signal occupies a known narrow bandwidth. Any interference outside of that bandwidth can be reduced with a temporal (as opposed to spatial) filter. If interference is expected to occupy the same bandwidth but comes from a different direction than the desired signal, a beamformer can be useful. Signals from a spatial array of sensors can be combined in such a way as to enhance the signal coming from one direction and reduce the signal from another. The combination of the signals, for the narrowband case, is usually a sum of weighted, phase-shifted signals. Differences in the weights affect the shape of the beam, and changes in phase are used to change or steer the beam to a specific direction without physically moving the array.

The beamformer weights are usually complex numbers and represent both changes in phase and magnitude. A typical narrowband array beamformer is shown in Figure 2.19.

Narrowband beamformers can have many elements in any geometrical pattern. The gains of typical narrowband beamformers can be chosen so as to produce a wide array of beamforms. The gains can also be updated adaptively.

Adaptive beamforming algorithms are considered in Chapter 7, in estimating robust beamforming techniques against many uncertainty factors. Adaptive minimum BER beamforming algorithms are described in Chapter 8.

2.5.2 DS/CDMA with Antenna Array

A K-user DS-CDMA communication system employing an M-element ULA at the base station is considered. The received continuous-time baseband signal at antenna element m can be modeled as follows [6]:

$$r^{(m)}(t) = \sum_{n=-\infty}^{\infty} \sum_{k=1}^{K} b_k(n) h_k^{(m)}(t - nT_b) + \sigma_\eta \eta^{(m)}(t) \tag{2.92}$$

where T_b is the bit interval, $b_k(n)$ is the nth data bit of the kth user, $h_k^{(m)}(t)$ is the effective signature waveform of the kth user at antenna element m, $\eta^{(m)}(t)$ is AWGN at antenna m with unit power spectral density, and σ_η is the noise power spectral density. It is assumed that the data symbols are independent, equally likely random variables from a finite alphabet, which are independent from $\eta^{(m)}(t)$. The effective signature waveform of user k at antenna element m can be modeled in the following form:

$$h_k^{(m)}(t) = \sum_{l=0}^{L-1} c_k(l) g_k^{(m)}(t - lT_c) \tag{2.93}$$

where L is the spreading factor, $T_c = T_b/L$ is the chip interval, $c_k(l)$ is the lth chip in the spreading sequence of the kth user, and $g_k^{(m)}(t)$ is the chip waveform of the kth user at antenna element m that has been distorted by the multipath propagation and front-end and transmit filters.

For M antennas, assuming the propagating signals are plane waves, multiple copies of the transmitted signal can be collected with different times of arrival due to reflection from the surrounding obstacles. The channel is assumed to be time invariant and elliptically modeled [28]. Therefore, the distorted chip waveform of the kth user can be modeled as follows:

$$\mathbf{g}_k(t) = \left[g_k^1(t) \; \ldots \; g_k^M(t)\right] = \sigma_k \sum_{f=0}^{F_k-1} \alpha_{k,f} \psi_k(t - \tau_{k,f}) \boldsymbol{a}_{k,f} \tag{2.94}$$

Here, σ_k is the amplitude of kth user signal, F_k is the number of multipath components that are associated with the kth user, $\alpha_{k,f}$ is the proportion of the kth user's amplitude, scattered in the fth path, assumed to be Rayleigh distributed, $\tau_{k,f}$ is the time delay of this path, $\psi_k(t)$ is the original chip waveform that has been filtered on receiver and transmitter filters.

$$\boldsymbol{a}_{k,f} = \boldsymbol{a}(\theta_{k,f}) = \left[1 \;\; e^{-j\left\{2\pi\frac{d}{\lambda}\sin\theta_{k,f}\right\}} \;\; \dots \;\; e^{-j\left\{2\pi\frac{d}{\lambda}(M-1)\sin\theta_{k,f}\right\}}\right]^T$$

is the array response vector to the fth path of the kth user, with $\theta_{k,f}$ the DOA of the impinging signal of the kth user's fth ray, and d is the spacing between the antenna elements, usually assumed to be $\lambda/2$ where λ is the wavelength.

Without loss of generality, we assume the desired user is the first user and the receiver is synchronized to that user, ($\tau_{1,0} = 0$). In general, the code sequence $\boldsymbol{c}_1 = [c_1(0) \dots c_1(L-1)]^T$ is known for the desired user. The multipath channel parameters of the desired user and multipath ray DOAs, and all parameters for the other users are assumed unknown.

Assuming the received signal is sampled at the chip rate, the output from each array element is placed into a vector with length $[(L + L_s - 1)N]$, where N is the number of bits and L_s is the multipath delay spread in chips. The sampled received signal for user 1 at antenna element m is as follows:

$$\boldsymbol{r}^{(m)}(n) = b_1(n)\boldsymbol{h}_1^{(m)} + \sum_{k=2}^{K} b_k(n)\boldsymbol{h}_k^{(m)} + \sigma_\eta \boldsymbol{\eta}^{(m)}(n) \tag{2.95}$$

where $\boldsymbol{h}_k^{(m)}$ and $\boldsymbol{\eta}^{(m)}(n)$ are the sampled signature waveform vector for user k and the sampled AWGN vector, respectively, at antenna element m with length $[(L + L_s - 1)N]$.

For M antennas, we expand (2.95) from vector form to matrix form as follows:

$$\boldsymbol{R}(n) = b_1(n)\boldsymbol{H}_1 + \sum_{k=2}^{K} b_k(n)\boldsymbol{H}_k + \sigma_\eta \boldsymbol{\Gamma}(n) \tag{2.96}$$

where $\boldsymbol{R}(n) = [(\boldsymbol{r}^{(1)}(n))^T \dots (\boldsymbol{r}^{(M)}(n))^T]^T$ are the sampled received signals at the antenna array (sampled array observations), $\boldsymbol{H}_k = [(\boldsymbol{h}_k^{(1)})^T \dots (\boldsymbol{h}_k^{(M)})^T]^T$ is the space-time signature matrix of the kth user, $\boldsymbol{\Gamma}(n) = [(\boldsymbol{\eta}^{(1)}(n))^T \dots (\boldsymbol{\eta}^{(M)}(n))^T]^T$ is the sampled noise matrix, and n is the sample index.

For convenience, a complete DS/CDMA system employing an antenna array at the receiver is illustrated in Figure 2.20. At the receiver side, the sampled received signal from each antenna element is multiplied by a matched filter at each antenna branch. The output signal after the matched filter bank, $\boldsymbol{x}(n) = [x^{(1)}(n) \ldots x^{(M)}(n)]^T$ with size $M \times 1$, is then multiplied by the complex-valued beamformer weight vector, $\boldsymbol{w} = [w_1 \ldots w_M]^T$ with size $M \times 1$ to produce the output statistics of the antenna array as follows:

$$y(n) = \boldsymbol{w}^H \boldsymbol{x}(n) \tag{2.97}$$

where $(.)^H$ stands for the Hermitian. The beamformer output $y(n)$ can be expressed as:

$$y(n) = \boldsymbol{w}^H \ (\overline{\boldsymbol{x}}(n) + \sigma_\eta \boldsymbol{u}(n)) = \overline{y}(n) + \overline{\eta}(n) \tag{2.98}$$

where, $\overline{\boldsymbol{x}}(n)$ is the noiseless matched filter output vector, $\overline{\eta}(n) = \boldsymbol{w}^H \boldsymbol{u}(n)$ is Gaussian with zero mean and variance $E[|\overline{\eta}(n)|^2] = 2{\sigma_\eta}^2 \boldsymbol{w}^H \boldsymbol{w}$, and $\boldsymbol{u}(n)$ is the noise vector after being multiplied by the matched filter.

The bit decision is made according to

$$\hat{b}(n) = \text{sgn}(\text{Re}\{y(n)\}) = \text{sgn}(y_R(n)) \tag{2.99}$$

where sgn(.) denotes the sign function, and $y_R(n) = \text{Re}\{y(n)\}$ is the real part of the beamformer output $y(n)$.

2.6 Simulation Software

Three sets of simulation software are provided with the book:

1. Multiuser detection based on DS/CDMA
2. Robust adaptive beamforming algorithms
3. Minimum BER beamforming based on MIMO/OFDM system.

The multiuser algorithms are based on the DS/CDMA simulation software system and the system model developed by Anair [1]. The algorithms presented in Chapters 3–6 are based on this system model. However, the reader can easily extend these algorithms to other models, such as MIMO/OFDM and even evolved versions of OFDM in 5G New Radio. The

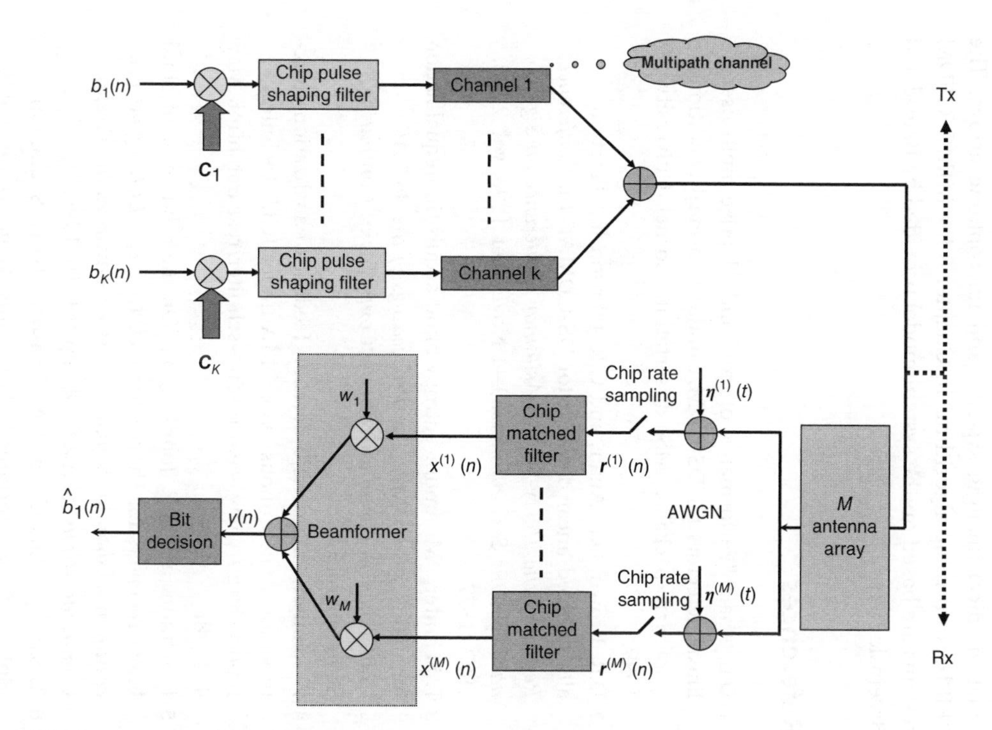

Figure 2.20 DS/CDMA system model with antenna array.

beamforming algorithms in Chapter 7 are based on a simplified model, as explained in that chapter [64]. Again, these algorithms can be easily extended to any wireless communication system or any other type of beamforming application such as radar, sonar, medical imaging systems and microphone arrays. The BER beamforming algorithms in Chapter 8 for MIMO/OFDM systems are based on the system model provided by Bansal and Brzeinski [54].

References

1 D.L. Anair, "Performance of fixed and adaptive multi-user linear detectors for DS-CDMA under non-ideal conditions: a software package," MSc dissertation, Cornell University, 1999.

2 D.R. Brown, D.L. Anair, and C.R. Johnson Jr., "Fractionally sampled linear detector for DS-CDMA," In: *Conference Record of the Thirty-Second Asilomar Conference on Signals, Systems & Computers,* 1–4 November 1998, vol. 2, pp. 1873–1877.

3 P. Schniter, "Minimum-entropy blind acquisition/equalization for uplink DS-CDMA." In: *Proceedings of the 36th Allerton Conference on Communications, Control, and Computing,* September 1998.

4 P. Schniter, "Linear and decision feedback equalization structures for asynchronous DS-CDMA under ICI," Technical report, http://www.ece.osu.edu/~schniter/research.html, May 4, 1998.

5 P. Schniter and C.R. Johnson, Jr., "On the robustness of blind linear receivers for short-code CDMA." *2nd IEEE Signal Processing workshop on Signal Processing Advances in Wireless Communications,* SPAWC '99, pp. 13–16, 1999.

6 T. Samir, S. Elnoubi, and A. Elnashar, "Block-Shanno minimum BER beamforming," *IEEE Trans. Vehic. Techn.,* vol. 57, no. 5, pp. 2981–2990, 2008.

7 J. Korhonen, *Introduction to 3G Mobile Communications,* 2nd edn, Artech House, 2003.

8 U. Madhow, "Blind adaptive interference suppression for direct sequence CDMA," *Proc. IEEE*, vol. 86, pp. 2049–2069, 1998.

9 J.S. Lee and L.E. Miller, *CDMA Systems Engineering Handbook*, Artech House, 1998.

10 M. Elnashar and A. El-Saidny, "Extending the battery life of smartphones and tablets: a practical approach to optimizing the LTE network," *IEEE Vehic. Techn. Mag.*, vol. 2, pp. 38–49, 2014.

11 J.P. Castro, *The UMTS Network Radio Access Technology*, John Wiley, 2001.

12 W. Konhauser, W. Mohr, and R. Prasad, *Third Generation Mobile Communication Systems*, Artech House, 2000.

13 Z. Tian, *Blind Multi-user Detection with Adaptive Space-time Processing for DS-CDMA Wireless Communications*, PhD thesis, George Mason University, 2000.

14 X. Zhengyuan, *Blind Channel Estimation and Multiuser detection for CDMA Communications*, PhD thesis, Steven Institute of Technology, 1999.

15 S. Chen, A.K. Samingan, and L. Hanzo, "Adaptive minimum error rate training for neural networks with application to multiuser detection in CDMA communication system" *IEEE Trans. Signal Process.*, vol. 49, pp. 1240–1247, 2001.

16 S. Chen, A.K. Samingan, and L. Hanzo, "Support vector machine multiuser receiver for DS-CDMA signals in multipath channels" *IEEE Trans. Neural Netw.*, vol. 12, pp. 604–611, 2001.

17 S. Chen, A.K. Samingan, B. Mulgrew, and L. Hanzo, "Adaptive minimum-BER linear multiuser detection for DS-CDMA signals in multipath channels," *IEEE Trans. Signal Process*, vol. 49, pp. 1240–1247, 2001.

18 S. Chen, B. Mulgrew, and L. Hanzo, "Adaptive least error rate algorithm for neural network classifiers" In: *Proceedings of the 2001 IEEE Signal Processing Society Workshop Neural Networks for Signal Processing* XI, 10–12 September 2001, pp. 223–232.

19 P. Arasaratnam, S. Zhu, and A.G. Constantinides, "Fast convergent multiuser constant modulus algorithm for use in multiuser DS-CDMA environment," *IEEE International Conference on Acoustics, Speech, and Signal Processing*, (ICASSP '02), 13–17 May 2002, vol. 3, pp. 2761–2764.

20 A. Bahai and B. Saltzberg, *Multicarrier Digital Communications: Theory and Applications of OFDM*. Kluwer Academic, 1999.

21 I. Koutsopoulos and L. Tassiulas, "Adaptive resource allocation in SDMA-based wireless broadband networks with OFDM signaling." In: *Proceedings of IEEE INFOCOM* 2002, vol. 3, pp. 1376–1385.

22 J. Campello de Souza, *Discrete Bit Loading for Multicarrier Modulation Systems*, PhD thesis, Stanford University. 1999.

23 P.S. Chow, J.M. Cioffi and J.A.C. Bingham, "A practical discrete multitone transceiver loading algorithm for data transmission over spectrally shaped channels," *IEEE Trans. Commun.*, vol. 43, no. 2, pp. 773–775, 1995.

24 E. Parzen, "On estimation of a probability density function and mode", *Ann. Math. Stat.*, vol. 33, pp. 1066–1076, 1962.

25 B.W. Silverman, *Density Estimation*. Chapman and Hall, 1996.

26 A.W. Bowman and A. Azzalini, *Applied Smoothing Techniques for Data Analysis*, Oxford University Press, 1997.

27 S. Chen, L. Hanzo, N.N. Ahmad, and A. Wolfgang, "Adaptive minimum bit error rate beamforming assisted QPSK receiver." In: *2004 IEEE International Conference on Communications*, 20–24 June 2004 vol. 6, pp. 3389–3393.

28 S. Chen, X.C. Yang, and L. Hanzo, "Space-time equalization assisted minimum bit-error ratio multiuser detection for SDMA systems." In: *61st IEEE Vehicular Technology Conference*, 30 May–1 June 2005, vol. 2, pp. 1220–1224.

29 R.J. Kozick and B.M. Sadler, "Maximum-likelihood array processing in non-Gaussian noise with Gaussian mixtures," *IEEE Trans. Signal Process*, vol. 48, no. 12, pp. 3520–3535, 2000.

30 R.J. Kozick and B.M. Sadler, "Robust subspace estimation in nonGaussian noise," in *Proceedings of IEEE Conference on Acoustics, Speech and Signal Processing* (ICASSP), Istanbul, Turkey, 2000, pp. 3818–3821.

31 T.C. Chuah , B.S. Sharif, and O.R. Hinton "Robust adaptive spread-spectrum receiver with neural-net preprocessing in non-Gaussian noise", *IEEE Trans. Neural Netw.*, vol. 12, no. 3, pp. 546–558, 2001.

32 K.J. Wang and Y. Yao, "New nonlinear algorithms for narrow-band interference suppression in CDMA spread-spectrum systems", *IEEE J. Select. Areas Commun.*, vol. 17, pp. 2148–2153, 1999.

33 M. Nakagami, *The m-Distribution—A General Formula of Intensity Distribution of Rapid Fading. Statistical Methods in Radio Wave Propagation.* Pergamon, 1960.

34 P. Lombardo, G. Fedele, and M.M. Rao, "MRC performance for binary signals in Nakagami fading with general branch correlation model," *IEEE Trans. Commun.*, vol. 47, pp. 44–52, 1999.

35 P.Y. Kam, "Bit error probability of MDPSK over the nonselective Rayleigh fading channel with diversity reception," *IEEE Trans. Commun.*, vol. 39, pp. 220–224, 1991.

36 L.-L. Yang and L. Hanzo "Performance of generalized multicarrier DS-CDMA over Nakagami-m fading channels", *IEEE Trans. Commun.*, vol. 50, pp. 956–956, 2002.

37 J.-M. Chaufray , P. Loubaton and P. Chevalier, "Consistent estimation of Rayleigh fading channel second-order statistics in the context of the wideband CDMA mode of the UMTS", *IEEE Trans. Signal Process.*, vol. 49, pp. 3055–3064, 2001.

38 J.R. Foerster and L.B. Milstein "Coded modulation for a coherent DS-CDMA system employing an MMSE receiver in a fading channel", *IEEE Trans. Commun.*, vol. 48, pp. 1909–1918, 2000.

39 Trigui, I., Laourine, A., Affes, S., and Stephenne, A. "Performance analysis of mobile radio systems over composite fading/shadowing channels with co-located interference", *IEEE Trans. Wireless. Commun.*, vol. 8, no. 7, pp. 3448–3453, 2009.

40 J.C. Lin , W.C. Kao , Y.T. Su, and T.H. Lee, "Outage and coverage considerations for microcellular mobile systems in a shadowed-Rician/shadowed-Nakagami environment", *IEEE Trans. Vehic. Techn.*, vol. 48, no. 1, pp. 66–75, 1999.

41 IXIA White paper, "SC-FDMA Single carrier FDMA in LTE," 2009.

42 H.G. Myung, J. Lim, and D.J. Goodman, "Single carrier FDMA for uplink wireless transmission". *IEEE Vehic. Techn. Mag.*, vol. 1, no. 3, pp. 30–38, 2006.

43 A. Osseiran, F. Boccardi, V.A Braun, et al., "Scenarios for the 5G mobile and wireless communications: the vision of the METIS project", *IEEE Commun. Mag.*, vol. 52, no. 5, pp. 26–35, 2014.

44 A. Osseiran, V. Braun, T. Hidekazu, et al., "The foundation of the mobile and wireless communications system for 2020 and beyond challenges, enablers and technology solutions." In: *VTC Spring 2013*, June 2–5, 2013.

45 H. Lin and P. Siohan, "An advanced multi-carrier modulation for future radio systems." In: *Acoustics, Speech and Signal Processing* (ICASSP), 2014.

46 W.Y. Zou and Y. Wu, "COFDM: An overview," *IEEE Trans. Broadcasting*, vol. 41, no. 1, pp. 1–8, 1995.

47 B. Le Floch, M. Alard, and C. Berrou, "Coded orthogonal frequency division multiplex", *Proc. IEEE*, vol. 83, pp. 982–996, 1995.

48 G. Cherubini, E. Eleftheriou, and S. Olcer, "Filtered multitone modulation for very high-speed subscriber lines," *IEEE J. Select. Areas Commun.*, vol. 20, no. 5, pp. 1016–1028, 2002.

49 A.M. Tonello, "A novel multi-carrier scheme: cyclic block filtered multitone modulation." In: Proceedings o of ICC 2013, Budapest, June 2013.

50 M. Fuhrwerk, J. Peissig, M. Schellmann, "Channel adaptive pulse shaping for OQAM-OFDM systems." In: Proceedings of the 22nd European Signal Processing Conference (EUSIPCO), 2014.

51 T. Strohmer and S. Beaver, "Optimal OFDM design for time-frequency dispersive channels," *IEEE Trans. Commun.*, vol. 51, no. 7, pp. 1111–1122, 2003.

52 Y.S. Cho, J. Kim, W.Y. Yang, and C.G. Kang, *MIMO-OFDM Wireless Communications with MATLAB.* Wiley, 2010.

53 A. Elnashar and M. Elsaidney, *Design, Deployment, and Performance of 4G-LTE Networks: A Practical Approach.* Wiley, 2014.

54 P. Bansal and A. Brzeinski, "Adaptive loading in MIMO/OFDM systems," Technical paper, 13 December 2001.

55 A. Elnashar, and M.A. El-Saidny, "Extending the battery life of smartphones and tablets: a practical approach to optimizing the LTE network," *IEEE Vehic. Technol. Mag.*, vol. 9, no. 2, pp. 38–49, 2014.
56 Y.S. Song, H.M. Kwon, and B.J. Min, "Computationally efficient smart antennas for CDMA wireless communications" *IEEE Trans. Vehic. Techn.*, vol. 50, no. 6, 2001.
57 A. Goldsmith, *Wireless Communications*, Cambridge University Press, 2005.
58 J.M.G. Linmartz, Wireless Communication, Baltzer Science Publishers.
59 M.Y. Alias, A.K. Samingan, S. Chen, and L. Hanzo, "Multiple antenna aided OFDM employing minimum bit error rate multiuser detection", *Electronics Lett.*, vol. 39, no. 24, pp. 1769–1770, 2003.
60 G. Huang, J. He, and Q. Zhang, "Research on adaptive subcarrier-bitpower allocation for OFDMA." In: *Proceedings of International Conference on Wireless Communications, Networking and Mobile Computing*, Sept. 2009, pp. 1–4
61 J. Litva and T.K. Lo, *Digital Beamforming in Wireless Communications*. Artech House, 1996.
62 T.E. Biedka, A General Framework for the Analysis and Development of Blind Adaptive Algorithms, PhD thesis, Virginia Tech, 2001.
63 W.L. Stutzman and G.A. Thiele, *Antenna Theory and Design*. John Wiley, 1981.
64 A. Elnashar, and M. Elsaidney, Practical Guide to LTE-A, VoLTE and IoT: Paving the way towards 5G, May 2018, Wiley.

3
Adaptive Detection Algorithms

3.1 Introduction

In this chapter, we will survey adaptive detection algorithms based on the DS/CDMA model. However, the adaptive techniques that are summarized in this survey can be easily extended to MIMO/OFDM and smart antenna arrays. The DS/CDMA model is the most complicated system model because of the need for multiuser interference cancellation and because the channel is frequency selective, as explained in Chapter 2. Despite the various advantages of the DS/CDMA system, it is interference limited due to multiuser interference and it cannot be easily extended to ultra-broadband systems, as explained in Chapter 2. A conventional DS/CDMA receiver treats each user separately, as a signal, with other users considered as noise or multiple access interference (MAI). A major drawback of such conventional DS/CDMA systems is the near–far problem: degradation in performance due to the sensitivity to the power of desired user against interference power. Reliable demodulation is impossible unless tight power control algorithms are used. The near–far problem can significantly reduce the capacity. Multiuser detection (MUD) algorithms have been developed to improve capacity dramatically over that achievable with conventional single-user detection techniques. MUD considers signals from all users, which lead us to joint detection. MUD reduces interference and hence leads to a capacity increase and alleviation of the near–far problem. Power control algorithms can be used but are not necessary.

Simplified Robust Adaptive Detection and Beamforming for Wireless Communications, First Edition. Ayman Elnashar.

Companion website: www.wiley.com/go/elnashar49

Among the first advances in the area of MUD was the development of the optimum MUD, in the sense of minimizing error probability or maximum likelihood [1, 2]. Although the optimal receiver has been shown to be the maximum likelihood sequence estimator (MLSE) [1], exponential complexity prohibits its practical realization. The key algorithmic structure of optimum MUD is that of a bank of matched filters, followed by a dynamic programming algorithm. More recently, there has been considerable interest in linear MUD with tolerable performance loss.

3.2 The Conventional Detector

The conventional detector for the received signal given in (2.3) is a bank of K correlators, as shown in Figure 3.1. The user signal is recovered by correlating the received signal with the user code. The correlation detector can also be implemented through a bank of matched filters (MFs) [3]. The outputs of the MF bank are sampled at the bit rate, which yields a "soft" estimate of the user data. The *sgn*(.) operator is applied to the soft estimate to get the "hard" data decision. The conventional detector neglects the presence of MAI and ISI and it is the simplest linear MUD.

It appears from Figure 3.1 that the conventional detector follows a single-user detection strategy; each user is detected

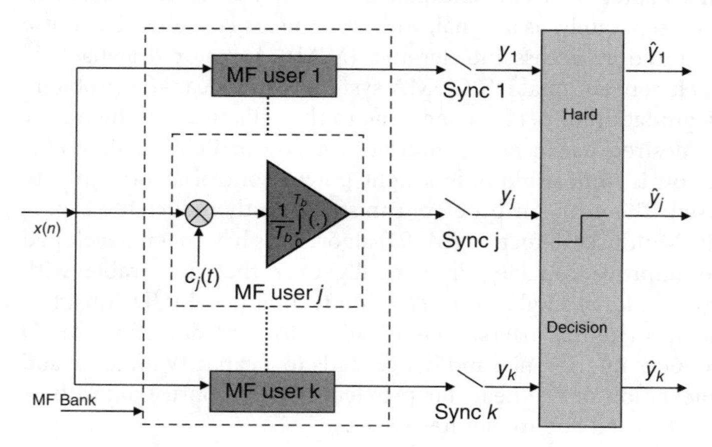

Figure 3.1 The conventional single user DS/CDMA detector: a bank of MFs.

in a standalone manner without considering the MAI from other users.[1] More specifically, single-user detector does not consider any other users in the system or ISI caused by channel dynamics and therefore it is not robust to asynchronism, frequency-selective fading channels, the near–far effect, or substantial cross-correlation between signature waveforms. In the IS-95 system, which uses the single-user detection strategy, alternative techniques are employed to mitigate the near–far effect and MAI: perfect power control, sectored/adaptive antennas, and source coding, among others.

3.3 Multiuser Detection

Multiuser detection techniques can substantially increase the capacity of CDMA systems, and a significant body of research has addressed such schemes. Verdu's seminal work proposed and analyzed the optimal multiuser detector, or the maximum likelihood detector [1]. Unfortunately, this optimal detector is too complex for practical implementation. In an asynchronous DS/CDMA, a search over 2^{NK} possible solutions is conducted. Alternatively, a maximum likelihood detector can be implemented using the Viterbi algorithm [4, 5]. However, the computational complexity is still an exponential function of the number of users. Therefore, over the last two decades, most research efforts have focused on finding suboptimal MUD receivers with low complexity. It worth noting that it is not recommended to develop further robust multiuser techniques for DS/CDMA for mobile broadband systems. Instead, OFDM and its variants, versions of which are being adopted now in 4G and 5G systems, should be developed further for wireless broadband systems (mobile and fixed). Therefore, the algorithms presented in this and the following three chapters can be extended for OFDM and its alternatives rather than developing more complicated receiver designs for DS/CDMA systems, which will not be used in future wireless systems except for some specific applications.

Figure 3.2 shows most of the suboptimal receivers and their different categories. Most of the proposed receivers can

1 The MAI is considered as background noise.

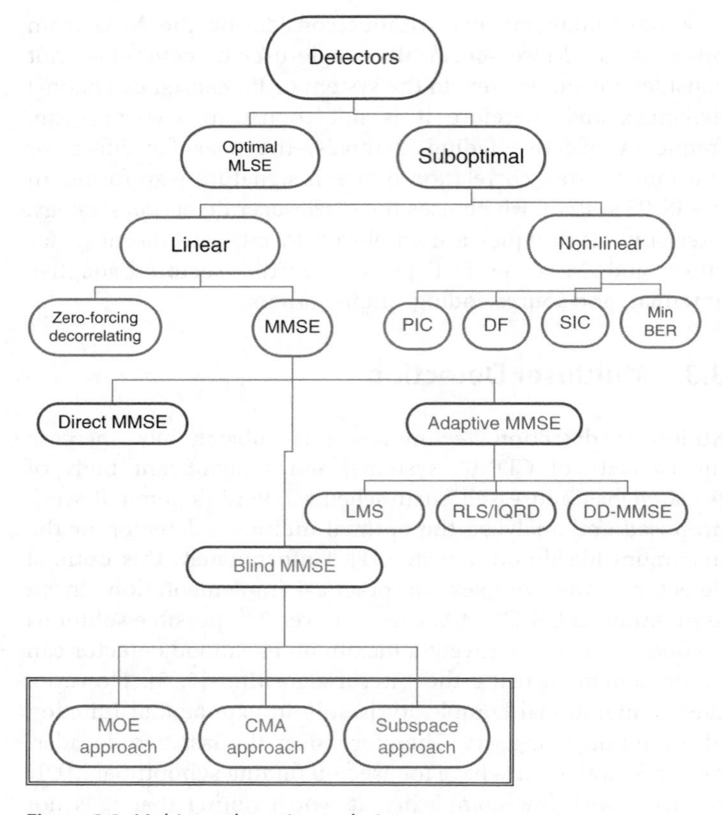

Figure 3.2 Multiuser detection techniques.

be classified into one of two categories: linear or nonlinear detectors. The most important nonlinear receivers are:

- subtractive interference cancellation detectors
- neural network-based detectors [6]
- decision-feedback detectors [7–9]
- minimum BER receivers [10–14].

The subtractive interference cancellation detectors can be divided into successive interference cancellation [15] and parallel interference cancellation [16] subcategories.

The linear detectors apply a linear mapping to the soft output of the conventional detector or directly to the received signal samples, to reduce the MAI seen by each user. The two most popular linear receivers are the minimum mean-squared error detector (MMSE) and the decorrelating or zero-forcing detector [3]. We focus our attention in this book on linear detection, where the linear detector output is a linear combination of the received chip sampled signals. Among different detection techniques, linear receivers are of great significance due to their ease of practical implementation.

3.3.1 Decorrelating Detector

The decorrelating detector (sometimes termed the zero-forcing (ZF) detector) was initially proposed by Schneider [17] and it has been extensively analyzed [18, 19]. The decorrelating detector projects the received signal orthogonally to the interference subspace. Thus, the near–far resistance of the decorrelating detector equals its asymptotic efficiency [18, 19]. The decorrelating detector removes all the MAI interference and therefore the power of each user does not need to be estimated or controlled. Unfortunately, the decorrelating receiver suffers from noise enhancement caused by noise constituent $\sigma_v^2 \boldsymbol{f}^H \boldsymbol{f}$ in the output SINR of the linear detector, as given in Chapter 2. Therefore, at low SNR, the detector performance will be dramatically affected. According to the general model proposed in Chapter 2, we can develop a close- form solution for the decorrelating detector as follows [20]:

$$\boldsymbol{f}_{ZF}^{(j)} = \left(\boldsymbol{H}\boldsymbol{H}^H\right)^{-1} \boldsymbol{H}\boldsymbol{e}_{j,\delta} \tag{3.1}$$

where $e_{k,\delta}$ is an all-zero vector with a one at the position corresponding to delay δ of the user of interest. It is important to note that the decorrelating detector can exist if and only if the matrix $(\boldsymbol{H}\boldsymbol{H}^H)$ is invertible. The ZF detector is capable of eliminating both MAI and ISI, but has some tradeoffs at low SNR, as stated above.

3.3.2 Minimum Mean-squared Error Detector

The MMSE [21] detector minimizes the mean-squared error (MSE) between the decision statistics and the desired symbol. The MMSE receiver is closely related to the decorrelating receiver but it maximizes the output SINR ratio. The performance of the MMSE receiver is close to the decorrelating receiver at high SNR. This means that the asymptotic efficiency and near–far resistance of the MMSE and ZF receivers are the same [22]. The MMSE receiver optimizes the tradeoff between noise enhancement and interference cancellation [21, 23]. The MMSE detector can be derived as follows:

$$\boldsymbol{f}^{(j)}_{MMSE} = \min_{f} \left\{ E\left(|y_j - \hat{y}_j|^2 \right) \right\} \tag{3.2}$$

$$\boldsymbol{f}^{(j)}_{MMSE} = \left[\boldsymbol{H}\boldsymbol{H}^H + \boldsymbol{I}_{N_f}\sigma_w^2 \right]^{-1} \boldsymbol{H}\boldsymbol{e}_{j,\delta} \tag{3.3}$$

For a synchronous system, the MMSE detector is given by:

$$\boldsymbol{f}^{(j)}_{MMSE} = \boldsymbol{R}^{-1}\boldsymbol{h}_j$$

In practical applications, the system matrix $\boldsymbol{R} = \boldsymbol{H}\boldsymbol{H}^T + \boldsymbol{I}_{N_f}\sigma_w^2$ is not known. Therefore, the inverse of received signal autocorrelation matrix $\boldsymbol{R}_{xx}^{-1}$, which can be estimated from the received data, can be used instead of the system matrix. The autocorrelation matrix $\boldsymbol{R}_{xx}$ can be obtained using one of the following methods:

Sample covariance method

$$\boldsymbol{R}_{xx}(n) = \frac{1}{N}\sum_{n=1}^{N} \boldsymbol{x}(n)\boldsymbol{x}^H(n) \tag{3.4}$$

where N is the number of snapshots.

Exponentially decaying data window

$$\boldsymbol{R}_{xx}(n) = \eta\boldsymbol{R}(n-1) + \boldsymbol{x}(n)\boldsymbol{x}^H(n) \tag{3.5}$$

where η is the usual forgetting factor, with $0 \ll \eta \leq 1$.

The second method can be used along with a recursive least squares (RLS) approach in an adaptive manner. The selection of the forgetting factor has a significant impact on the performance of the RLS algorithm. The value of the forgetting factor controls the tradeoff between the stability and the tracking ability. In a system identification setting, both the detector length and a leakage phenomenon affect the selection of the forgetting factor. A variable forgetting factor may be proposed for optimum estimation of the autocorrelation matrix.

A new robust RLS adaptive filtering algorithm that uses *a priori* error-dependent weights has been proposed [24].

Robustness against impulsive noise is achieved by choosing the weights on the basis of the norms of the crosscorrelation vector and the input-signal autocorrelation matrix. The proposed algorithm also uses a variable forgetting factor that leads to fast tracking [24]. It is robust with respect to impulsive noise as well as long bursts of impulsive noise, in the sense that it converges back to the steady state much faster than during the initial convergence. The proposed algorithm also tracks sudden system disturbances. Simulation results show that the proposed algorithm achieves improved robustness and better tracking than conventional RLS and recursive least-M estimate algorithms [25]. This is an interesting topic to be considered along with the algorithms presented in this book.

3.3.3 Adaptive Detection

Adaptive interference suppression is analogous to adaptive equalization of time-invariant channels by virtue of the analogy between MAI and ISI. Applications of these methods to DS-CDMA were proposed by a number of different authors at approximately the same time [26–29]. The adaptive linear receiver is based on the MMSE criterion [30, 31]. These receivers require only a training sequence and synchronization to the required user, and can be implemented recursively using the traditional recursive RLS [32] or least mean-square (LMS) algorithms [33]. The adaptive receiver can be switched to decision-directed mode after the training phase to efficiently handle the bandwidth resources. However, decision-directed adaptation is still vulnerable to sudden channel variations. As a consequence, transmitters are requested to send flash training sequences [30]. In addition, training is wasteful, as a part of the available bandwidth is used for training before the actual data transmission.

3.3.4 Blind Detection

Considerable research efforts have been carried out on the blind detection problem. In order to recover from failure of decision-directed adaptation without requiring the transmitter to send a fresh training sequence, it has proven necessary to develop blind adaptive techniques that do not require a training sequence. Our primary focus in this book is on blind receivers that require only the signature waveform of the required user and the timing information of that user (the receiver is to be synchronized to the required user, but the multipath channel is unknown). We provide here a brief discussion of blind multiuser receivers that use only second order statistics (SOS) of the received signal. Different SOS receiver designs use subspace methods [34–36], constrained optimization [37–52], or the constant modulus algorithm [53–57]. We will summarize these three methods in the remainder of this chapter. These techniques are mainly derived from beamforming algorithms and therefore what will be presented here can be easily extended to beamforming applications or MIMO/OFDM.

As evident from the number of references, constrained optimization is a most interesting approach and it has been widely explored for robust adaptive detection and beamforming for wireless communications as well as other applications. The majority of the presented and developed algorithms in this book are based on constrained optimization approaches, either with single or multiple constraints. Multiple cost functions will be exploited, including output power, MSE, BER, and constant modulus types. The constrained optimization approach can be combined with any of these cost functions. The reader can consider using the developed algorithms with different cost functions or with any SOS algorithm. The author's papers [56–60] give details of similar techniques.

3.3.4.1 Constrained Optimization

Constrained optimization methods have received considerable attention as a means to derive blind multiuser receivers as well as for robust adaptive beamforming with low complexity. Direct estimation of the linear receiver/beamforming parameters can be realized by minimizing some inverse filtering criterion. Depending on the criterion used, appropriate constraints using hypothesized nominals are needed so that the trivial all-zero solution is excluded. A well-known cost function for constrained optimization is the output energy or the variance. The first minimum output energy (MOE) detector was developed by Honig *et al.* [38]. The receiver's output variance is minimized subject to appropriate constraints, which depend on the multipath structure of the signal of interest. The output energy of the linear detector presented in Chapter 2 is given by:

$$E\left\{|y(n)|^2\right\} = E\left\{|\boldsymbol{f}^H \boldsymbol{x}(n)|^2\right\} = \boldsymbol{f}^H \boldsymbol{R}_{xx} \boldsymbol{f} \tag{3.6}$$

where $\boldsymbol{R}_{xx}(n) = E\{\boldsymbol{x}(n)\boldsymbol{x}^H(n)\}$ is the received signal autocovariance matrix.

3.3.4.1.1 Minimum Output Energy with Single Constraint

MOE detection has been proposed as a blind adaptive technique for multiuser detection in DS-CDMA systems. It assumes that the channel of the desired user is known to the receiver. The detector corresponds to the solution of a constrained optimization problem employing a single linear constraint constructed from knowledge of the desired user's effective signature vector $\boldsymbol{h}^j_\delta$ at known delay δ.

The minimum output energy optimization problem can be expressed as:

$$\min_{\boldsymbol{f}} \boldsymbol{f}^H \boldsymbol{R}_{xx} \boldsymbol{f} \text{ } subject\ to.\ \boldsymbol{f}\boldsymbol{h}^1_\delta = 1$$
$$or\ subject\ to\ \boldsymbol{f}\boldsymbol{C}^j_\delta \cdot \boldsymbol{g}_j = 1 \tag{3.7}$$

The optimal solution is then given by:[2]

$$\boldsymbol{f} = \frac{\boldsymbol{R}_{xx}^{-1}\boldsymbol{h}_1}{\boldsymbol{h}_1^H \boldsymbol{R}_{xx}^{-1}\boldsymbol{h}_1} \tag{3.8}$$

$$\boldsymbol{f} = \frac{\boldsymbol{R}_{xx}^{-1}\boldsymbol{C}_1\boldsymbol{g}_1}{\boldsymbol{g}_1^H \boldsymbol{C}_1^H \boldsymbol{R}_{xx}^{-1}\boldsymbol{C}_1\boldsymbol{g}_1} \tag{3.9}$$

Generally, the MOE detector is a scaled version of the MMSE detector. Since scaling does not affect the output SINR, MOE has the same optimal performance as MMSE when the channel model is accurate. From another viewpoint, since the constraint freezes the contribution of the desired signal to the output, the MOE receiver can only suppress the sum of the noise and interference energies at the output. This is precisely the quantity being minimized by the MMSE receiver, except that the latter also optimizes the scaling of the desired vector contribution so

---

2 For simplicity, the user of interest – User 1 – is used as the timing reference.

as to track the desired user symbol, rather than any arbitrary scalar multiple of $\boldsymbol{s}_1(n)$.

The MOE detector requires that the effective signature vector of the desired user h_1 is known to the receiver. In the AWGN propagation channel, this reduces to knowledge of the spreading codes of the desired user, where $\boldsymbol{h}_1 = \alpha.\boldsymbol{c}_1$. Disregarding the scaling effect, the MOE detector can be designed by minimizing the energy in $y(n)$ subject to the constraint that the inner product of the weight vector with the desired user's code be a constant value, chosen to be one. Therefore:

$$\min_{\boldsymbol{f}} \boldsymbol{f}^H \boldsymbol{R}_{xx} \boldsymbol{f} \text{ subject to } \boldsymbol{f} \boldsymbol{c}_1 = 1 \tag{3.10}$$

Therefore, the optimal MOE solution in the AWGN channel is given by:

$$\boldsymbol{f} = \frac{\boldsymbol{R}_{xx}^{-1} \boldsymbol{c}_1}{\boldsymbol{c}_1^H \boldsymbol{R}_{xx}^{-1} \boldsymbol{c}_1} \tag{3.11}$$

In an AWGN environment with no multipath, this approach provides a blind solution with MMSE performance. Unfortunately, however, the imposed constraint is very sensitive to signal mismatch and inter-chip interference, making it unsuitable for systems with multipath distortion, signature mismatch, and asynchronous transmission. The MOE is similar to the MVDR beamforming algorithms that will be developed in Chapter 7. The user signature or the user code is replaced by the desired DOA steering vector. The mismatch error is the main concern in multiuser detection and beamforming and therefore the focus of this book is to develop robust adaptive detection algorithms against mismatch errors, uncertainty in DOA, and so on.

3.3.4.1.2 Minimum Output Energy with Multiple Constraints

In practical systems, there are always additional multipath components and other types of channel distortions. In addition, the effective signature waveform vector $\boldsymbol{h}_1$ and the channel vector $\boldsymbol{g}_1$ are generally unavailable to the receiver or become inaccurate. Therefore, the single constraint is not applicable any longer. The nominal constraint is very sensitive to signal mismatch, and generally cannot converge to correct detection.

In the temporal domain, MOE detection has been extended to handle the multipath case by using alternative constraint formulations. A multiple-constraint approach has been used to handle the multipath effect. This can be done by forcing the receiver response to all delayed copies of the signal of interest to zero. With these additional constraints, the MOE energy becomes applicable to the multipath environment, but this approach does not exploit the signal energy that is contained in the delayed copies of the signal of interest and hence does not offer optimal performance.

The optimal MOE linear detector can be obtained by minimizing the output energy of the receiver subject to certain constraints. To avoid cancellation of the signal of interest scattered over different multipaths during the minimization of the detector output energy, we can generally impose a set of linear constraints of the form $\boldsymbol{C}_1{}^H\boldsymbol{f} = \boldsymbol{g}$, where $\boldsymbol{g}$ is the parameter vector to be optimized as well. Therefore, the optimal MOE detector can be obtained by solving the following constrained minimization problem:

$$\min_{\boldsymbol{f}} \boldsymbol{f}^H \boldsymbol{R}_{xx} \boldsymbol{f} \ s.t. \ \boldsymbol{C}_1{}^H \boldsymbol{f} = \boldsymbol{g} \tag{3.12}$$

The closed-form optimal solution to the constrained optimization problem (3.12) can be obtained using the Lagrange multiplier methodology. This yields:

$$\boldsymbol{f} = \boldsymbol{R}_{xx}^{-1}\boldsymbol{C}_1(\boldsymbol{C}_1^H\boldsymbol{R}_{xx}^{-1}\boldsymbol{C}_1)^{-1}\boldsymbol{g} \tag{3.13}$$

The optimal receiver output power after the interference has been suppressed can be obtained by substituting from (3.13) into (3.6). This produces:

$$\xi_{\min} = \boldsymbol{g}^H(\boldsymbol{C}_1^H\boldsymbol{R}_{xx}^{-1}\boldsymbol{C}_1)^{-1}\boldsymbol{g} \tag{3.14}$$

3.3.4.1.3 Channel Vector Estimation

Max/Min Approach The optimum constrained vector $\boldsymbol{g}$ (channel vector estimator) can be obtained by maximizing the minimum output energy (3.14) after the interference has been rejected. That is,

$$\max_{\|\boldsymbol{g}\|=1} \boldsymbol{f}^H \boldsymbol{R}_{xx} \boldsymbol{f} = \max_{\|\boldsymbol{g}\|=1} \boldsymbol{g}^H(\boldsymbol{C}_1^H\boldsymbol{R}_{xx}^{-1}\boldsymbol{C}_1)^{-1}\boldsymbol{g} \tag{3.15}$$

Therefore, the optimal max/min constrained vector is the eigenvector of $(C_1^H R_{xx}^{-1} C_1)^{-1}$ corresponding to maximum eigenvalue, or the eigenvector of $C_1^H R_{xx}^{-1} C_1$ corresponding to the minimum eigenvalue. That is:

$$g = \max_{\|g\|=1} g^H (C_1^H R_{xx}^{-1} C_1)^{-1} g \tag{3.16}$$

or

$$g = \min_{\|g\|=1} g^H (C_1^H R_{xx}^{-1} C_1) g \tag{3.17}$$

The performance of this method tends to be close to the MMSE detector at high SNR in the presence of multipath fading [39].

Improved Cost Function The constrained MOE detector with a max/min approach exhibits some performance loss because the optimal constraint vector is a perturbed channel vector. The perturbation depends on the background noise; it cannot be neglected, especially at low SNR. This problem can be mitigated by developing a new constrained method to improve channel estimation.

An improved cost function [61] is based on a relation between the constrained optimization approach and the subspace approach. The new cost function in this case is as follows:

$$\tilde{g} = \min_{\|g\|=1} \tilde{g}^H (C_1^H R_{xx}^{-1} C_1 - \gamma . C_1^H C_1) \tilde{g} \tag{3.18}$$

where γ is the reciprocal of largest eigenvalue of R_{xx}. Therefore, the channel vector $\tilde{g}$ can be estimated as the eigenvector corresponding to the smallest eigenvalue of the cost function (3.18). With this estimate, the detector can be constructed based on the MMSE criterion

$$f_{IMOE} = R_{xx}^{-1} C_1 \tilde{g} \tag{3.19}$$

Modified Cost Function The constraint vector has been shown to be a biased estimate of the channel vector [39]. The bias, in terms of the vector norm, is proportional to the noise power. Therefore, there is some performance loss from the MOE receiver compared with the MMSE receiver. Detailed study of the noise effect has been performed [39]. In order to mitigate the noise effect and obtain an *unbiased* channel estimate, a new cost function removes the noise contribution from the data covariance matrix [62]. That is:

$$\overline{\boldsymbol{R}}_{xx} = \boldsymbol{R}_{xx} - \alpha\sigma^2 \boldsymbol{I}_{N_f} \tag{3.20}$$

where σ^2 is the noise power and the parameter α quantifies the extent of the noise mitigation; it should satisfy $0 \leq \alpha < 1$ such that the modified covariance matrix is positive definite.

Following the max/min approach described above, $\overline{\boldsymbol{g}}$ can be obtained as follows:

$$\overline{\boldsymbol{g}} = \min_{\|\overline{\boldsymbol{g}}\|=1} \overline{\boldsymbol{g}}^H (\boldsymbol{C}_1^H \overline{\boldsymbol{R}}_{xx}^{-1} \boldsymbol{C}_1) \overline{\boldsymbol{g}} \tag{3.21}$$

The noise power can be estimated from the smallest eigenvalue of the estimated data covariance matrix. Once the optimal channel vector is obtained, the MMOE detector is forced to take an MMSE-like form:

$$\boldsymbol{f}_{MMOE} = \boldsymbol{R}_{xx}^{-1} \boldsymbol{C}_1 \overline{\boldsymbol{g}} \tag{3.22}$$

Capon Method The Capon method is based on the estimation of the channel vector using a generalized eigenvalue decomposition problem. A procedure reminiscent to the Capon estimation method has been proposed [42]. This optimizes the MOE power as a function of the effective signature waveform subject to the constraint $\|\boldsymbol{h}\| = 1$. That is,

$$\widehat{\boldsymbol{g}} = \max_{h=C_1\widehat{g}} \frac{\boldsymbol{h}_1^H \boldsymbol{h}_1}{\boldsymbol{h}_1^H \boldsymbol{R}_{xx}^{-1} \boldsymbol{h}_1} \tag{3.23}$$

If $\boldsymbol{h}_1$ is replaced by $\boldsymbol{h}_1 = \boldsymbol{C}_1 \widehat{\boldsymbol{g}}$, we get the following optimization problem:

$$\widehat{\boldsymbol{g}} = \min_{\widehat{g}} \frac{\widehat{\boldsymbol{g}}^H \boldsymbol{C}_1^H \boldsymbol{R}_{xx}^{-1} \boldsymbol{C}_1 \widehat{\boldsymbol{g}}}{\widehat{\boldsymbol{g}}^H \boldsymbol{C}_1^H \boldsymbol{C}_1 \widehat{\boldsymbol{g}}} \tag{3.24}$$

Therefore, Equation (3.24) is equivalent to a generalized eigenvalue problem involving the pensile of $(\boldsymbol{C}_1^H \boldsymbol{R}_{xx}^{-1} \boldsymbol{C}_1, \boldsymbol{C}_1^H \boldsymbol{C}_1)$.

Power Method It has been verified that the channel estimation error $\boldsymbol{g}_{moe} - \frac{g_1}{\|g_1\|}$ is proportional to the noise power σ^2, which is the smallest eigenvalue of $\boldsymbol{R}_{xx}$ [39, 62]. Therefore, if we boost the power of $\boldsymbol{R}_{xx}$ (POR) in the MOE cost function, the minimum eigenvalue of the resulting matrix will decrease exponentially. As a consequence, the channel estimation error is expected to decrease as well [63]. Following this approach, POR channel estimation can be formulated as follows

$$\breve{\boldsymbol{g}} = \min_{\|\boldsymbol{g}\|=1} \breve{\boldsymbol{g}}^H (\boldsymbol{C}_1^H \boldsymbol{R}_{xx}^{-m} \boldsymbol{C}_1) \breve{\boldsymbol{g}} \tag{3.25}$$

where m is a positive integer. Once the channel vector $\breve{\boldsymbol{g}}$ has been obtained, the POR receiver takes an MMSE-like form:

$$\boldsymbol{f}_{POR} = \boldsymbol{R}_{xx}^{-1} \boldsymbol{C}_1 \breve{\boldsymbol{g}} \tag{3.26}$$

Further discussion on this method and the adaptive implementation of a POR receiver is provided in the literature [63].

3.3.4.1.4 SINR Analysis

For any linear receiver $\boldsymbol{f}$, the output SINR can be expressed as follows:

$$SINR = \frac{\boldsymbol{f}^H \boldsymbol{R}_s \boldsymbol{f}}{\boldsymbol{f}^H \boldsymbol{R}_n \boldsymbol{f}} \tag{3.27}$$

where

$$\boldsymbol{R}_{xx} = \boldsymbol{R}_s + \boldsymbol{R}_n \text{ (received data covariance matrix)} \tag{3.28a}$$

$$\boldsymbol{R}_s = \boldsymbol{h}_1 \boldsymbol{h}_1^H \text{ (signal contribution)} \tag{3.28b}$$

$$\boldsymbol{R}_{xx} = \boldsymbol{H}\boldsymbol{H}^H + \sigma^2 \boldsymbol{I} \text{ (MAI + ISI + noise contribution)} \tag{3.29}$$

Therefore,

$$SINR = \frac{\boldsymbol{f}^H \boldsymbol{h}_1 \boldsymbol{h}_1^H \boldsymbol{f}}{\boldsymbol{f}^H \boldsymbol{R}_{xx} \boldsymbol{f} - \boldsymbol{f}^H \boldsymbol{h}_1 \boldsymbol{h}_1^H \boldsymbol{f}} \tag{3.30}$$

The autocovariance data matrix can be estimated from the received data using (3.4) or (3.5). The SINR of MMSE detector can be acquired by plugging the MMSE detector (3.7) into (3.27), so that:

$$SINR_{mse} = \frac{\boldsymbol{h}_1^H \boldsymbol{R}_{xx}^{-1} \boldsymbol{h}_1}{1 - \boldsymbol{h}_1^H \boldsymbol{R}_{xx}^{-1} \boldsymbol{h}_1} \tag{3.31}$$

For the MOE detector given in (3.13), the SINR can be reformulated as follows [39]:

$$SINR_{MOE} = \frac{\|g^H g_1\|^2}{\left\|g^H (C_1^H R_{xx}^{-1} C_1)^{-1} g_1\right\| - \|g^H g_1\|^2} \tag{3.32}$$

For the MOE detectors given in (3.19), (3.22) and (3.26), the SINR can be formulated as follows:

$$SINR_{MOE} = \frac{\left\|g^H (C_1^H R_{xx}^{-1} C_1) g_1\right\|^2}{\left\|g^H (C_1^H R_{xx}^{-1} C_1)^{-1} g_1\right\| - \left\|g^H (C_1^H R_{xx}^{-1} C_1) g_1\right\|^2} \tag{3.33}$$

where g takes the corresponding optimum value of the certain channel estimation method.

Further analysis of the SINR of the MOE detector and its connection to the non-blind MMSE and subspace methods is provided in the literature [39, 61–63].

3.3.4.1.5 *Partition Linear Interference Cancellation Structure*

The partition linear interference cancellation (PLIC) structure has been implemented in adaptive array processing and beamforming algorithms, referred to as the generalized sidelobe canceller (GSC) [64, 65]. Schodorf and Williams [66, 67] adopted this structure for the linear MUD problem and it was further employed in other studies [52, 66, ,68, 69]. The essence of the PLIC structure approach is to convert a *constrained* optimization problem to an *unconstrained* optimization problem. This is accomplished by dividing the detector vector into two orthogonal components via a projection matrix $P_c = C_1(C_1^H C_1)^{-1} C_1^H$ associated with the constraint equation (3.12):

A non-adapting part f_c ($N_f \times 1$) satisfies the constraints:

$$f_c = P_c f_{opt} \tag{3.34}$$

$$f_c = C_1(C_1^H C_1)^{-1} g \tag{3.35}$$

Note that f_c is not data dependent and therefore it represents the non-adapting portion of the solution. The channel vector g is acting as the constrained vector, and must be optimized using any channel estimation technique.

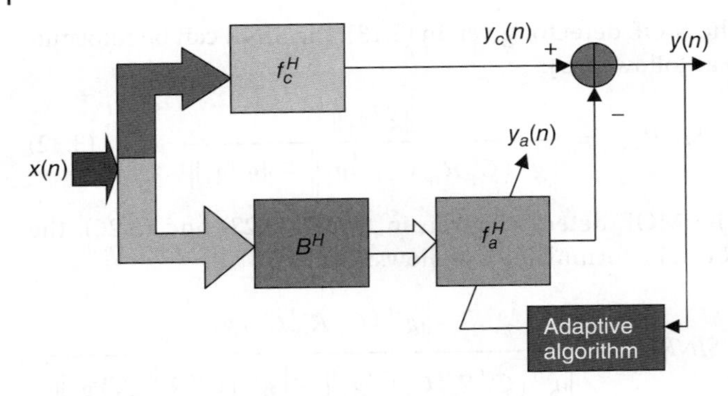

Figure 3.3 Block diagram of PLIC structure.

The adapting part $\boldsymbol{f}_a$ $(N_a \times 1, N_a = N_f - N_g)$ is orthogonal to $\boldsymbol{f}_c$ and can be adapted without constraints to mitigate the interference using any unconstrained update scheme, such as the LMS or RLS algorithms, until a given tolerance is reached.

The solution vector decomposition using the PLIC structure is depicted in Figure 3.3. The constraint equation is satisfied by the non-adapting weight vector $\boldsymbol{f}_c$ in the upper branch, while the lower branch weights are adapted to suppress MAI. The weight vector is then divided into two orthogonal parts and can be expressed as follows,

$$\boldsymbol{f} = \boldsymbol{f}_c - \boldsymbol{B}\boldsymbol{f}_a \tag{3.36}$$

In order to prevent cancellation of the desired source, a blocking matrix $\boldsymbol{B}$ is inserted to ensure orthogonality between upper and lower branches, and this satisfies:

$$\boldsymbol{B}^H \boldsymbol{C}_1 = \boldsymbol{0} \tag{3.37}$$

and

$$\boldsymbol{B}^H \boldsymbol{B} = \boldsymbol{I} \tag{3.38}$$

Equation (3.37) ensures that no components from the received signal matching the constraints are contained in the lower branch.

The optimal weight vector can be found by solving the unconstrained minimization problem:

$$\min_{\boldsymbol{f}_a} (\boldsymbol{f}_c - \boldsymbol{B}\boldsymbol{f}_a)^H \boldsymbol{R}_{xx} (\boldsymbol{f}_c - \boldsymbol{B}\boldsymbol{f}_a) \tag{3.39}$$

The Lagrange method can be invoked to solve this problem and the optimum weight vector is then given by [52]:

$$\boldsymbol{f}_a = (\boldsymbol{B}^H \boldsymbol{R}_{xx} \boldsymbol{B})^{-1} \boldsymbol{B}^H \boldsymbol{R}_{xx} \boldsymbol{f}_c \tag{3.40}$$

A short proof of equivalence between GSC and the linearly constrained minimum variance receiver is provided by Breed and Strauss [70].

3.3.5 Constant Modulus Approach

The constant modulus (CM) criterion has become popular approach in the design of blind linear estimators of sub-Gaussian independent identically distributed processes transmitted through unknown linear channels in the presence of unknown additive interference. Godared [55] and Treichler *et al.* [71] proposed the CM criterion as an alternative to minimizing MSE. The CM approach minimizes the dispersion of the receiver output about the dispersion constant $r = E\{|S|^4\}/E\{|S|^2\}$. The CM cost function can be formulated as follows [56]:

$$J(\boldsymbol{f}) \triangleq E\{(|\boldsymbol{f}^H \boldsymbol{x}|^2 - r)^2\} \tag{3.41}$$

The locations in receiver parameter space of the local minima of $J(\boldsymbol{f})$ are referred to as CM receivers. In BPSK signals, the dispersion constant $r = E\{|S|^4\}/E\{|S|^2\} = 1$. The CM algorithm (CMA) seeks to minimize the cost function defined by the CM criterion (3.41). In most applications, a CM receiver is obtained from the stochastic gradient algorithm. The gradient of $J(\boldsymbol{f})$ is given by [56]:

$$\frac{\partial}{\partial \boldsymbol{f}} J(\boldsymbol{f}) \triangleq 2E\{(|\boldsymbol{f}^H \boldsymbol{x}|^2 - r) y \boldsymbol{x}\} \tag{3.42}$$

The CMA involves a stochastic gradient update of the receiver coefficients by removing the expectation operator in (3.41) and correcting $\boldsymbol{f}$ by a small amount in the opposite direction. This yields:

$$\boldsymbol{f}(k) \triangleq \boldsymbol{f}(k-1) - \mu\{(|y(k)|^2 - r) y(k) \boldsymbol{x}(k)\} \tag{3.43}$$

The existence of multiple CM minima makes it difficult for CM-minimizing schemes to generate estimates of the desired source rather than interference in multiuser environments.

Sufficient conditions for local convergence of gradient descent minimization of the CM cost function are provided by Schniter and Johnson [72]. The sufficient conditions are expressed in terms of the statistical properties of initial estimates, specificall, CM cost, kurtosis, and SINR. Schniter also proposed a near–far resistance initialization procedure for, and applications of, the CMA to uplink DS-CDMA [73]. Their approach was based on pre-whitening of the received signal. Unfortunately, pre-whitening is costly to implement in an adaptive manner [74].

A new statistical criterion to separate linear mixtures of signals in multiuser communications consists of minimizing a cost function that contains a sum of CM criteria and a set of cross-correlation terms to prevent the same signal from being extracted at several outputs simultaneously [75–77]. This method is almost applicable in MIMO channels, where all signals are required to be extracted simultaneously. Unfortunately, this is not the case in most practical cases, especially in the downlink. In addition, error accumulation can occurred due to cross-correlation terms. Some alternative approaches, similar to the constrained optimization approach, have been proposed [56, 78–80]. More specifically, and similar to the MOE receiver, to avoid the cancellation of the signal of interest when scattered in different multipaths during the minimization of the dispersion of the receiver output instead of the output variance, we can generally impose a set of linear constraints of the form $\boldsymbol{C}_1{}^H\boldsymbol{f} = \boldsymbol{g}$, where $\boldsymbol{g}$ is the constraint vector. Consequently, the so-called linear constraint CMA (LCCMA) detector can be obtained by solving the following constrained minimization problem [56]:

$$\min_{f} J_1(\boldsymbol{f}) \triangleq E\{(|\boldsymbol{f}^H\boldsymbol{x}|^2 - r)^2\} \;\; s.t. \;\boldsymbol{C}_1{}^H\boldsymbol{f} = \boldsymbol{g} \tag{3.44}$$

The constrained optimization problems given in (3.44) can be converted to unconstrained minimization problems using the PLIC structure as well. The optimal weight vector $\boldsymbol{f}_a$ can be found by plugging (3.36) into (3.44) and solving using the stochastic gradient method as follows:

$$\boldsymbol{f}_a(k) \triangleq \boldsymbol{f}_a(k-1) - \mu\{e(n)y(k)\boldsymbol{z}(k)\} \tag{3.45}$$

where $e(n) = |y(k)|^2 - r$ is the output error, $z(k) = B^H x(n)$, and μ is the algorithm step-size, which controls the speed of convergence. To increase the convergence speed, one might go for variable $\mu(\mathrm{V}\ \mu - \mathrm{CMA})$. In the study of Migues and Castedo [80], the variable step-size is selected as follows:

$$\mu(n) = \alpha \min\left\{ \left| \frac{y(n)-1}{e(n)y(n)\rho(n)} \right|, \left| \frac{y(n)+1}{e(n)y(n)\rho(n)} \right| \right\} \tag{3.46}$$

where $\alpha < 1$ is chosen to reduce the algorithm misadjustment noise and $\rho(n) = x^H(n)x(n)$ is the current estimate of the autocovariance data matrix.

3.3.6 Subspace Approach

Subspace-based high-resolution methods play an important role in sensor array processing, spectrum analysis, and general parameter analysis [34]. Several recent studies have addressed the use of subspace-based methods for the MUD problem [34–36]. Subspace methods are attractive because of the closed-form identification. They can also be adaptively implemented using subspace tracking algorithms [51, 81–83]. On the other hand, the approach may not be robust against modeling errors, especially when the system matrix is close to being singular. In addition, these methods depend on eigendecomposition, which is computationally expensive. Moreover, we need to estimate channel response g_1 and hence in time-varying channels we need to adaptively adjust the detector parameters. Furthermore, the subspace approach requires the estimation of signal and noise subspaces and rank. Unfortunately, in most practical cases the signal and noise dimensions are not known or inaccurate and hence the subspace approach is not appropriate for dynamic systems. In this book, the subspace approach is used only for channel vector estimation using subspace tracking algorithms with an MOE detector. In this approach, neither the signal and noise subspaces nor the eigendecomposition are estimated.

For the sake of comprehensiveness, the subspace approach is briefly summarized. The autocorrelation matrix of the received signal can be expressed as follows:

$$R_{xx} = E\{x^H(n)x(n)\} \triangleq HH^H + \sigma_w^2 I_{N_f} \tag{3.47}$$

By performing eigendecomposition of the autocorrelation matrix $\boldsymbol{R}_{xx}$, we get

$$\boldsymbol{R}_{xx} = \boldsymbol{U}\boldsymbol{\Lambda}\boldsymbol{U}^H = [\boldsymbol{U}_s\boldsymbol{U}_n]\begin{bmatrix}\boldsymbol{\Lambda}_s & 0\\ 0 & \boldsymbol{\Lambda}_n\end{bmatrix}[\boldsymbol{U}_s\boldsymbol{U}_n]^H \tag{3.48}$$

where

$$\boldsymbol{U} = [\boldsymbol{U}_s\ \boldsymbol{U}_n] \tag{3.49}$$

$$\boldsymbol{\Lambda} = diag(\boldsymbol{\Lambda}_s, \boldsymbol{\Lambda}_n) \tag{3.50}$$

$$\boldsymbol{\Lambda}_s = \begin{pmatrix}\lambda_1 & & \\ & \ddots & \\ & & \lambda_k\end{pmatrix} \tag{3.51}$$

contains the largest k eigenvalues of $\boldsymbol{R}_{xx}$ in descending order,

$$\boldsymbol{U}_s = [\boldsymbol{u}_1 \cdots \boldsymbol{u}_k] \tag{3.52}$$

contains the corresponding eigenvectors,

$$\boldsymbol{\Lambda}_n = \sigma_w^2\boldsymbol{I}_{N_f-k} \tag{3.53}$$

and

$$\boldsymbol{U}_n = [\boldsymbol{u}_{k+1} \cdots \boldsymbol{u}_{N_f}] \tag{3.54}$$

contain the $N_f - k$ orthogonal eigenvectors that correspond to the eigenvalue σ^2. The range space of $\boldsymbol{U}_s$ is called the *signal subspace* and the *noise subspace* is spanned by $\boldsymbol{U}_n$.

The decorrelating detector and MMSE detector can be obtained blindly, respectively, as follows [35]:

$$d_1 = \frac{1}{[\boldsymbol{h}_1^H\boldsymbol{U}_s(\Lambda_s - \sigma_w^2\boldsymbol{I})^{-1}\boldsymbol{U}_s^H\boldsymbol{h}_1]}\boldsymbol{U}_s(\Lambda_s - \sigma_w^2\boldsymbol{I})^{-1}\boldsymbol{U}_s^H\boldsymbol{h}_1 \tag{3.55}$$

$$m_1 = \frac{1}{[h_1^H\boldsymbol{U}_s\Lambda_s^{-1}\boldsymbol{U}_s^H h_1]}\boldsymbol{U}_s\Lambda_s^{-1}\boldsymbol{U}_s^H h_1 \tag{3.56}$$

If the dimension of the signal subspace is unknown, the true desired signal vector may have a significant component in the estimated noise subspace, thereby leading to possible errors in timing estimation due to a "miss" of the correct timing values. On the other hand, overestimation of the signal dimension means that nominals corresponding to incorrect timing have a larger component in the signal subspace, thereby making false alarms more likely [30]. Therefore, the effect of estimation errors on subspace-based receivers deserves detailed investigation.

3.4 Simulation Results

The performance and stability of the detectors can be analyzed by numerical simulations. DS-CDMA waveforms, with BPSK data and spreading modulation by a raised-cosine pulse shaping filter with cut-off factor $\alpha = 0.22$ are considered. The number of users is 5 or 10, with processing gain of 31; that is, the chip duration $T_c = T/31$. Orthogonal Gold codes are employed as the signature sequences and the channel length (maximum delay spread) is assumed to be ten delayed components. Five multipath components, with equal, random, or custom power distributions, are considered and the multipath delays are randomly distributed. The detector length is 31 chips and is assumed to be synchronized to the required user. This means that the user delay $1 \leq \delta \leq N_s$ is known. Performance evaluations of different detectors are considered in the next subsections. All simulation results can be easily obtained using the first software package for MUD, as explained in Chapter 2.

3.4.1 Linear Detectors

The performance of linear detectors is evaluated in terms of MSE and BER. The MSE is simulated against iterations. User powers are assumed to be equal except that the user of interest is assumed to be 10 dB less than the other users to simulate the near–far effect. The SNR is 30 dB. Figure 3.4 illustrates the MSE for MF, MMSE, and decorrelating detectors versus snapshot. The MMSE detector exhibits better performance than the decorrelating (ZF) and the MF detectors. However, the MSE is not a precise measure as it is plotted at a certain SNR. Therefore, for an adequate comparison, the average BER of the user of interest is simulated against the SNR. Figure 3.5 illustrates the BER performance of the three detectors versus SNR. The SNR changes from 0 to 60 dB and the received signal is estimated at every SNR. The figure underscores the superiority of the MMSE detector. The decorrelating detector exhibits serious degradation and it is

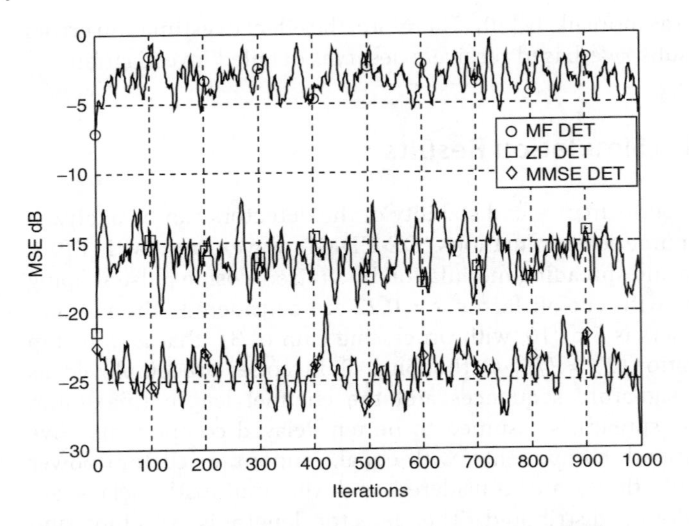

Figure 3.4 MSE for MF, ZF, and MMSE detectors versus snapshot.

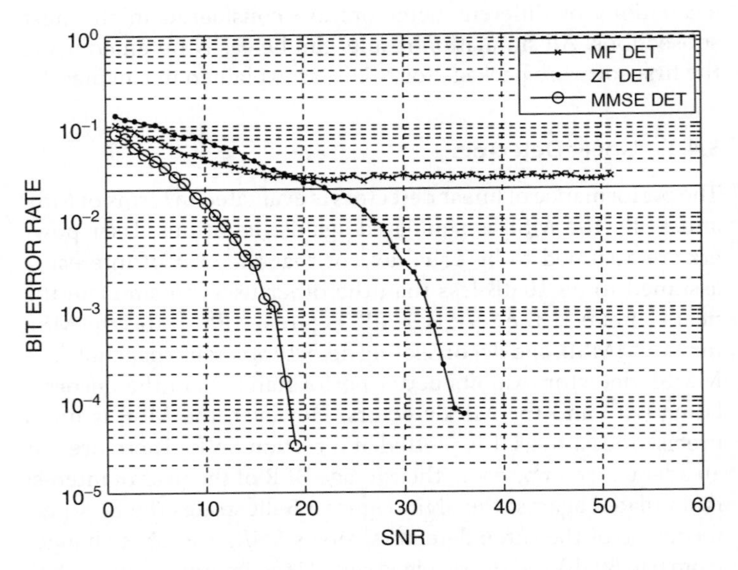

Figure 3.5 Average BER versus SNR for the MF, ZF, and MMSE detectors.

even inferior to the MF detector at low SNRs due to its noise enhancement shortcomings.

3.4.2 MOE Detectors

3.4.2.1 MOE Detector with Single Constraint

Firstly, the performance of the MOE detector with a single constraint is analyzed. Three scenarios are simulated based on the above-mentioned system. The performance is evaluated in terms of the MSE of the required user, which is User 1. The first scenario is the ideal system, where the channel is an AWGN channel and user powers are equal. The second scenario employs a multipath fading channel with equal power. In the third scenario the near–far effect is added via reduction of the required user power to 10 dB less than the other users. Figure 3.6 illustrates the MOE with single constraint for the three scenarios. It is evident that the MOE detector with single constraint cannot handle multipath or near–far effects and it gives reasonable results only in AWGN channels without multipath or near–far effects.

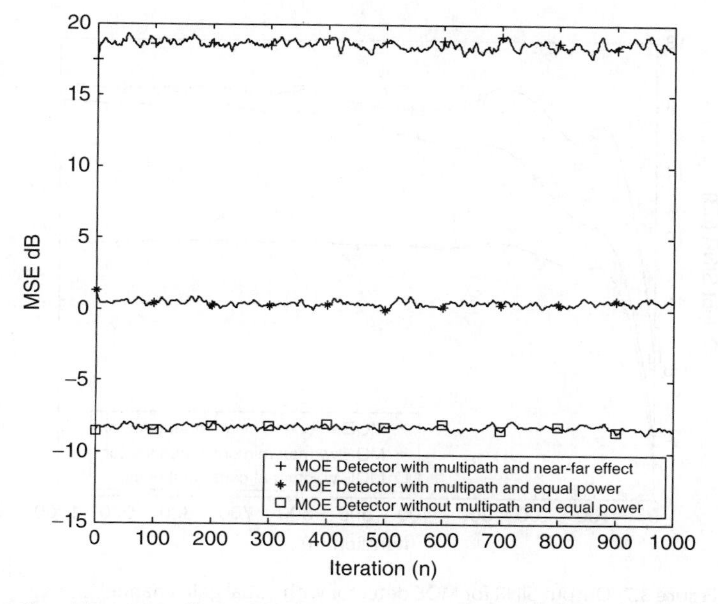

Figure 3.6 MSE of MOE with single constraint for three different scenarios.

3.4.2.2 MOE Detector with Multiple Constraints

The performance of the MOE with multiple constraints is evaluated in terms of the SINR and the BER versus snapshot. Three adaptive MOE detectors are simulated based on the RLS algorithm. The first detector channel is similar to the equal-gain combining system where the channel parameters are equal and normalized such that $\mathbf{g} = [\alpha \ \cdots \ \alpha]^T/\|\mathbf{g}\|$ where $\alpha = 1/N_g$. The second detector channel is adapted using the max/min approach and is initialized with the channel of the first detector. The optimum channel, which is the true channel, is simulated with the third detector. The SNR of the system is set to 30 dB and the detector vectors are initialized with all-zero vectors. The forgetting factor is set to one.

Figures 3.7 and 3.8 show the output SINR and BER for these detectors, demonstrating that the MOE detector with constant constraint vector has a degraded performance compared to the optimized constraint detector. The max/min optimized constraint detector exhibits comparable performance to the optimum constraint detector at high SNR. The

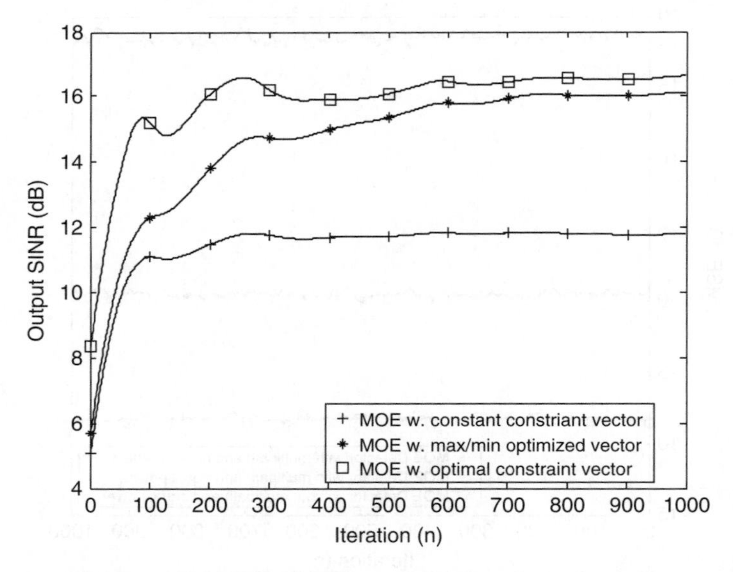

Figure 3.7 Output SINR for MOE detector with equal gain channel, optimized channel, and optimum channel.

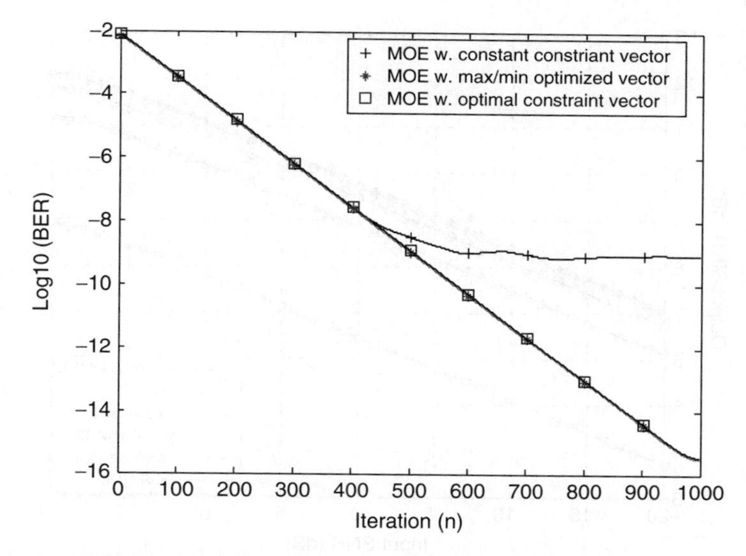

Figure 3.8 BER for MOE detector with equal gain channel, optimized channel, and optimum channel.

minor degradation in the output SINR between the optimized constraint detector and the true channel detector is a result of the max/min channel being a biased estimate of the actual channel.

3.4.3 Channel Estimation Techniques

It is interesting now to compare channel estimation techniques. The MOE detector with five channel estimators summarized in Section 3.2.1 is compared with the non-blind MMSE detector. SINR rather than SNR is adopted here as the performance measure. For the sake of a reasonable comparison, the SINR is computed using the intuitive equation (3.30). The system proposed above is used for this simulation and the actual system matrix is used. The detectors developed here are non-adaptive detectors. Figures 3.9 and 3.10 show the output SINR against the input SNR for perfect power (equal user power) and a −10 dB near–far effect (that is, the required user has power 10 dB less than other users)

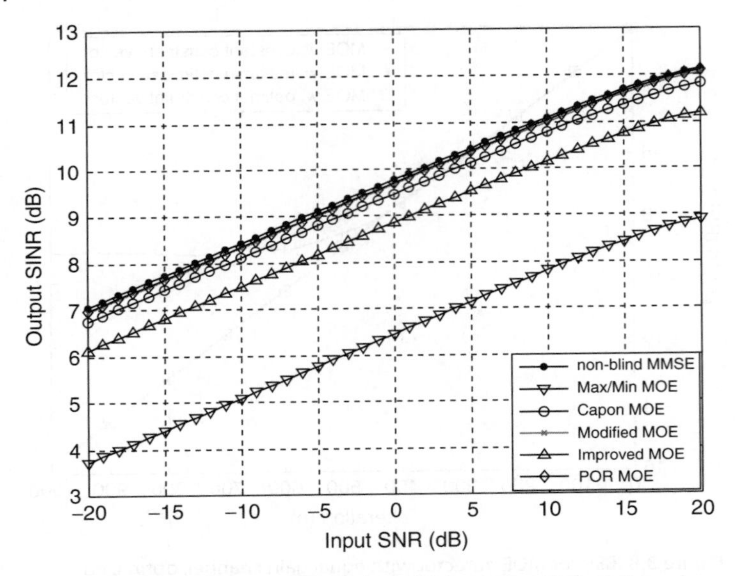

Figure 3.9 Output SINR versus SNR for MOE detector with five channel estimators and the MMSE detector with perfect power control.

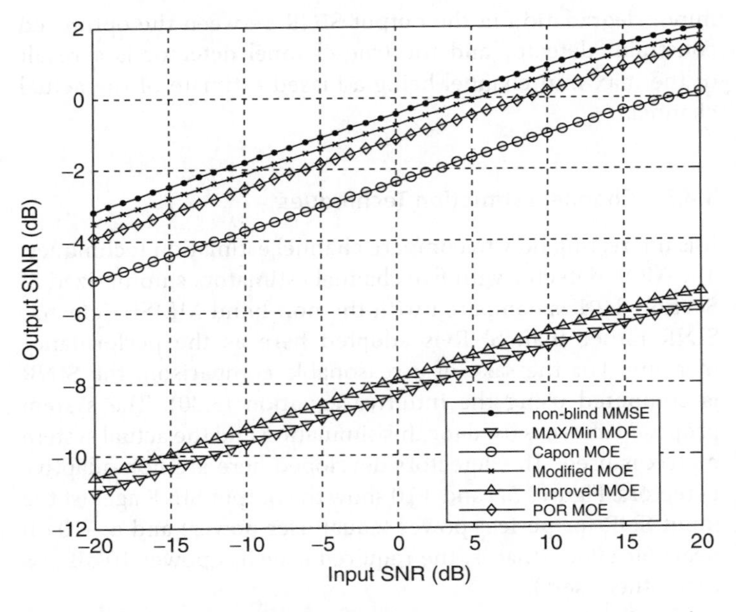

Figure 3.10 Output SINR versus SNR for MOE detector with five channel estimators and the MMSE detector with −10 dB near–far effect.

- The max/min method has the worst performance.
- The improved method is ranked second after the max/min method. Its performance is dramatically affected by near–far effect (Figure 3.9).
- The Capon method is ranked third. However, its performance is slightly affected by the near–far effect.
- The modified method offers a little improvement over the POR method and appears to be better than other channel estimation techniques. This is, not surprisingly, due to the noise cancellation offered by this method. However, in practical implementations the noise power may be unknown or inaccurate. In addition, the noise cancellation by negative diagonal loading of the autocorrelation matrix may affect its positive definiteness. Therefore, the true performance comparison should be done through adaptive implementation and using the estimated autocorrelation matrix from the received data. This performance comparison will be examined in Chapter 4.
- The non-blind MMSE detector has the optimum performance.

3.4.4 LCCMA Detector

Finally, computer simulations are conducted using the LCCMA detector. Again, the above system is invoked and an adaptive step-size is employed using Equation (3.46). Two scenarios were simulated in this experiment. The first is equal power system and the second incorporates the near–far effect, with the required user set to 10 dB less than other users. The evaluation is performed in terms of output SINR and BER. Two LCCMA detectors were simulated. In the first detector, the received signal is pre-whitened using the debiased technique. The second detector is adapted without weighting. Figures 3.11 and 3.12 illustrate the equal power and near–far effect systems, respectively. It is evident from these figures that preweighting is a mandatory technique for LCCMA detector convergence under the near–far effect. Further analysis of the performance and near–far resistance of the LCCMA detector will be provided in Chapter 6.

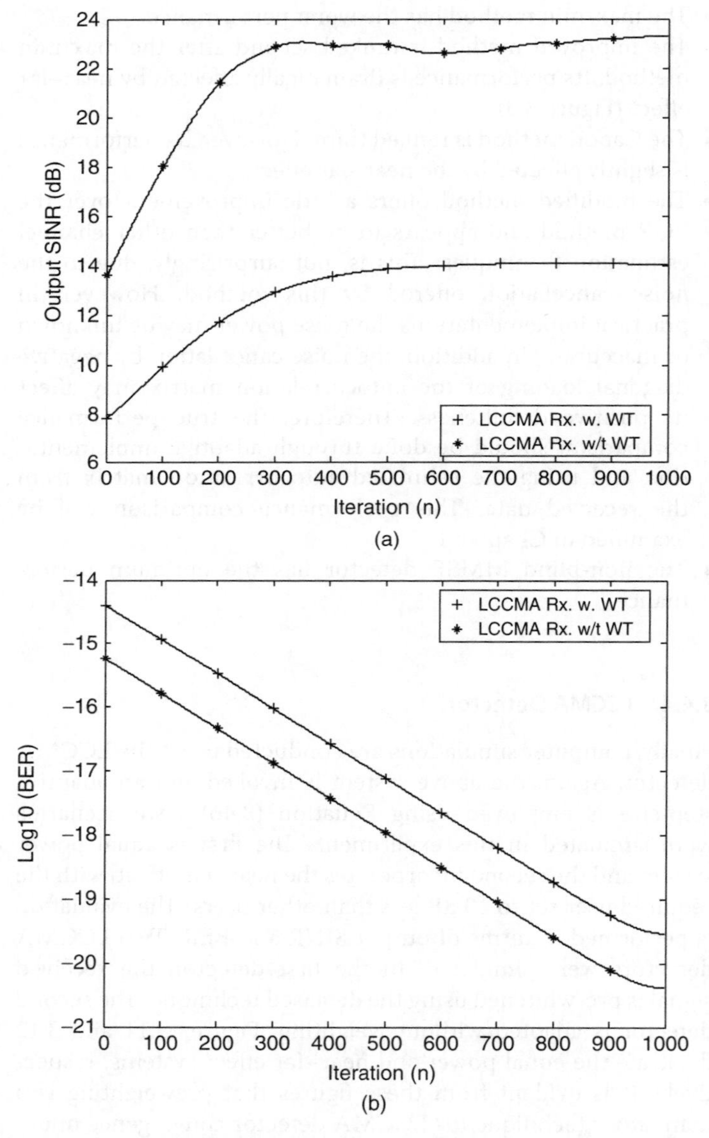

Figure 3.11 Adaptive LCCMA detector with/without whitening under equal power: (a) BER and (b) output SINR.

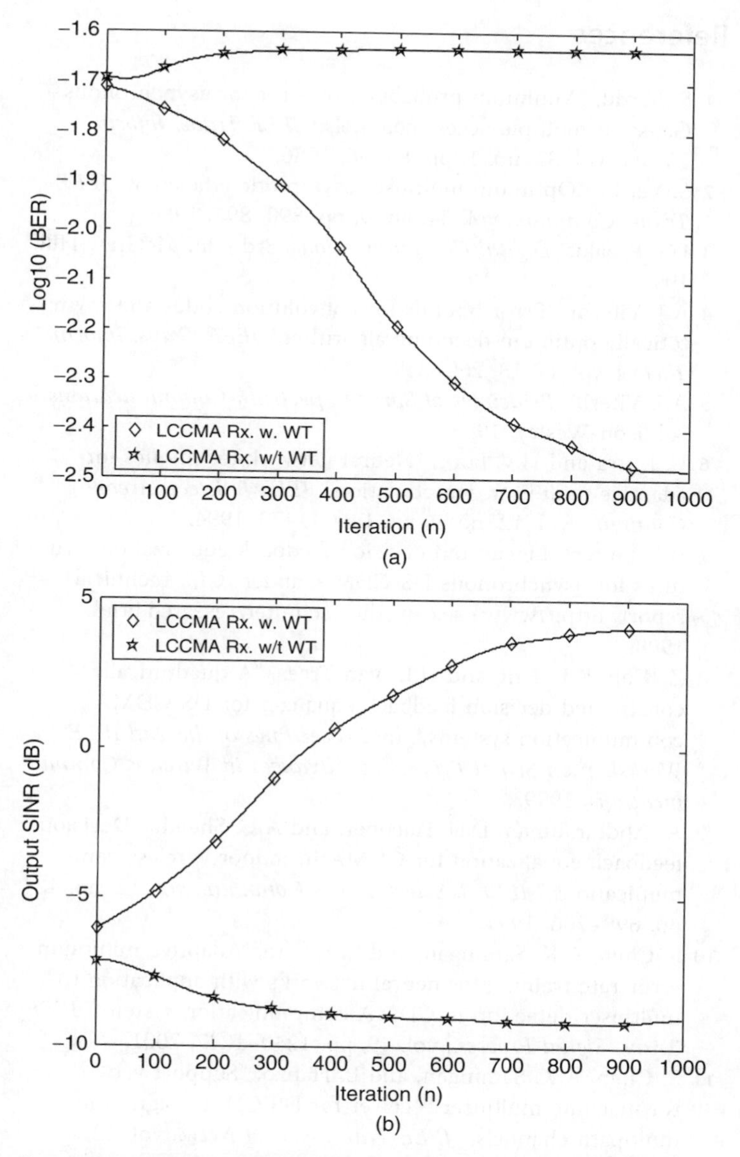

Figure 3.12 Adaptive LCCMA detector with/without whitening under near–far effect: (a) BER and (b) output SINR.

References

1 S. Verdu, "Minimum probability of error for asynchronous Gaussian multiple-access channels," *IEEE Trans. Inform. Theory*, vol. 32, no. 1, pp. 85–96, 1986.
2 S. Verdu, "Optimum multiuser asymptotic efficiency," *IEEE Trans. Commun.*, vol. 34, no. 9, pp. 890–897, 1986.
3 J.G. Proakis, *Digital Communications*, 3rd edn. McGraw-Hill, 1995.
4 A.J. Viterbi, "Error bounds for convolution codes and asymptotically optimum decoding algorithm," *IEEE Trans. Inform. Theory*, vol. IT-13, 260–269.
5 A.J. Viterbi, *Principles of Spread Spectrum Communications*, Addison-Wesley, 1995.
6 U. Mitra and H.V. Poor, "Neural network techniques for adaptive multiuser demodulation," *IEEE J. Select. Areas Commun.*, vol. 12. no. 9, pp. 1460–1470, 1994.
7 P. Schniter, "Linear and decision feedback equalization structures for asynchronous DS-CDMA under ICI," Technical report, http://www.ece.osu.edu/~schniter/research.html, 1998.
8 Z. Tian, K.L. Bell, and H.L. Van Trees, "A quadratically constrained decision feedback equalizer for DS-CDMA communication systems," In: *Proceedings of the 2nd IEEE Workshop on Signal Processing Advances in Wireless Communications*, 1999.
9 A. Abdulrahman, D.D. Falconer, and A.U. Sheikh, "Decision feedback equalization for CDMA in indoor wireless communications," *IEEE J. Select. Areas Commun.*, vol. 12, no. 4, pp. 698–706, 1994.
10 S. Chen, A.K. Samingan, and L. Hanzo, "Adaptive minimum error rate training for neural networks with application to multiuser detection in CDMA communication system," *IEEE Trans. Signal Process.*, vol. 49, pp. 1240–1247, 2001.
11 S. Chen, A.K. Samingan, and L. Hanzo, "Support vector machine multiuser receiver for DS-CDMA signals in multipath channels," *IEEE Trans. Neural Netw.*, vol. 12, pp. 604–611, 2001.
12 S. Chen, A.K. Samingan, B. Mulgrew, and L. Hanzo, "Adaptive minimum-BER Linear multiuser detection for

DS-CDMA signals in multipath channels," *IEEE Trans. Signal Process.*, vol. 49, pp. 1240–1247, 2001.

13 S. Chen, B. Mulgrew, and L. Hanzo, "Adaptive least error rate algorithm for neural network classifiers." In: *Proceedings of the 2001 IEEE Signal Processing Society Workshop Neural Networks for Signal Processing* XI, pp. 223–232, 10–12 September 2001.

14 S. Chi, J. Choi, H.J. Im, and B. Choi, "A novel adaptive beamforming algorithm for antenna array CDMA systems with strong interferes," *IEEE Trans. Vehic. Techn.*, vol. 51, no. 5, pp. 808–816, 2002.

15 A.J. Viterbi, "Very low rate convolutional codes for maximum theoretical performance of spread-spectrum multiple-access channels," *IEEE J. Select. Areas Commun.*, vol. 8, no. 4, pp. 641–649, 1990.

16 M.K. Varanasi and B. Aazhang, "Multistage detection in asyncrounous code-division multiple-access communications," *IEEE Trans. Commun.*, vol. 38, no. 4, pp. 675–682, 1990.

17 K.S. Schneider, "Optimum detection of code division multiplexed signals," *IEEE Trans. Aerospace Elect. Sys.*, vol. AES-15, pp. 181–185, 1979.

18 R. Lupas, S. Verdu, "Linear multiuser detectors for synchronous code-division multiple-access channels," *IEEE Trans. Inform. Theory*, vol. 35, no. 1, pp. 123–136, 1989.

19 R. Lupas and S. Verdu, "Near–far resistance of multiuser detectors for asynchronous channels," *IEEE Trans. Comm.*, vol. 38, no. 4, pp. 496–508, 1990.

20 D.L. Anair, "*Performance of fixed and adaptive multi-user linear detectors for DS-CDMA under non-ideal conditions: a software package*," MSc dissertation, Cornell University, 1999.

21 Z. Xie, R.T. Short, and C.K. Rushforth, "A family of suboptimum detectors for coherent multi-user communications," *IEEE J. Select. Areas Commun.*, vol. 8, no. 4, pp. 683–690, 1990.

22 H.V. Poor and S. Verdu, "Probability of error in MMSE multiuser detection," *IEEE Trans. Info. Theory*, vol. 43, no. 3, pp. 858–871, 1997.

23 M. Honig, U. Madhow, S. Verdu, "Blind adaptive multiuser detection," *IEEE Trans. Info. Theory*, vol. 41, no. 4, pp. 944–960, 1995.
24 M.Z.A. Bhotto and A. Antoniou, "Robust recursive least-squares adaptive-filtering algorithm for impulsive-noise environments," *Signal Process. Lett.*, vol. 18, no. 3, pp. 185–188, 2011.
25 Y. Zou, S.C. Chan, and T.S. Ng, "A recursive least M-estimate (RLM) adaptive filter for robust filtering in impulsive noise," *IEEE Signal Process. Lett.*, vol. 7, no. 11, pp. 324–326, 2000.
26 S.L. Miller, "An adaptive direct-sequence code-division-multiple-access receiver for multiuser interference rejection," *IEEE Trans. Commun.*, vol. 43, no. 2/3/4, pp. 1746–1755, 1995.
27 P.B. Rapajic and B.S. Vuctic, "Adaptive receiver structures for asynchronous CDMA systems," *IEEE J. Select. Areas Commun.*, vol. 12, no. 4, pp. 685–697, 1994.
28 U. Madhow and M.L. Hoing, "MMSE interference suppression for direct-sequence spread-spectrum CDMA," *IEEE Trans. Commun.*, vol. 42, no. 12, pp. 3178–3188, 1994.
29 A.N. Barbosa and S.L. Miller, "Adaptive detection of DS/CDMA signals in fading channels," *IEEE Trans. Commun.*, vol. 46, no. 1, pp. 115–124, 1998.
30 U. Madhow, "Blind adaptive interference suppression for direct sequence CDMA," *Proc. IEEE*, vol. 86, pp. 2049–2069, 1998.
31 T.J. Lim and S. Roy, "Adaptive filters in multiuser (MU) CDMA detection," *Wireless Networks*, vol. 4, pp. 307–318, 1998.
32 J.M. Cioffi and T. Kailath, "Fast recursive least squares transversal filters for adaptive filtering," *IEEE Trans. Acoust. Speech Signal Process*, vol. 32, no. 2, pp. 304–337, 1984.
33 V. Krishnamurthly, G. Yin, and S. Singh, "Adaptive step-size algorithms for blind interference suppression in DS/CDMA systems," *IEEE Trans. Signal Process.*, vol. 49, no. 1, pp. 190–201, 2001.
34 X. Wang and H.D. Poor, "Blind adaptive multiuser detection in multipath CDMA channels based on subspace tracking," *IEEE Trans. Signal Process.*, vol. 46, pp. 3030–3044, 1998.

35 X. Wang and H.D. Poor, "Blind equalization and multiuser detection in dispersive CDMA channels," *IEEE Trans. Commun.*, vol. 46, pp. 91–103, 1998.

36 X. Wang and H.V. Poor, *Wireless Communication Systems: Advanced Techniques for Signal Reception*, Prentice Hall, 2002.

37 M. Hoing and M.K. Tsatsanis, "Adaptive techniques for multiuser CDMA receivers," *IEEE Signal Process. Mag.*, pp. 49–61, 2000.

38 M. Honig, U. Madhow, and S. Verdu, "Blind adaptive multiuser detection," *IEEE Trans. Info. Theory*, vol. 41, no. 4, pp. 944–960, 1995.

39 M.K. Tsatsanis and Z. Xu, "Performance analysis of minimum variance CDMA receivers," *IEEE Trans. Signal Processing.*, vol. 46, no. 11, pp. 3014–3022, 1998.

40 M.K. Tsatsanis, "Inverse filtering criteria for CDMA systems," *IEEE Trans. Signal Process.*, vol. 45, no. 1, pp. 102–112, 1997.

41 M.K. Tsatsanis and Z. Xu, "Constrained optimization methods for direct blind equalization," *IEEE J. Select. Areas Commun.*, vol. 17, no. 3, pp. 424–433, 1999.

42 M.K. Tsatsanis and Z. Xu, "On minimum output energy CDMA receives in presence of multipath." In: *Conference on Information Sciences and Systems (CISS'97)*, pp. 377–381, 1997.

43 M.L. Honig and M.K. Tsatsanis, "Adaptive techniques for multiuser CDMA receivers," *IEEE Trans. Signal Process.*, vol. 17, no. 3, pp. 49–61, 2000.

44 Z. Tian, *Blind Multi-user Detection with Adaptive Space-Time Processing for DS-CDMA Wireless Communications*, PhD dissertation, George Mason University, 2000.

45 Z. Tian, K.L. Bell, and H.L. Van Trees, "A quadratically constrained decision feedback equalizer for DS-CDMA communication systems." In: *Proceedings of the 2nd IEEE Workshop on Signal Processing Advances in Wireless Communications*, 1999.

46 Z. Tian, K.L. Bell, and H.L. Van Trees, "A recursive least squares implementation for LCMP beamforming under

quadratic constraint" *IEEE Trans. Signal Processing.*, vol. 49, no. 6, pp. 1138–1145, 2001.

47 Z. Tian, K.L. Bell, and H.L. Van Trees, "A RLS implementation for adaptive beam-forming under quadratic constraint." In *9th IEEE Workshop on Statistical Signal and Array Processing*, Portland, Oregon, September 1998.

48 Z. Tian, K.L. Bell, and H.L. Van Trees, "Quadratically constrained RLS Filtering for adaptive beamforming and DS-CDMA multi-user detection." In: *Adaptive Sensor Array Processing Workshop*, MIT Lincoln Lab, Lexington, MA, 1999.

49 Z. Tian, K.L. Bell, and H.L. Van Trees, "Robust RLS implementation of the minimum output energy detector with multiple constraints." In: *Center of Excellence in C3I*, George Mason University, USA, July 15, 2000.

50 Z. Tian, K.L. Bell, and H.L. Van Trees, "Robust constrained linear receivers for CDMA wireless systems," *IEEE Trans. Signal Process.*, vol. 49, no. 7, pp. 1510–1522, 2001.

51 A. Elnashar, S. Elnoubi, and H. Elmikati "Performance analysis of blind adaptive MOE multiuser receivers using inverse QRD-RLS algorithm," *IEEE Trans. Circ. Systems I*, vol. 55, no. 1, pp. 398–411, 2008.

52 A. Elnashar, S. Elnoubi, and H. Elmikati, "Low-complexity robust adaptive generalized sidelobe canceller detector for DS/CDMA systems," *Int. J. Adapt. Cont. Signal Process.*, vol. 23, no. 3, pp. 293–310, 2008.

53 D.N. Godard, "Self-recovering equalization and carrier tracking in two-dimensional data communication systems," *IEEE Trans. Commun.*, vol. COMM-28, pp. 1867–1875, 1980.

54 J.K. Tugnait and T. Li. "Blind detection of asynchronous CDMA signals in multipath channels using code-constrained inverse filter criterion," *IEEE Trans. Signal Process.*, vol. 49, no. 7, pp. 1300–1309, 2001.

55 J.R. Treichler and B.G. Agee, "A new approach to multipath correction of constant modulus signals," *IEEE Trans. Acoust., Speech, Signal Process.*, vol. 31, pp. 459–471, 1983.

56 A. Elnashar, S. Elnoubi, and H. Elmikati, "Sample-by-sample and block-adaptive robust constant modulus-based algorithms," *IET Signal Process.*, vol. 6, no. 8, pp. 805–813, 2012.

57 A. Elnashar, S. Elnoubi, and H. Elmikati "Performance analysis of blind adaptive MOE multiuser receivers using inverse QRD-RLS algorithm," *IEEE Trans. Circuits Systems I*, vol. 55, no. 1, pp. 398–411, 2008.

58 A. Elnashar, S. Elnoubi, and H. Elmikati "Further study on robust adaptive beamforming with optimum diagonal loading," *IEEE Trans. Antennas Propag.*, vol. 54, no. 12, pp. 3647–3658, 2006.

59 A. Elnashar, "On efficient implementation of robust adaptive beamforming based on worst-case performance optimization" *IET Signal Process.*, vol. 2, no. 4, pp. 381–393, 2008.

60 T. Samir, S. Elnoubi, and A. Elnashar, "Block-Shanno minimum BER beamforming," *IEEE Trans. Vehic. Techn.*, vol. 57, no. 5, pp. 2981–2990, 2008.

61 Z. Xu, "Improved constraint for multipath mitigation in constrained MOE multiuser detection" *J. Commun. Netw.*, vol. 3, no. 3, pp. 249–256, 2001.

62 Xu Zhengyuan, "Blind Channel Estimation and Multiuser Detection for CDMA Communications," PhD dissertation, Steven Institute of Technology, 1999.

63 Z. Xu, L. Ping, and X. Wang "Blind multiuser detection: from MOE to subspace methods," *IEEE Trans. Signal Process.* vol. 52, no. 2, pp. 510–524, 2004.

64 K.M. Buckly and L.J. Griffiths, "An adaptive generalized sidelobe canceller with derivative constraint," *IEEE Trans. Antennas Propag.*, vol. AP34, pp. 311–319, 1986.

65 S.-J. Yu and J.-H. Lee, "The statistical performance of eigenspace-based adaptive array processors," *IEEE Trans. Antennas Propag.*, vol. 44, no. 5, pp. 665–671, 1996.

66 J.B. Schodorf and D.B. Williams, "A constrained optimization approach to multi-user detection," *IEEE Trans. Signal Processing.*, vol. 45, no. 1, pp. 258–262, Jan. 1997.

67 J.B. Schodorf and D.B. Williams, "Array processing techniques for multiuser detection," *IEEE Trans. Commun.*, vol. 45, no. 11, pp. 1375–1378, Nov. 1997.

68 S.N. Batalama, M.J. Medley, and D.A. Pados, "Robust adaptive recovery of spread-spectrum signals with short data records," *IEEE Trans. Commun.*, vol. 48, no. 10, pp. 1725–1731, 2000.

69 S.N. Batalama, M.J. Medley, and I.N. Psaromiligkos, "Adaptive robust spread-spectrum receivers," *IEEE Trans. Commun.*, vol. 47, no. 6, pp. 905–917, 1999.
70 B.R. Breed, J. Strauss, "A short proof of the equivalence of LCMV and GSC beamforming," *IEEE Signal Process. Lett.*, vol. 9, no. 6, pp. 168–169, 2002.
71 J.R. Treichler and B.G. Agee, "A new approach to multipath correction of constant modulus signals," *IEEE Trans. Acoust., Speech, Signal Process*, vol. 31, pp. 459–471, 1983.
72 P. Schniter and C.R. Johnson, Jr., "Sufficient conditions for the local convergence of constant modulus algorithms," *IEEE Trans. Signal Process.*, vol. 48, pp. 2785–2796, 2000.
73 P. Schniter, "Minimum-entropy blind acquisition/equalization for uplink DS-CDMA." In: *Proceedings of the 36th Allerton Conference on Communications, Control, and Computing*, 1998.
74 J.K. Tugnait and B. Huang, "On a whitening approach to partial channel estimation and blind equalization in FIR/IIR multiple-input multiple-output channels," *IEEE Trans. Signal Process*, vol. 48. no. 3, pp. 832–845, 2000.
75 D.J. Brooks, S. Lambotharan, and J.A. Chambers, "Optimum delay and mean square error using a mixed crosscorrelation and constant modulus algorithm," *IEEE Proc. Commun.*, vol. 147, no. 1, pp. 18–22, 2000.
76 Y. Luo, J.A. Chambers, and S. Lambotharan, "Global convergence and mixing parameter selection in the cross-correlation constant modulus algorithm for the multiuser environment," *IEE Proc.-Vis. Image Signal Process.*, vol. 148, no. 1, pp. 9–20, 2001.
77 Y. Luo, J.A. Chambers, "Steady-state mean-square error analysis of the cross-correlation and constant modulus algorithm in a MIMO conventional system," *IEE Proc.-Vis. Image Signal Process.*, vol. 149, no. 4, pp. 196–203, 2002.
78 C. Xu and G. Feng, "Non-canonically constrained CMA for blind multiuser detection," *Electr. Lett.*, vol. 36, no. 2, pp. 171–172, 2000.
79 C. Xu and G. Feng, and K.S. Kwak, "A modified constrained constant modulus approach to blind adaptive multiuser

detection," *IEEE Trans. Commun.*, vol. 49, no. 9, pp. 1642–1648, 2001.

80 J. Migues and L. Castedo, "A linearly constrained constant modulus approach to blind adaptive multiuser interference suppression," *IEEE Commun. Letters*, vol. 2, no. 8, pp. 217–219, 1998.

81 Y. Hua, Y. Xiang, T. Chen, K.A. Meriam, and Y. Miao, "Natural power method for fast subspace tracking." In: *Neural Networks for Signal Processing IX, 1999. Proceedings of the 1999 IEEE Signal Processing Society Workshop*, pp. 176–185, 23–25 August 1999

82 P. Strobach, "Square-root QR inverse iteration for tracking the minor subspace," *IEEE Trans. Signal Process.*, vol. 48, no. 11, pp. 2994–2999, 2000.

83 Z. Fu and E.M. Dowling, "Conjugate gradient projection subspace tracking," *IEEE Trans. Signal Process.*, vol. 45, no. 6, pp. 1664–1668, 1997.

4

Robust RLS Adaptive Algorithms

4.1 Introduction

A linear receiver can be designed by minimizing some inverse filtering criterion [1–3]. Appropriate constraints are used to avoid the trivial all-zero solution. A well-known cost function for the constrained optimization problem is the variance or the power of the output signal. A minimum output energy (MOE) detector has been developed for multiuser detection based on the constrained optimization approach [3]. In an additive white Gaussian noise (AWGN) environment with no multipath effects, this detector provides a blind solution with MMSE performance. Unfortunately, the approach experiences performance degradation in the presence of signal mismatch, inter-chip interference, and multipath propagation [4]. An improved constrained optimization approach to handle multipath fading uses only the main multipath component [4]; other delayed components are forced to zero. This approach is not the optimal solution. Although it can handle the multipath case, but it does not maximize the signal-to-interference noise ratio (SINR). An optimal solution [5] obtains the constrained vector by a max/min approach. The theoretical performance of this method tends to be close to the optimal non-blind MMSE receiver at high signal-to-noise ratios (SNR) in the presence of multipath fading. Adaptive implementation algorithms for this method have been developed [6–8]. Unfortunately, this method requires eigenvalue decomposition, which adds higher complexity. Additionally, performance degradation is incurred because of noise-induced channel estimation error.

Simplified Robust Adaptive Detection and Beamforming for Wireless Communications, First Edition. Ayman Elnashar.

Companion website: www.wiley.com/go/elnashar49

An improved constrained MOE detector for handling the noise effect with low SNR requires minimum eigenvector and corresponding eigenvalue estimation for a new cost function plus the maximum or minimum eigenvalue of the received signal autocovariance matrix, respectively [9, 10]. In these approaches, the contribution of the noise is partially subtracted from the cost function [9] or completely eliminated if its power can be estimated [10]. If noise power is not perfectly known, as is the case in practical or severe fading channels, we can only perform subtraction in an *ad-hoc* manner by choosing a suitable close candidate from an estimate [11]. These approaches are expected to perform well if the original channel matrix is used instead of the estimated autocorrelation matrix. It has been shown [6] that the channel estimation error is inversely proportional to the input SNR. Motivated by this result, Xu *et al.* [11] proposed a power of $\boldsymbol{R}_{xx}$ (POR) method to raise the power of the data covariance matrix $\boldsymbol{R}_{xx}$ in the MOE cost function to a positive integer m in order to virtually increase the SNR from an estimation perspective. It has been shown [11] that an MOE receiver with MMSE-like form built on the estimated channel parameters when $m \geq 2$ asymptotically converges to the optimal MMSE receiver in terms of output SINR, thus eliminating the penalty in the MOE detector. However, the POR method carries a high computational burden due to the power increase of the autocorrelation or covariance matrix.

Most detectors developed in previous approaches are based on the well-known RLS algorithm. In the conventional RLS algorithm, the calculation of the Kalman gain requires inversion of the autocovariance matrix of the received signal. If the data matrix is ill-conditioned, or in the worst-case has rank less than the number of the weight vector elements, convergence of the conventional RLS algorithm will rapidly become impossible. Moreover, it has another shortcoming, namely that it generally does not lend itself to efficient hardware implementation [12]. Furthermore, finite-precision implementations of RLS filters have sometimes been observed to be numerically unstable [13]. A well-known approach for overcoming these shortcomings is the rotation-based QR-RLS algorithm. The QR decomposition transforms the original RLS problem into a problem that uses only transformed data values by Cholesky factorization of the original least-squares data matrix [14]. The QRD-RLS algorithm, which is also referred to as the Givens-rotation or CORDIC-based RLS algorithm is the most promising RLS algorithm because it possesses desirable properties for VLSI implementations, such as regularity, and it can be mapped on CORDIC arithmetic-based processors [15]. This algorithm exhibits a high degree of parallelism, and can be mapped to triangular systolic arrays for efficient parallel implementation. In addition, a fast low-complexity implementation in both fixed-order and an order-recursive lattice-type is also available [16–19].

Poor and Wang investigated the use of QRD-RLS algorithm for MUD of DS/CDMA systems [14]. Unfortunately, the QRD-RLS algorithm suffers from a major drawback, namely back-substitution. This is a costly operation to be performed in systolic array structures. Alexander and Ghirnikar proposed a new computationally efficient algorithm for recursive least squares filtering, based on an IQRD algorithm [13]. In the IQRD-RLS updating method, the least-squares weight vector can be calculated without back-substitution. Furthermore, the IQRD method employs orthogonal rotation operations to recursively update the filter, and thus preserves the inherent stability properties of QR approaches to RLS filters. The IQRD-RLS algorithm has been adopted for beamforming applications [20–22] and multiuser detection [23–25].

In this chapter, linearly constrained IQRD-RLS algorithms with multiple constraints are developed and implemented for MUD in DS/CDMA systems. As explained before, the same algorithms can be extended to MVDR beamforming algorithms. Two approaches are considered: with a constant constrained vector and with an optimized constrained vector. Three IQRD-based detectors are developed

- a direct form MOE detector based on the IQRD update method with fixed constraints
- an MOE detector in PLIC structure, also based on the IQRD-RLS algorithm [23]
- an optimal MOE algorithm built on the IQRD update method and a subspace tracking algorithm for tracking the channel vector [23, 26, 27].

The constrained vector (estimated channel vector) is obtained using the max/min approach with IQRD-RLS based subspace tracking algorithms that are analyzed and tested for channel vector tracking. Recently proposed subspace tracking algorithms are tested and analyzed for channel estimation:

1. the fast orthogonal PAST (OPAST) algorithm [28–30]
2. the normalized orthogonal Oja (NOOja) algorithm [31–34]
3. a fast subspace tracking algorithm based on the Lagrange multiplier methodology and the IQRD algorithm [23].

Moreover, a new strategy for combining max/min channel estimation technique with the robust quadratic constraint technique is proposed. This is anchored in the direct form algorithm. Specifically, a robust MOE detector based on max/min approach and a quadratic inequality constraint (QIC) on the weight vector norm to overcome noise enhancement at low SNR is developed. A direct-form solution is introduced for the quadratic constraint detector, with a variable loading (VL) technique employed to satisfy the QIC. Thus, the IQRD algorithm acts as a core to the proposed receivers, which facilitates real-time implementation through systolic implementation. However, the same algorithms can be easily implemented using fast and robust RLS-based algorithms [35–39].

Systolic array implementation is the most promising technology that motivated the QR-RLS algorithms [40–46]. A systolic array implementation for RLS computations is discussed in the literature [47, 48]. The possibility of systolic array implementations for the presented IQRD-RLS receivers is demonstrated. A systolic implementation for the optimal MOE w. max/min approach is established [23]. Finally, computer simulations will demonstrate that the second detector is a good alternative to the first, which was proposed in [12] for moving-jammer suppression in beamforming applications. Moreover, we will demonstrate that the third detector, which is based on subspace tracking, yields the best SINR and lowest BER compared to the non-blind MMSE detector. A comparison of the complexity involved in existing channel estimation techniques and in recently proposed MOE receivers anchored in the IQRD-RLS algorithm is performed. An alternative adaptive implementation for the generalized eigenvalue problem is also presented.

4.2 IQRD-RLS Algorithm

The IQRD-RLS algorithm is used to update the inverse Cholesky factor $\boldsymbol{R}^{-H}(n)$ of the autocovariance matrix of the received vector $\boldsymbol{x}(n)$ where:

$$\boldsymbol{R}_{xx} = \boldsymbol{R}^{H}(n)\boldsymbol{R}(n) \tag{4.1}$$

A rotation matrix $\boldsymbol{P}(n)$, which successively annihilates the elements of intermediate vector $\boldsymbol{a}(n) = \frac{\boldsymbol{R}^{-H}(n-1)\boldsymbol{x}(n)}{\sqrt{\lambda}}$ against $\frac{\boldsymbol{R}^{-H}(n-1)}{\sqrt{\lambda}}$ into a related Kalman gain value $b(n)$ using a sequence of Givens rotations, can be used to update the inverse Cholesky factor according to the following equations:

$$\begin{bmatrix} \boldsymbol{R}^{-H}(n) \\ \boldsymbol{j}^{H}(n) \end{bmatrix} = \boldsymbol{P}(n) \begin{bmatrix} \frac{\boldsymbol{R}^{-H}(n-1)}{\sqrt{\lambda}} \\ 0^{T} \end{bmatrix} \tag{4.2}$$

where

$$\boldsymbol{P}(n) \begin{bmatrix} \boldsymbol{a}(n) \\ 1 \end{bmatrix} = \begin{bmatrix} 0 \\ b(n) \end{bmatrix} \tag{4.3}$$

and

$$\boldsymbol{j}(n) = \frac{\boldsymbol{R}^{-H}(n-1)\boldsymbol{a}(n)}{\sqrt{\lambda}b(n)}$$

The Kalman gain can be calculated by scaling the vector $\boldsymbol{j}(n)$ by $b(n)$, where the vector $\boldsymbol{j}(n)$ is computed using Givens rotations during the inverse Cholesky factor update. The derivation of the IQRD algorithm is provided in Appendix 4.A and further discussion can be found in the literature [12, 13]. For the sake of comparison, the QR decomposition for MIMO-OFDM applications is provided in Appendix 4.B [41].

Givens rotation is the most commonly used method in performing QRD and IQRD updating. The generic formula for these rotations requires explicit square-root (sqrt) computations, a step which constitutes a computational bottleneck and is quite undesirable from the practical VLSI circuit design point of view [49]. More specifically, in VLSI circuit design, the sqrt operation is expensive, because it takes up much space on a VLSI chip or make take many cycles to complete the calculation. Therefore, it is beneficial to avoid or minimize sqrt operations, and much effort has been spent on minimizing or even eliminating it from the Givens rotation method. A unified systematic approach for sqrt-free Givens rotations is provided in the paper by Hsieh *et al.* [49].

A VLSI architecture of QR decomposition for 4×4 MIMO-OFDM systems was proposed by Huang and Tsai [41], who combine both complex Givens rotation and real Givens rotation algorithms to reduce the required CORDIC operations. Moreover, they eliminate the delay elements required in the conventional complex Givens rotation systolic array and improve hardware utilization at the real Givens rotation stage through use of the time-sharing technique. A comparison of different Givens rotations and other techniques is provided in Appendix 4.B and Table 4.B.1.

4.3 IQRD-Based Receivers with Fixed Constraints

4.3.1 Direct-form MOE Detector

Starting from the optimal MOE detector in (3.13) and using (4.1), the optimal MOE detector is then given by:

$$f = [\boldsymbol{R}^H(n)\boldsymbol{R}(n)]^{-1}\boldsymbol{C}_1\{\boldsymbol{C}_1^H[\boldsymbol{R}^H(n)\boldsymbol{R}(n)]^{-1}\boldsymbol{C}_1\}^{-1}\boldsymbol{g} \tag{4.4}$$

Two new matrices are defined here: $\boldsymbol{\Delta}(n) = [\boldsymbol{R}^H(n)\boldsymbol{R}(n)]^{-1}\boldsymbol{C}_1$ and $\boldsymbol{\Pi}(n) = \boldsymbol{C}_1^H\boldsymbol{\Delta}(n)$, and hence the optimal weight vector can be expressed as:

$$\boldsymbol{f}(n) = \boldsymbol{\Delta}(n)\boldsymbol{\Pi}^{-1}(n)\boldsymbol{g} \tag{4.5}$$

The recursive algorithm is developed by deriving recursive update equations for matrices $\boldsymbol{\Delta}(n)$ and $\boldsymbol{\Pi}^{-1}(n)$. The optimal weight vector $\boldsymbol{f}(n)$ can then be updated recursively.

Based on the update equation (4.2) of the inverse Cholesky factor and multiplying each side with its Hermitian transpose, the following update equation is obtained [50]:

$$\boldsymbol{R}^{-1}(n)\boldsymbol{R}^{-H}(n) = \frac{\boldsymbol{R}^{-1}(n-1)\boldsymbol{R}^{-H}(n-1)}{\lambda} - \boldsymbol{j}(n)\boldsymbol{j}^H(n) \tag{4.6}$$

Postmultiplying (4.6) by $\boldsymbol{C}_1$, a recursive formula for $\boldsymbol{\Delta}(n)$ matrix is derived:

$$\boldsymbol{\Delta}(n) = \lambda^{-1}\boldsymbol{\Delta}(n-1) - \boldsymbol{j}(n)\boldsymbol{\pi}^H(n) \tag{4.7}$$

where

$$\boldsymbol{\pi}(n) = \boldsymbol{C}_1^H \boldsymbol{j}(n) \tag{4.8}$$

Premultiplying (4.7) by $\boldsymbol{C}_1^H$, a recursive formula for the matrix $\boldsymbol{\Pi}(n)$ is obtained as follows:

$$\boldsymbol{\Pi}(n) = \lambda^{-1}\boldsymbol{\Pi}(n-1) - \boldsymbol{\pi}(n)\boldsymbol{\pi}^H(n) \tag{4.9}$$

By applying the matrix inversion lemma [5] to (4.9),

$$\boldsymbol{\Pi}^{-1}(n) = \lambda\boldsymbol{\Pi}^{-1}(n-1) + \frac{\lambda^2\boldsymbol{\Pi}^{-1}(n-1)\boldsymbol{\pi}(n)\boldsymbol{\pi}^H(n)\boldsymbol{\Pi}^{-1}(n-1)}{1-\lambda\boldsymbol{\pi}^H(n)\boldsymbol{\Pi}^{-1}(n-1)\boldsymbol{\pi}(n)} \tag{4.10}$$

Therefore, the recursive formulas for both $\boldsymbol{\Delta}(n)$ and $\boldsymbol{\Pi}^{-1}(n)$ matrices have been derived, and consequently the recursive implementation is completed. A recursive formula for the optimum adaptive detector $\boldsymbol{f}(n)$ can be obtained by substituting from (4.7) and (4.10) into (4.5). Equations (4.5), (4.7), and (4.10) form the basis for the direct-form MOE algorithm with fixed constraints (direct-MOE-IQRD).

A recursive update equation for $\boldsymbol{f}(n)$ has been developed for beamforming applications [12]. Therefore, the algorithms developed in this chapter can be easily extended to beamforming applications. Also, the QR decomposition algorithm has been exploited in a MIMO-OFDM system [41], and consequently IQRD algorithms can also be used with the MIMO-OFDM model presented in Chapter 2.

4.3.2 MOE Detector based on IQRD-RLS and PLIC

This algorithm is based on the PLIC structure combined with the IQRD-RLS algorithm for weight vector update. The PLIC structure is presented in Chapter 3 and demonstrated in Figure 4.1. The IQRD algorithm is used to update the inverse Cholesky factor $\boldsymbol{R}_z^{-H}(n)$ of the autocovariance matrix of the blocked data

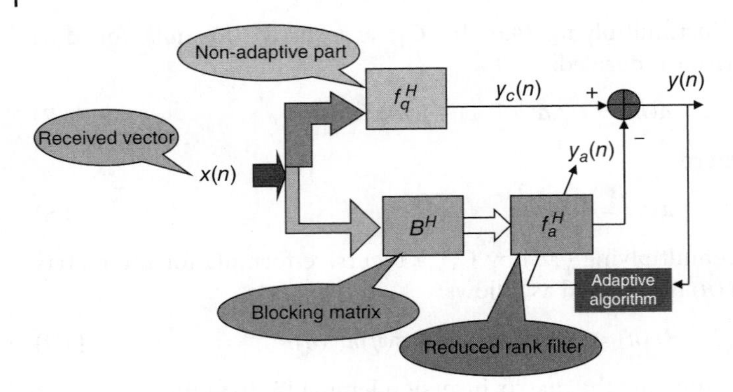

Figure 4.1 Block diagram of PLIC structure.

vector $\boldsymbol{z}(n) = \boldsymbol{B}^H\boldsymbol{x}(n)$ (the lower branch of the PLIC structure). A rotation matrix $\boldsymbol{P}_z(n)$, which successively annihilates the elements of the intermediate vector $\overline{\boldsymbol{a}}(n) = \frac{\boldsymbol{R}_z^{-H}(n-1)\boldsymbol{z}(n)}{\sqrt{\lambda}}$ into a related Kalman gain value $\overline{b}(n)$, can be used to update the inverse Cholesky factor. The Kalman gain can be calculated again by scaling the vector $\boldsymbol{u}(n)$ by $\overline{b}(n)$, where $\boldsymbol{u}(n)$ is computed using Givens rotations, during inverse Cholesky factor update. The MOE with PLIC structure, implemented using the IQRD-RLS algorithm (referred to as PLIC-MOE-IQRD), is summarized in Box 4.1.

Box 4.1 Summary of PLIC-MOE-IQRD receiver with fixed constraints.

Initialization

- $\boldsymbol{R}_z^{-H}(0) = \varepsilon \boldsymbol{I}_{N_f-N_h}$
- select $\boldsymbol{g}$ to be constant vector
- $\boldsymbol{f}_a^H(0) = \boldsymbol{0}_{N_f-N_g}$

For $n = 1,2,\cdots$, **do**

- Take N_f samples from the received vector $\boldsymbol{x}(n)$
- Calculate the output of blocking matrix $\boldsymbol{z}(n) = \boldsymbol{B}^H\boldsymbol{x}(n)$
- Calculate the output $y(n)$ of the detector using

$$y(n) = y_c(n) - y_a(n) = \boldsymbol{f}_c^H\boldsymbol{x}(n) - \boldsymbol{f}_a^H(n)\boldsymbol{z}(n)$$

- Update the inverse Cholesky *factor* $\boldsymbol{R}_z^{-H}(n)$ using Givens rotations according to:

$$- \begin{bmatrix} \boldsymbol{R}_z^{-H}(n) \\ \boldsymbol{u}^H(n) \end{bmatrix} = \boldsymbol{P}_z(n) \begin{bmatrix} \dfrac{\boldsymbol{R}_z^{-H}(n-1)}{\sqrt{\lambda}} \\ \boldsymbol{0}^T \end{bmatrix}$$

  – where $P_z(n) \begin{bmatrix} \overline{\boldsymbol{a}}(n) \\ 1 \end{bmatrix} = \begin{bmatrix} \boldsymbol{0}^T \\ \overline{b}(n) \end{bmatrix}$ with $\boldsymbol{P}_z(n)$ the rotation matrix and $\overline{\boldsymbol{a}}(n) = \dfrac{\boldsymbol{R}_z^{-H}(n-1)\boldsymbol{z}(n)}{\sqrt{\lambda}}$

- Calculate the Kalman gain using: $\boldsymbol{k}(n) = \dfrac{\boldsymbol{u}(n)}{\overline{b}(n)}$
- Update $\boldsymbol{f}_a(n)$ as follows: $\boldsymbol{f}_a(n) = \boldsymbol{f}_a(n-1) + \boldsymbol{k}(n)y^*(n)$
- Calculate the weight vector using: $\boldsymbol{f}(n) = \boldsymbol{f}_c - \boldsymbol{B}\boldsymbol{f}_a(n)$

End

4.4 IQRD-based Receiver with Optimized Constraints

Consider now the optimal MOE receiver in direct form (3.13) and with the min/max approach [5, 6]. The optimized constrained vector $\boldsymbol{g}(n)$ can be derived as follows:

$$\max_{\|g\|=1} \boldsymbol{f}^H_{\max/\min} \boldsymbol{R}^H(n)\boldsymbol{R}(n)\boldsymbol{f}_{\max/\min} \tag{4.11}$$

Substituting from (4.5) into (4.11), it can be shown that the optimal solution for $\boldsymbol{g}$ is $\boldsymbol{v}_1$. The principal eigenvector corresponding to the maximum eigenvalue of the matrix $\boldsymbol{\Pi}^{-1}(n)$ or the minor eigenvector $\boldsymbol{\gamma}_1$ corresponding to the minimum eigenvalue of matrix $\boldsymbol{\Pi}(n)$, and then the optimal min/max weight vector, can be respectively updated as follows:

$$\boldsymbol{f}_{\max/\min} = \beta_1 \boldsymbol{\Delta}(n)\boldsymbol{v}_1 \tag{4.12}$$

$$\boldsymbol{f}_{\max/\min} = \frac{1}{\alpha_1}\boldsymbol{\Delta}(n)\boldsymbol{\gamma}_1 \tag{4.13}$$

where β_1 is the maximum eigenvalue and α_1 is the minimum eigenvalue of $\boldsymbol{\Pi}^{-1}(n)$ and $\boldsymbol{\Pi}(n)$, respectively.

In (4.7), a recursive formula for the update of $\boldsymbol{\Delta}(n)$ matrix is derived. Therefore, a complete recursive implementation for the max/min optimal weight vector using the IQRD method can be obtained by recursively updating the eigenvector (principal or minor) of the matrix ($\boldsymbol{\Pi}^{-1}(n)$ or $\boldsymbol{\Pi}(n)$), respectively, using the IQRD-based subspace tracking algorithm.

To track the principal component of the target matrix $\boldsymbol{\Pi}^{-1}(n)$, it must be reformulated in the proper form as $\boldsymbol{\Psi}(n) = \sum_{i=1}^{n} \lambda^{n-i}\boldsymbol{d}(n)\boldsymbol{d}^H(n)$ or as the well-known recursive formula

$$\boldsymbol{\Psi}(n) = \lambda\boldsymbol{\Psi}(n-1) + \boldsymbol{d}(n)\boldsymbol{d}^H(n) \tag{4.14}$$

where λ is the usual forgetting factor, with $0 < \lambda \leq 1$. By comparing the update equation (4.10) of $\boldsymbol{\Pi}^{-1}(n)$ with (4.14),

$$\boldsymbol{d}(n) = \frac{\lambda\boldsymbol{\Pi}^{-1}(n-1)\boldsymbol{\pi}(n)}{\sqrt{1-\lambda\boldsymbol{\pi}^H(n)\boldsymbol{\Pi}^{-1}(n-1)\boldsymbol{\pi}(n)}} \tag{4.15}$$

Any orthogonal subspace tracking algorithm can be employed for tracking the principal component of $\boldsymbol{\Pi}^{-1}(n)$ and hence the optimal max/min weight vector can be recursively updated. Such algorithms are the OPASTd algorithm and the NOOja algorithm. The deflated OPAST algorithm (OPASTd) is very effective and has low computational complexity and it sequentially tracks principal components. The orthogonal versions from the subspace tracking algorithms are selected to ensure the orthogonality of the constrained vector during the update process. These subspace tracking algorithms are summarized in Appendix 4.C.

Tracking the minor component involves less computational complexity than tracking the principal component, as the matrix $\boldsymbol{\Pi}(n)$ will be used instead of its inverse. But due to the instability in tracking the minor components [51], an alternative approach is adopted here for tracking the minor component of the matrix $\boldsymbol{\Pi}(n)$. A similar approach can be found in the literature [23, 52]. The new approach depends on the matrix itself instead of its update form. Let us start by formulating the following real valued

Lagrangian function:

$$\boldsymbol{\Psi}_n(\boldsymbol{g},\zeta) = \boldsymbol{g}^H(n-1)\boldsymbol{\Pi}(n)\boldsymbol{g}(n-1) + \frac{1}{2}\zeta(n-1)(1-\boldsymbol{g}^H(n-1)\boldsymbol{g}(n-1)) \quad (4.16)$$

The optimum channel vector $\boldsymbol{g}(n)$ can be obtained by minimizing the above cost function and hence the channel vector can be updated as follows:

$$\boldsymbol{g}(n) = \boldsymbol{g}(n-1) - \mu\nabla_g\boldsymbol{\Psi}(\boldsymbol{g},\zeta) \quad (4.17)$$

From (4.16) the gradient vector can be estimated as:

$$\nabla_g(n) = \boldsymbol{\Pi}(n)\boldsymbol{g}(n-1) - \zeta(n)\boldsymbol{g}(n-1) \quad (4.18)$$

The step-size μ should be set to a value that guarantees algorithm convergence. Further improvement to this method can be achieved by updating the step-size on the basis of the optimal calculation. The optimal step-size can be estimated using an approach similar to the one introduced by Attallah and Abed-Meraim [34]. Substituting from (4.17) into (4.16) and incorporating the variable step-size $\mu(n)$, instead of a fixed step-size. into (4.17) and omitting the second term for a moment, the cost function update is obtained as follows:

$$\boldsymbol{\Psi}_n(\boldsymbol{g},\zeta) = \boldsymbol{\Psi}_{n-1}(\boldsymbol{g},\zeta) - 2\mu(n-1)\boldsymbol{g}^H(n-1)\boldsymbol{\Pi}(n)\nabla_{\boldsymbol{g}}^H \times\boldsymbol{\Psi}_{n-1}(\boldsymbol{g},\zeta) + \mu^2(n-1)\nabla_g^H\boldsymbol{\Psi}_{n-1}(\boldsymbol{g},\zeta)\boldsymbol{\Pi}(n)\nabla_g \times\boldsymbol{\Psi}_{n-1}(\boldsymbol{g},\zeta) \quad (4.19)$$

Therefore, the optimal step-size can be obtained by differentiating (4.19) with respect to the adaptive step-size $\mu(n)$ and setting the result to zero. Assuming that $\boldsymbol{\Psi}_n(\boldsymbol{g},\zeta)$ is independent from $\mu(n)$ then the optimal step size can be obtained as follows:

$$\mu_{opt}(n) = \frac{\alpha\boldsymbol{g}^H(n-1)\boldsymbol{\Pi}(n)\nabla_g}{\nabla_g{}^H\boldsymbol{\Pi}(n)\nabla_g + \eta} \quad (4.20)$$

where α,η are two positive constants, the former added to improve the numerical stability of the algorithm and the latter to avoid dividing by zero, respectively. The second term in (4.16) will not affect (4.20) as it will be equal to zero if the orthogonality of the channel vector during the update process is guaranteed.

Assuming the previous estimated channel vector $\mathbf{g}(n-1)$ satisfies the normalization constraint, the updated value $\mathbf{g}(n)$ must satisfy the normalization constraint (that is, $\mathbf{g}^H(n)\mathbf{g}(n) = 1$). By substituting from the update equation (4.17), an expression for the Lagrange multiplier $\zeta(n)$ is obtained as a solution for the following quadratic equation:

$$a\zeta^2(n) - 2b\zeta(n) + c \tag{4.21}$$

where

$$a = \mu_{opt}(n-1) \tag{4.22a}$$

$$b = 1 + \mu_{opt}(n-1)\mathbf{g}^H(n-1)\boldsymbol{\Pi}(n)\mathbf{g}(n-1) \tag{4.22b}$$

$$c = \mu_{opt}(n-1)\mathbf{g}^H(n-1)\boldsymbol{\Pi}^2(n)\mathbf{g}(n-1) + 2\mathbf{g}^H(n-1)\boldsymbol{\Pi}(n)\mathbf{g}(n-1) \tag{4.22c}$$

Therefore, the Lagrange multiplier $\zeta(n)$ can be expressed as:

$$\zeta(n) = \frac{-b \pm \sqrt{b^2 - ac}}{a} \tag{4.23}$$

Indeed, it is easily verified that $b^2 > ac$. As a consequence, the roots of the quadratic equation (4.21) are both real. The smaller root is selected to ensure better algorithm stability. Finally, a recursive implementation for (4.13) can be found by injecting (4.7) and (4.17) into (4.13).

Note that, the Lagrange method addressed in this section can be used for tracking the principal component of matrix $\boldsymbol{\Pi}^{-1}(n)$ by replacing the matrix $\boldsymbol{\Pi}(n)$ with $\boldsymbol{\Pi}^{-1}(n)$ in the cost function (4.16). This cost function is then maximized rather than minimized. Therefore, a recursive update equation for (4.12) can be obtained as well. The final optimum MOE detector with minor subspace tracking (referred to as direct-MOE-IQRD w. max/min) is summarized in Box 4.2.

Box 4.2 Summary of MOE receiver with optimized constraints

Initialization

- $\mathbf{R}^{-H}(0) = \mathbf{I}_{N_f}$ lower triangular matrix
- $\boldsymbol{\Delta}(0) = \mathbf{R}^{-1}(0)\mathbf{R}^{-H}(0)\mathbf{C}_1$
- $\boldsymbol{\Pi}^{-1}(0) = [\mathbf{C}_1{}^H\boldsymbol{\Delta}(0)]^{-1}$

- select $\boldsymbol{g}(0)$ to be any orthogonal vector
- $\mu_{opt}(0) = 0.001$
- $\boldsymbol{f}_{\max/\min}(0) = \boldsymbol{\Delta}(0)\boldsymbol{g}(0)$

For $n = 1,2,\cdots$, **do**

- Update the inverse Cholesky factor $\boldsymbol{R}^{-H}(n)$, similar to first algorithm
- Update the $\boldsymbol{\Delta}(n)$ matrix as follows: $\boldsymbol{\Delta}(n) = \lambda^{-1}\boldsymbol{\Delta}(n-1) - \boldsymbol{j}(n)\boldsymbol{\pi}(n)$
- Update the $\boldsymbol{\Pi}(n)$ matrix as follows: $\boldsymbol{\Pi}(n) = \dfrac{1}{\lambda}\boldsymbol{\Pi}(n-1) - \boldsymbol{\pi}^H(n)\boldsymbol{\pi}(n)$
- $a = \mu_{opt}(n-1)$
- $b = 1 + \mu_{opt}(n-1)\boldsymbol{g}^H(n-1)\boldsymbol{\Pi}(n)\boldsymbol{g}(n-1)$
- $c = \mu_{opt}(n-1)\boldsymbol{g}^H(n-1)\boldsymbol{\Pi}^2(n)\boldsymbol{g}(n-1) + 2\boldsymbol{g}^H(n-1)\boldsymbol{\Pi}(n)\boldsymbol{g}(n-1)$
- $\zeta(n) = \dfrac{-b - \sqrt{b^2 - \mu_{opt}(n-1)c}}{\mu_{opt}(n-1)}$ (Lagrange multiplier)
- $\nabla_g(n) = \boldsymbol{\Pi}(n)\boldsymbol{g}(n-1) - \zeta(n)\boldsymbol{g}(n-1)$ (Gradient vector)
- $\mu_{opt}(n) = \dfrac{\alpha \boldsymbol{g}^H(n)\boldsymbol{\Pi}(n)\nabla_g}{\nabla_g^H\boldsymbol{\Pi}(n)\nabla_g + \eta}$ (optimum step-size)
- $\boldsymbol{g}(n) = \boldsymbol{g}(n-1) - \mu_{opt}(n)\nabla_g(n)$ (estimated channel vector)
- $\boldsymbol{f}_{\max/\min}(n) = \dfrac{1}{\zeta(n)}\boldsymbol{\Delta}(n)\boldsymbol{g}(n)$ (weight vector)

End

4.5 Channel Estimation Techniques

In this section, the channel estimation techniques discussed early in Chapter 3 will be implemented in adaptive fashion based on the IQRD-RLS algorithm.

4.5.1 Noise Cancellation Schemes

In these techniques, the contribution of the noise is partially subtracted from the cost function [10], or completely eliminated if the noise power can be estimated [9].

4.5.1.1 Adaptive Implementation of Improved cost function

The cost function given in (3.18) can be easily implemented using the IQRD-RLS algorithm as follows:

$$\boldsymbol{\varphi}(n) = \boldsymbol{C}_1^H\boldsymbol{R}_{xx}^{-1}(n)\boldsymbol{C}_1 - \gamma.\boldsymbol{C}_1^H\boldsymbol{C}_1 \tag{4.24}$$

where γ is the reciprocal of largest eigenvalue of $\boldsymbol{R}_{xx}(n)$. Substituting from (4.9) into (4.24), we get

$$\boldsymbol{\varphi}(n) = \boldsymbol{\Pi}(n) - \gamma.\boldsymbol{C}_1^H \boldsymbol{C}_1 \tag{4.25}$$

Therefore, the channel vector $\tilde{\boldsymbol{g}}$ can be estimated as the eigenvector corresponding to the smallest eigenvalue of the cost function (4.25). The proposed Lagrange method is the most suitable technique for tracking the minor component of the cost function (4.25). With this estimate, the detector can be constructed based on the MMSE criterion:

$$\boldsymbol{f}_{IMOE}(n) = \boldsymbol{\Delta}(n)\tilde{\boldsymbol{g}}(n) \tag{4.26}$$

Therefore, the algorithm summarized in Box 4.2 can be used to estimate this detector, except that the cost function $\boldsymbol{\Pi}(n)$ is replaced by $\boldsymbol{\varphi}(n)$. In addition, the largest eigenvector of $\boldsymbol{R}_{xx}(n)$ is estimated.

4.5.1.2 Adaptive Implementation of Modified Cost Function

The cost function given in (3.20), which is constructed by removing the noise contribution from the data covariance matrix, is estimated as follows:

$$\overline{\boldsymbol{R}}_{xx}(n) = \boldsymbol{R}_{xx}(n) - \alpha\sigma_w^2 \boldsymbol{I}_{N_f} \tag{4.27}$$

where σ_w^2 is the noise power and the parameter α quantifies the extent of the noise mitigation and should satisfy $0 \leq \alpha < 1$ such that the modified covariance matrix is positive definite. The main drawback of this cost function is that it is not easy to get the inverse of $\overline{\boldsymbol{R}}_{xx}(n)$. Following an approach similar to [53], we get:

$$\overline{\boldsymbol{R}}_{xx}^{-1}(n) = [\boldsymbol{I}_{N_f} - \alpha\sigma^2 \boldsymbol{R}_{xx}^{-1}(n)]^{-1} \boldsymbol{R}_{xx}^{-1}(n) \tag{4.28}$$

We expand the bracketed term using Taylor series as:

$$[\boldsymbol{I}_{N_f} - \alpha\sigma^2 \boldsymbol{R}_{xx}^{-1}(n)]^{-1} = \boldsymbol{I}_{N_f} + \alpha\sigma^2 \boldsymbol{R}_{xx}^{-1}(n) + \cdots \tag{4.29}$$

Using Taylor series approximation, we get

$$[\boldsymbol{I}_{N_f} - \alpha\sigma^2 \boldsymbol{R}_{xx}^{-1}(n)]^{-1} \approx \boldsymbol{I}_{N_f} + \alpha\sigma^2 \boldsymbol{R}_{xx}^{-1}(n) \tag{4.30}$$

Then

$$\overline{\boldsymbol{R}}_{xx}^{-1}(n) = [\boldsymbol{I}_{N_f} + \alpha\sigma^2 \boldsymbol{R}_{xx}^{-1}(n)] \boldsymbol{R}_{xx}^{-1}(n) \tag{4.31}$$

This approximation is valid only when the following inequality is guaranteed [53]:

$$\alpha\sigma^2 \|\boldsymbol{R}_{xx}^{-1}(n)\| < 1 \tag{4.32}$$

where $\|.\|$ is any valid matrix norm used to evaluate the closeness of approximation in (4.30). Generally, the approximation in (4.30) holds for small $\alpha\sigma^2$. Therefore, at high SNR we can set α close to 1, whereas at low SNR we set it to a small value.

Following the max/min approach described above, the modified channel vector $\overline{\boldsymbol{g}}$ can be obtained as follows:

$$\overline{\boldsymbol{g}} = \min_{\|\overline{g}\|=1} \overline{\boldsymbol{g}}^H (\boldsymbol{C}_1^H \overline{\boldsymbol{R}}_{xx}^{-1} \boldsymbol{C}_1) \overline{\boldsymbol{g}} \tag{4.33}$$

Substituting from (4.31) into (4.33), we get

$$\overline{\boldsymbol{g}} = \min_{\|\overline{g}\|=1} \overline{\boldsymbol{g}}^H ([\boldsymbol{C}_1^H + \alpha\sigma^2 \boldsymbol{\Delta}^H(n)]\boldsymbol{\Delta}(n)) \overline{\boldsymbol{g}} \tag{4.34}$$

Equation (4.34) can be further simplified as follows:

$$\overline{\boldsymbol{g}} = \min_{\|\overline{g}\|=1} \overline{\boldsymbol{g}}^H (\overline{\boldsymbol{\Delta}}(n)\boldsymbol{\Delta}(n)) \overline{\boldsymbol{g}} \tag{4.35}$$

where

$$\overline{\boldsymbol{\Delta}}(n) = [\boldsymbol{C}_1^H + \alpha\sigma^2 \boldsymbol{\Delta}^H(n)] \tag{4.36}$$

The noise power can be estimated from the smallest eigenvalue of the estimated data covariance matrix. The channel vector can be estimated also using the proposed Lagrange method. Once the optimized channel vector is obtained, the MMOE detector is forced to take a MMSE-like form:

$$\boldsymbol{f}_{MMOE}(n) = \boldsymbol{\Delta}(n)\overline{\boldsymbol{g}}(n) \tag{4.37}$$

Finally, the $\boldsymbol{f}_{MMOE}(n)$ detector can be obtained using the pseudocode in Box 4.2 with minor modifications.

4.5.2 Adaptive Implementation of POR Method

Using the POR method described in Chapter 3 with power of 2, the channel can be estimated as follows:

$$\breve{\boldsymbol{g}} = \min_{\|g\|=1} \breve{\boldsymbol{g}}^H (\boldsymbol{C}_1^H \boldsymbol{R}_{xx}^{-2} \boldsymbol{C}_1) \breve{\boldsymbol{g}} \tag{4.38}$$

The power integer m is set to 2, because with $m > 2$ the computational complexity increases exponentially and the possibility of systolic implementation will be complicated. By rearranging (4.38), we get:

$$\breve{\boldsymbol{g}} = \min_{\|g\|=1} \breve{\boldsymbol{g}}^H(\boldsymbol{C}_1^H \boldsymbol{R}_{xx}^{-H} \boldsymbol{R}_{xx}^{-1} \boldsymbol{C}_1)\breve{\boldsymbol{g}} \tag{4.39}$$

and hence

$$\breve{\boldsymbol{g}} = \min_{\|g\|=1} \breve{\boldsymbol{g}}^H(\Delta^H(n)\Delta(n))\breve{\boldsymbol{g}} \tag{4.40}$$

Once the channel vector $\breve{\boldsymbol{g}}$ is obtained, the POR receiver takes also MMSE-like form

$$\boldsymbol{f}_{POR}(n) = \Delta(n)\breve{\boldsymbol{g}}(n) \tag{4.41}$$

Similar to the noise cancellation approaches, the POR receiver can be implemented using the code in Box 4.2 as well.

4.5.3 Adaptive Implementation of Capon Method

The Capon method is based on the estimation of the channel vector using a generalized eigenvalue decomposition problem. Equation 3.24 is equivalent to a generalized eigenvalue problem involving the pensile of $(\boldsymbol{C}_1^H \boldsymbol{R}_{xx}^{-1} \boldsymbol{C}_1, \boldsymbol{C}_1^H \boldsymbol{C}_1)$. The generalized eigenvalue involves solving the matrix equation

$$\boldsymbol{C}_1^H \boldsymbol{R}_{xx}^{-1} \boldsymbol{C}_1 \hat{\boldsymbol{g}} = \boldsymbol{C}_1^H \boldsymbol{C}_1 \hat{\boldsymbol{g}} \lambda \tag{4.42}$$

where $\hat{\boldsymbol{g}}$ is the principal generalized eigenvector and λ is the largest generalized eigenvalue. A simple recursive solution for the generalized eigenvalue problem is available [54]. Despite, the simplicity of this approach, its simplicity is based on replacing the covariance data matrix with the current estimate (that is, $\boldsymbol{R}_{xx} = x(n)x^H(n)$) without recursive estimation. The approach adopted here is similar to the one proposed by Rao *et al.* [55]. The generalized eigenvector $\hat{\boldsymbol{g}}$ is a stationary point of the function:

$$J(\hat{\boldsymbol{g}}) = \frac{\hat{\boldsymbol{g}}^H \boldsymbol{C}_1^H \boldsymbol{R}_{xx}^{-1} \boldsymbol{C}_1 \hat{\boldsymbol{g}}}{\hat{\boldsymbol{g}}^H \boldsymbol{C}_1^H \boldsymbol{C}_1 \hat{\boldsymbol{g}}} \tag{4.43}$$

Then

$$\frac{\partial J(\widehat{\boldsymbol{g}})}{\partial \widehat{\boldsymbol{g}}} = \frac{\widehat{\boldsymbol{g}}^H \boldsymbol{C}_1^H \boldsymbol{R}_{xx}^{-1} \boldsymbol{C}_1 \widehat{\boldsymbol{g}}}{\widehat{\boldsymbol{g}}^H \boldsymbol{C}_1^H \boldsymbol{C}_1 \widehat{\boldsymbol{g}}} = 0 \tag{4.44}$$

and hence

$$\boldsymbol{C}_1^H \boldsymbol{R}_{xx}^{-1} \boldsymbol{C}_1 \widehat{\boldsymbol{g}} = \frac{\widehat{\boldsymbol{g}}^H \boldsymbol{C}_1^H \boldsymbol{R}_{xx}^{-1} \boldsymbol{C}_1 \widehat{\boldsymbol{g}}}{\widehat{\boldsymbol{g}}^H \boldsymbol{C}_1^H \boldsymbol{C}_1 \widehat{\boldsymbol{g}}} \boldsymbol{C}_1^H \boldsymbol{C}_1 \widehat{\boldsymbol{g}} \tag{4.45}$$

Premultiplying (4.45) by $\boldsymbol{\Pi}^{-1}(n)$, we get

$$\widehat{\boldsymbol{g}}(n) = \beta \boldsymbol{\Pi}^{-1}(n)(n)\mathbb{C}\widehat{\boldsymbol{g}}(n-1) \tag{4.46}$$

where

$$\beta = \frac{\widehat{\boldsymbol{g}}^H(n-1)\boldsymbol{\Pi}(n)\widehat{\boldsymbol{g}}(n-1)}{\widehat{\boldsymbol{g}}^H(n-1)\mathbb{C}\widehat{\boldsymbol{g}}(n-1)} \tag{4.47}$$

and

$$\mathbb{C} = \boldsymbol{C}_1^H \boldsymbol{C}_1 \tag{4.48}$$

Equation (4.46) is used to update the channel vector $\widehat{\boldsymbol{g}}(n)$. It is interesting to note that the factor β will not affect the channel estimation. The Capon detector takes the following MMSE-like form:

$$\boldsymbol{f}_{Capon}(n) = \boldsymbol{\Delta}(n)\widehat{\boldsymbol{g}}(n) \tag{4.49}$$

For clarity, the final Capon detector is summarized in Box 4.3.

Box 4.3 Summary of Capon receiver with optimized constraints

Initialization

- $\boldsymbol{R}^{-H}(0) = \boldsymbol{I}_{N_f}$ lower triangular matrix
- $\boldsymbol{\Delta}(0) = \boldsymbol{R}^{-1}(0)\boldsymbol{R}^{-H}(0)\boldsymbol{C}_1$
- $\boldsymbol{\Pi}^{-1}(0) = [\boldsymbol{C}_1{}^H \boldsymbol{\Delta}(0)]^{-1}$
- $\mathbb{C} = \boldsymbol{C}_1^H \boldsymbol{C}_1$
- select $\boldsymbol{g}(0)$ to be any orthogonal vector
- $\boldsymbol{f}_{opt}(0) = \boldsymbol{\Delta}(0)\boldsymbol{g}(0)$

(Continued)

Box 4.3 (Continued)

For $n = 1,2,\cdots,$**do**

- Update the inverse Cholesky factor $\boldsymbol{R}^{-H}(n)$, similar to first algorithm
- Update the $\boldsymbol{\Delta}(n)$ matrix as follows: $\boldsymbol{\Delta}(n) = \lambda^{-1}\boldsymbol{\Delta}(n-1) - \boldsymbol{j}(n)\boldsymbol{\pi}(n)$
- Update the $\boldsymbol{\Pi}(n)$ matrix as follows: $\boldsymbol{\Pi}(n) = \frac{1}{\lambda}\boldsymbol{\Pi}(n-1) - \boldsymbol{\pi}^H(n)\boldsymbol{\pi}(n)$
- Update the $\boldsymbol{\Pi}^{-1}(n)$ matrix as follows: $\boldsymbol{\Pi}^{-1}(n) = \lambda\boldsymbol{\Pi}^{-1}(n-1)$ $+\dfrac{\lambda^2\boldsymbol{\Pi}^{-1}(n-1)\boldsymbol{\pi}(n)\boldsymbol{\pi}^H(n)\boldsymbol{\Pi}^{-1}(n-1)}{1-\lambda\boldsymbol{\pi}^H(n)\boldsymbol{\Pi}^{-1}(n-1)\boldsymbol{\pi}(n)}$
- $\beta = \dfrac{\widehat{\boldsymbol{g}}^H(n-1)\boldsymbol{\Pi}(n)\widehat{\boldsymbol{g}}(n-1)}{\widehat{\boldsymbol{g}}^H(n-1)\mathbb{C}\widehat{\boldsymbol{g}}(n-1)}$
- $\widehat{\boldsymbol{g}}(n) = \beta\boldsymbol{\Pi}^{-1}(n)\mathbb{C}\widehat{\boldsymbol{g}}(n-1)$ (estimated channel vector)
- $\boldsymbol{f}_{Capon}(n) = \boldsymbol{\Delta}(n)\widehat{\boldsymbol{g}}(n)$ (weight vector)

End

4.6 New Robust Detection Technique

The max/min approach embraced early in this chapter is notorious for exhibiting performance loss at low SNR [11, 23]. This can be further clarified from its principle: "the maximization of the power after the interference has been mitigated."

$$\max_{\boldsymbol{f}_{\max/\min}} \boldsymbol{f}^H_{\max/\min}\boldsymbol{R}_{xx}\boldsymbol{f}_{\max/\min} \tag{4.50}$$

where $\boldsymbol{f}_{\max/\min}$ is the max/min optimal robust detector update. The received signal correlation matrix $\boldsymbol{R}_{xx}$ can be expressed as in (3.37):

$$\boldsymbol{R}_{xx} = E\{\boldsymbol{x}^H(n)\boldsymbol{x}(n)\} = \boldsymbol{H}\boldsymbol{H}^T + \boldsymbol{R}_n \tag{4.51}$$

where $\boldsymbol{H}$ stands for the system matrix, as estimated in Chapter 2, and $\boldsymbol{R}_n = \sigma_w^2\boldsymbol{I}_{N_f}$ denotes the component of the autocorrelation matrix due to random noise. Substituting (4.51) into (4.50), we get:

$$\max_{\boldsymbol{f}_{\max/\min}} \left(\boldsymbol{f}^H_{\max/\min}\boldsymbol{H}\boldsymbol{H}^T\boldsymbol{f}_{\max/\min} + \sigma_w^2\boldsymbol{f}^H_{\max/\min}\boldsymbol{f}_{\max/\min}\right) \tag{4.52}$$

Therefore, the maximization process includes maximization of the noise constituent $\sigma_w^2\boldsymbol{f}^H_{\max/\min}\boldsymbol{f}_{\max/\min}$. In order to mitigate the effect of noise enhancement, a quadratic constraint can be imposed on the weight vector norm. In addition to this, the proposed detector may gain other merits of the quadratic constraint, such as robustness to pointing errors and random perturbation in detector parameters, and small training sample size [56, 57]. The quadratic inequality constraint will be discussed in detail in Chapter 5. For the sake of completeness, we will employ it here as a cooperative approach to handle the noise enhancement at low SNR. To start the formulation of this constraint, we directly impose the quadratic constraint on the maximization portion of the max/min optimization process. This yields:

$$\max_{\|\boldsymbol{g}\|=1}\left\{\min_{\boldsymbol{f}_{\max/\min}}\left\{\boldsymbol{f}^H_{\max/\min}\boldsymbol{R}_{xx}\boldsymbol{f}_{\max/\min}\right\} s.t.\;\; \boldsymbol{C}_1{}^H\boldsymbol{f}_{\max/\min}=\boldsymbol{g}\right\}$$
$$s.t.\;\; \boldsymbol{f}^H_{\max/\min}\boldsymbol{f}_{\max/\min}\le\tau \tag{4.53}$$

The solution to this optimization problem can be obtained by solving the inner optimization problem using the direct-MOE-IQRD algorithm and then solving the outer optimization problem using the Lagrange method. As a result, the max/min optimum detector with robustness against noise enhancement can be expressed as:

$$\hat{\boldsymbol{f}}_{\max/\min}=(\boldsymbol{R}_{xx}+\nu I)^{-1}\boldsymbol{C}_1\boldsymbol{g}_{\max/\min} \tag{4.54}$$

The max/min channel vector $\boldsymbol{g}_{\max/\min}$ can be obtained using any method from Section 4.5. Again, the Taylor series approximation adopted by Ganz *et al.* [53] can be used to obtain the diagonal loading term ν as follows:

$$\hat{\boldsymbol{f}}_{\max/\min}=(\boldsymbol{I}+\nu\boldsymbol{R}_{xx}^{-1})^{-1}\boldsymbol{R}_{xx}^{-1}\boldsymbol{C}_1\boldsymbol{g}_{\max/\min} \tag{4.55}$$

$$\hat{\boldsymbol{f}}_{\max/\min}\approx(\boldsymbol{I}-\nu\boldsymbol{R}_{xx}^{-1})\boldsymbol{f}_{\max/\min} \tag{4.56}$$

$$\hat{\boldsymbol{f}}_{\max/\min}\approx\boldsymbol{f}_{\max/\min}-\nu\tilde{\boldsymbol{f}}_{\max/\min} \tag{4.57}$$

where

$$\tilde{\boldsymbol{f}}_{\max/\min}=\boldsymbol{R}_{xx}^{-1}f_{\max/\min} \tag{4.58}$$

Therefore, by utilizing the quadratic constraint in (4.53) with the update equation (4.57), we obtain the diagonal loading term υ using a simple quadratic equation. This is similar to approaches found elsewhere [23, 57]. It is interesting to highlight the drawback of the Taylor series approximation (4.56), namely that the diagonal loading term υ will not be precisely calculated in this approximation and even real diagonal loading cannot be guaranteed. Nevertheless, consolidating the channel estimation and the quadratic constraint technique gives an extremely robust detector, especially at low SNR. This will be illustrated in Section 4.8. The quadratic constraint was proposed by Tian *et al.* [57], but their paper provided no direct-form implementation, and even the direct form they proposed is unsuitable for channel tracking because the inverse matrix update is a projection version of the inverse of the autocorrelation matrix. The detector is summarized in Box 4.4, with an IQRD implementation and channel estimation using the original max/min approach. The total multiplication complexity of this robust detector is about $O(3N_fN_g + 2N_g^2 + 2N_f^2 + 4N_g + 9N_f)$. The detailed complexity analysis will be provided in Section 4.9. It is mandatory to evaluate the robust detector in terms of its SINR or BER, and also from a computational complexity point of view. The focus in this book is on simplified robust detection algorithms that can be practically realized by techniques such as CORDIC and systolic arrays.

Box 4.4 Summary of MOE receiver with max/min approach and quadratic constraint

Initialization

- $\boldsymbol{R}^{-H}(0) = \boldsymbol{I}_{N_f}$ lower triangular matrix
- $\boldsymbol{\Delta}(0) = \boldsymbol{R}^{-1}(0)\boldsymbol{R}^{-H}(0)\boldsymbol{C}_1$, $\boldsymbol{\Pi}^{-1}(0) = [\boldsymbol{C}_1{}^H\boldsymbol{\Delta}(0)]^{-1}$
- select $\boldsymbol{g}(0)$ to be any orthogonal vector
- $\boldsymbol{f}_{\max/\min}(0) = \hat{\boldsymbol{f}}_{\max/\min}(0) = \boldsymbol{\Delta}(0)\boldsymbol{g}(0)$, $\mu_{opt}(0) = 0.001$

- **For** $n = 1,2,\cdots,$ **do:**

- Update the inverse Cholesky factor $\boldsymbol{R}^{-H}(n)$ using $\begin{bmatrix} \boldsymbol{R}^{-H}(n) \\ \boldsymbol{j}^H(n) \end{bmatrix}$

$$= \boldsymbol{P}(n)\begin{bmatrix} \frac{\boldsymbol{R}^{-H}(n-1)}{\sqrt{\lambda}} \\ \boldsymbol{0}^T \end{bmatrix}; 6N_f$$

- Where $\boldsymbol{P}(n)\begin{bmatrix} \boldsymbol{a}(n) \\ 1 \end{bmatrix} = \begin{bmatrix} 0 \\ b(n) \end{bmatrix}$ and $\boldsymbol{P}(n)$ is the rotation matrix

$$\boldsymbol{a}(n) = \frac{\boldsymbol{R}^{-H}(n-1)\boldsymbol{x}(n)}{\sqrt{\lambda}}$$

- Update the autocorrelation matrix as follows $\boldsymbol{R}_{xx}^{-1}(n) = \frac{\boldsymbol{R}_{xx}^{-1}(n-1)}{\lambda} - \boldsymbol{j}(n)\boldsymbol{j}^{H}(n)$, $N_f^{\,2}$
- Update the $\boldsymbol{\Delta}(n)$ matrix as follows $\boldsymbol{\Delta}(n) = \lambda^{-1}\boldsymbol{\Delta}(n-1) - \boldsymbol{j}(n)\boldsymbol{\pi}^{H}(n)$; $2N_fN_g$
- Update the $\boldsymbol{\Pi}(n)$ matrix as follows $\boldsymbol{\Pi}(n) = \frac{1}{\lambda}\boldsymbol{\Pi}(n-1) - \boldsymbol{\pi}^{H}(n)\boldsymbol{\pi}(n)$; $N_g^{\,2}$

Channel vector tracking

- $a = \mu_{opt}(n-1)$
- $b = 1 + \mu_{opt}(n-1)\boldsymbol{g}^{H}(n-1)\boldsymbol{\Pi}(n)\boldsymbol{g}(n-1)$; $N_g^{\,2} + N_g$
- $c = \mu_{opt}(n-1)\boldsymbol{g}^{H}(n-1)\boldsymbol{\Pi}^{2}(n)\boldsymbol{g}(n-1) + 2\boldsymbol{g}^{H}(n-1)\boldsymbol{\Pi}(n)\boldsymbol{g}(n-1)$; N_g
- $\zeta(n) = \dfrac{-b - \sqrt{b^2 - \mu_{opt}(n-1)c}}{\mu_{opt}(n-1)}$ (Lagrange multiplier)
- $\nabla_g(n) = \boldsymbol{\Pi}(n)\boldsymbol{g}(n-1) - \zeta(n)\boldsymbol{g}(n-1)$ (gradient vector)
- $\mu_{opt}(n) = \dfrac{\alpha\boldsymbol{g}^{H}(n)\boldsymbol{\Pi}(n)\nabla_g}{\nabla_g^{H}\boldsymbol{\Pi}(n)\nabla_g + \eta}$; $N_g^{\,2} + 2N_g$ (optimum step-size)
- $\boldsymbol{g}(n) = \boldsymbol{g}(n-1) - \mu_{opt}(n)\nabla_g(n)$ (estimated channel vector)

- $\boldsymbol{f}_{\max/\min}(n) = \boldsymbol{\Delta}(n)\boldsymbol{g}(n)$; N_fN_g (max/min detector of first algorithm)

Robust detector with QI constraint (second algorithm)

- if $\|\boldsymbol{f}_{\max/\min}(n)\|^2 > \tau$ (QIC)
  - $\tilde{\boldsymbol{f}}_{\max/\min}(n) = \boldsymbol{R}_{xx}^{-1}(n)\boldsymbol{f}_{\max/\min}(n)$; $N_f^{\,2}$
  - $a_1 = \boldsymbol{f}_{\max/\min}^{\;H}(n)\boldsymbol{f}_{\max/\min}(n)$; N_f
  - $b_1 = -2\mathrm{Re}\{\boldsymbol{f}_{\max/\min}^{\;H}(n)\tilde{\boldsymbol{f}}_{\max/\min}(n)\}$; N_f
  - $c_1 = \tilde{\boldsymbol{f}}_{\max/\min}^{H}(n)\tilde{\boldsymbol{f}}_{\max/\min}(n) - \tau$; N_f
  - $\nu = \dfrac{-b_1 \pm \mathrm{Re}\left\{\sqrt{b_1^{\,2} - 4a_1c_1}\right\}}{2a_1}$
  - $\widehat{\boldsymbol{f}}_{\max/\min} \approx \boldsymbol{f}_{\max/\min} - \nu\tilde{\boldsymbol{f}}_{\max/\min}$ (robust detector)
- else
  - $\widehat{\boldsymbol{f}}_{\max/\min} = \boldsymbol{f}_{\max/\min}$
  - $\nu = 0$
- end if

End

4.7 Systolic Array Implementation

Systolic array implementation of IQRD-based algorithms has important merits. The approach would be very beneficial for real-time applications and parallel processing [40–47]. An advantage of the systolic array over other architectures is that the individual processing element computational complexity does not grow when the number of constraints N_g is increased. The IQRD update method is selected for this purpose, because the direct QR method requires back-substitution for updating the detector, and hence two opposite-direction calculations (triangular update and back substitution) must be performed. Unfortunately, these two operations cannot be implemented in parallel fashion on triangular arrays. Therefore, one of the most important merits of such methods will be lost.

A systolic array implementation of the first algorithm can be found in the paper of Chern and Chang [12] where it was implemented for beamforming applications, as shown in Figure 4.2. The implementation can be tailored easily to multiuser detection or MIMO-OFDM detectors. The LC-IQRD-RLS beamforming algorithm and its systolic array implementation are shown in Box 4.5 and Figure 4.3, respectively. The second algorithm can be implemented pipelined as shown in Figure 4.4;

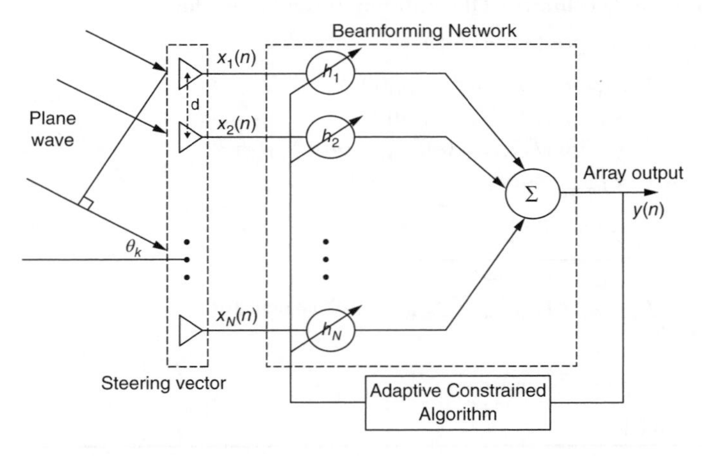

Figure 4.2 Configuration of linearly constrained adaptive-array beamformer.

Box 4.5 LC-IQRD-RLS algorithm

Initialization

- $\mathbf{R}^{-1}(0) = \delta^{-1}\mathbf{I}$, δ = small positive constant
- $\Gamma(0) = \mathbf{R}^{-1}(0)\mathbf{R}^{-H}(0)\mathbf{C}$
- $\mathbf{h}(0) = \Gamma(0)[\mathbf{C}^H\Gamma(0)]^{-1}\mathbf{f}$

For n = 1,2, …, **do**

1. Compute the intermediate vector $\mathbf{z}(n)$

$$\mathbf{z}(n) = \frac{\mathbf{R}^{-H}(n-1)\mathbf{x}(n)}{\sqrt{w}}$$

2. Evaluate the rotations that define $\mathbf{P}(n)$ which annihilates vector $\mathbf{z}(n)$ and compute the scalar variable $t(n)$

$$\mathbf{P}(n)\begin{bmatrix}\mathbf{z}(n)\\ 1\end{bmatrix} = \begin{bmatrix}\mathbf{0}\\ t(n)\end{bmatrix}$$

3. Update the lower triangular matrix $\mathbf{R}^{-H}(n)$ and compute the vector $\mathbf{g}(n)$ and $\alpha(n) = \mathbf{g}^H(n)\mathbf{C}$

$$\mathbf{P}(n)\begin{bmatrix}w^{-1/2}\mathbf{R}^{-H}(n-1)\\ \mathbf{0}^T\end{bmatrix} = \begin{bmatrix}\mathbf{R}^{-H}(n)\\ \mathbf{g}^H(n)\end{bmatrix}$$

Update following equations and intermediate inverse matrix:

- $$\Gamma(n) = w^{-1}\Gamma(n-1) - \mathbf{g}(n)\alpha(n)$$
  $$\mathbf{q}(n) = \frac{\sqrt{w}\Phi^{-1}(n-1)\alpha^H(n)}{1 - w\alpha(n)\Phi^{-1}(n-1)\alpha^H(n)}$$
- $\Phi^{-1}(n) = w[\mathbf{I} + \sqrt{w}\mathbf{q}(n)\alpha(n)]\Phi^{-1}(n-1)$

Updating the LS weight vector:

$$\mathbf{h}(n) = \mathbf{h}(n-1) - \rho(n)\varepsilon(n, n-1)$$

With

$$\rho(n) = \mathbf{k}(n) - \frac{\sqrt{w}}{t(n)}\Gamma(n)\mathbf{q}(n)$$
$$\varepsilon(n, n-1) = \mathbf{x}^H(n)\mathbf{h}(n-1)$$

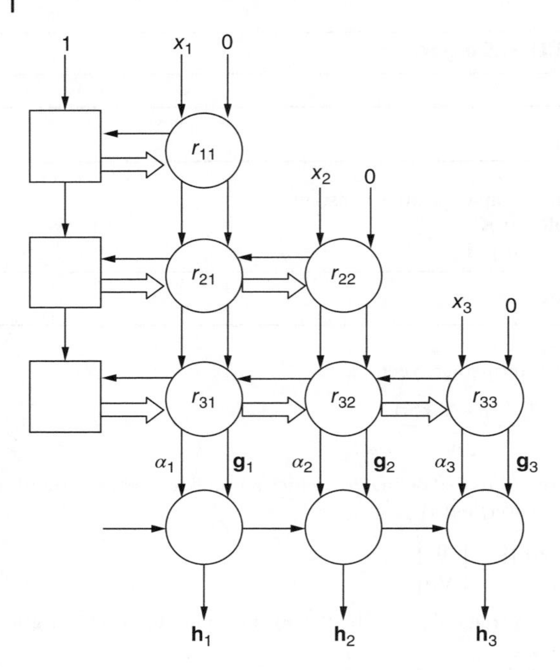

Figure 4.3 Systolic array implementation and processing cells for inverse updating of the LC-IQRD-RLS beamforming.

two approaches PLIC and IQRD-RLS systolic array are combined. The PLIC structure is used to convert the constrained optimization problem to unconstrained adaptive algorithm. IQRD-RLS algorithm is used to update the adaptive portion from the weight vector. The cells used in this implementation are three type cells as described in Figure 4.3. The internal cells are used to update inverse Cholesky factor and each cell contains the corresponding $r_{i,j}$ and concurrently generate the intermediate vector $\boldsymbol{a}(n)$. The boundary cells use the intermediate vector to generate the rotation matrices $\boldsymbol{P}_z(n)$ which are used to update the inverse Cholesky factor. The final cells are used to update the detector parameters and feedback again to the PLIC structure. A recursive formulation for the processing cells in adaptive fashion is also formulated in Figure. 4.4.

The main difference between the first and second schemes from a systolic implementation point of view is as follows: the second algorithm is based on systolic implementation of the unconstrained vector update, with a PLIC structure employed to convert the constrained optimization to an unconstrained one.

Figure 4.3 (*Continued*)

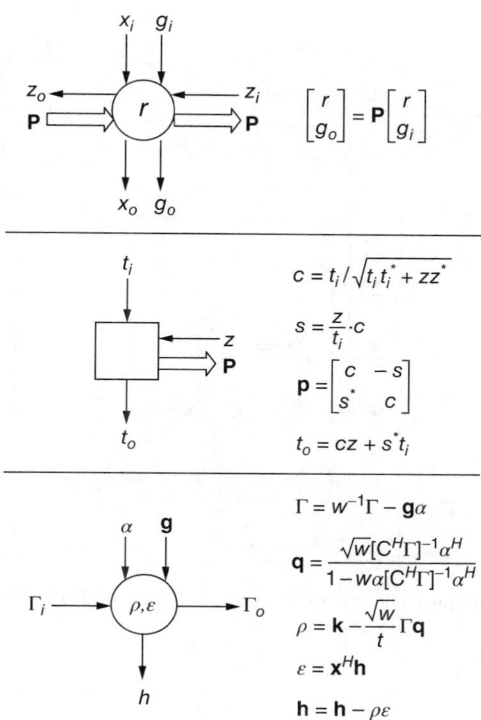

x_i g_i
z_o
P
r
z_i
P
$\begin{bmatrix} r \\ g_o \end{bmatrix} = \mathbf{P}\begin{bmatrix} r \\ g_i \end{bmatrix}$
x_o g_o
t_i
z
P
t_o
$c = t_i/\sqrt{t_i t_i^* + zz^*}$
$s = \frac{z}{t_i}\cdot c$
$\mathbf{p} = \begin{bmatrix} c & -s \\ s^* & c \end{bmatrix}$
$t_o = cz + s^* t_i$
α $\mathbf{g}$
Γ_i
ρ, ε
Γ_o
h
$\Gamma = w^{-1}\Gamma - \mathbf{g}\alpha$
$\mathbf{q} = \frac{\sqrt{w}[C^H\Gamma]^{-1}\alpha^H}{1 - w\alpha[C^H\Gamma]^{-1}\alpha^H}$
$\rho = \mathbf{k} - \frac{\sqrt{w}}{t}\Gamma\mathbf{q}$
$\varepsilon = \mathbf{x}^H\mathbf{h}$
$\mathbf{h} = \mathbf{h} - \rho\varepsilon$

Therefore, the number of constraints will not affect the systolic architecture. In the multiple-constraints case, the final row in the systolic implementation of the direct-MOE-IQRD algorithm is replicated for every constrained vector [12]. Therefore, for multiple constraints, the second scheme is recommended due to the simplicity of its systolic implementation.

The direct-MOE-IQED with max/min algorithm can be implemented as shown in Figure 4.5. The IQRD systolic implementation is used to update the inverse Cholesky factor and to generate the $\boldsymbol{j}(n)$ vector using Givens rotations. The octagonal cell grid with dimension $N_g \times N_f$ is used to update the $\boldsymbol{\Delta}(n)$ matrix and to generate the vector $\boldsymbol{\pi}(n)$ from the last column in the grid, where each cell contains the corresponding $c_{i,j}$ and $\delta_{i,j}$. The vector $\boldsymbol{\pi}(n)$ is used to update $\boldsymbol{\Delta}(n)$ matrix in the octagonal grid and to update the matrix $\boldsymbol{\Pi}(n)$ into a hexagonal cell grid with dimension $N_g \times N_g$. These two steps can be performed

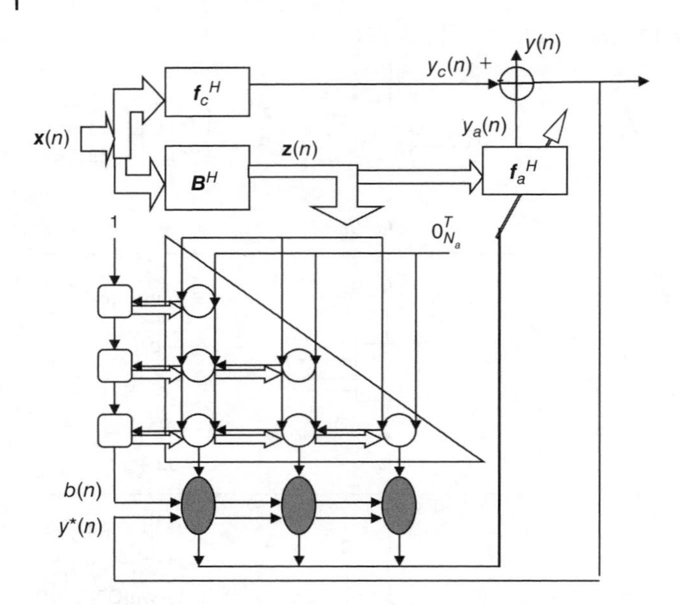

Figure 4.4 Systolic array implantation for PLIC-MOE-IQRD algorithm.

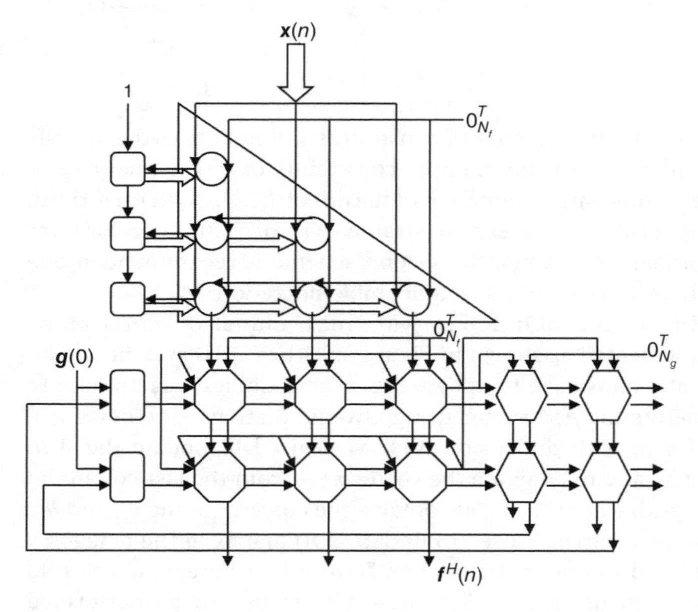

Figure 4.5 Systolic array implementation for direct-MOE-IQRD with max/min.

concurrently. After the update of the matrix $\boldsymbol{\Pi}(n)$, the hexagonal grid generates the matrix multiplication $\boldsymbol{\Pi}(n)\boldsymbol{g}(n-1)$, which is used in the rectangular cells to generate the updated channel vector $\boldsymbol{g}(n)$. The optimal weight vector can be obtained after the $\boldsymbol{\Delta}(n)$ update from the last row in the octagonal cell grid. The definitions for the three new cells used in this systolic implementation are given in Figure 4.6. The definitions of the cells used in IQRD update are similar to the ones used in the second algorithm, except the dimension is N_f instead of N_a.

Finally, some observations may be made concerning the parallel implementation of the third algorithm, which would be very beneficial in real-time applications. The next update of channel vector $\boldsymbol{g}(n+1)$ and the next update of vector $\boldsymbol{\pi}(n+1)$ can be performed concurrently with the estimate of the detector $\boldsymbol{f}(n)$.

4.8 Simulation Results

Five DS/CDMA synchronous users in a multipath Rayleigh fading channel with five multipath components are simulated. Orthogonal Gold codes are employed as the signature waveforms, and the channel length (maximum delay spread) is assumed to be ten delayed components. The multipath delays are randomly distributed. The detector is assumed to be synchronized to the required user. Users' powers are assumed to be equal, except the required user, which is assumed to be 10 dB less than the other users to simulate the near–far effect. Code length is equal to 31 chips and detector length is assumed to span one symbol length and therefore the detector length is 31. The SNR is 30 dB. The Matlab package provided with the book can generate this system or there is already ready a "mat" file with the package, which can be used for simulation.

4.8.1 Experiment 1

In this experiment, the performance of the three detectors presented in this chapter – direct-MOE-IQRD, PLIC-MOE-IQRD and direct-MOE-IQRD w. max/min – are investigated and compared with the direct-MOE-IQRD detector with optimum (actual) channel. The fixed constraints detectors are

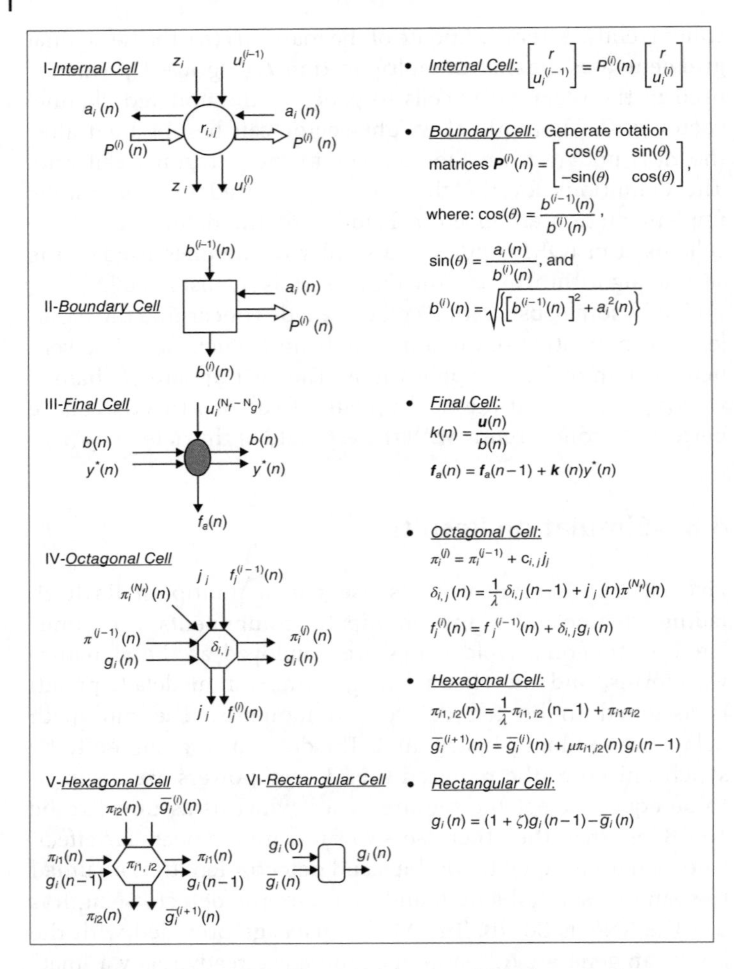

Figure 4.6 Definitions of cells used in the systolic implementations.

initialized with $\boldsymbol{g} = [\alpha \ \cdots \ \alpha]^T / \|\boldsymbol{g}\|$, where $\alpha = 1/N_g$, $\varepsilon = 1$, and $\boldsymbol{f}_a = \mathbf{0}^T_{N_f - N_g}$. The direct-MOE-IQRD w. max/min detector is initialized with $\boldsymbol{g} = [\alpha \ \cdots \ \alpha]^T / \|\boldsymbol{g}\|$, $\alpha = 0.9$ and $\eta = 0.4$. The forgetting factor for all algorithms is set to one. The initializations are the best for all algorithms, so as to give fair comparisons and benchmarking. Two performance measures are adopted here: the output SINR versus snapshot and BER versus snapshot.

Figures 4.7 and 4.8 show these measures for the four detectors. It is evident from the figures that the PLIC-MOE-IQRD algorithm performs similarly to the direct-MOE-IQRD algorithm. The direct-MOE-IQRD w. max/min algorithm has SINR performance close to that of the optimal channel detector, and the BER performance is almost as good.

4.8.2 Experiment 2

It is interesting now to investigate the subspace tracking algorithms. Four subspace tracking algorithms are analyzed as follows:

- The first algorithm is for tracking the minor component using the Lagrange method, and is referred to as MOE-IQRD w. Lagrange (MC).
- The second algorithm is for tracking the principal component using the NOOja algorithm, and is referred to as MOE-IQRD w. NOOja (PC).
- The third algorithm is for tracking the principal component using the Lagrange method, and is referred to as MOE-IQRD w. Lagrange (PC).
- The fourth algorithm is for tracking the principal component with the OPASTd algorithm, and is referred to as MOE-IQRD w. OPASTd (PC).

The Lagrange method, when used to track the minor component, is initialized with $\alpha = 0.95$ and $\eta = 0.4$, and it is initialized with $\alpha = 0.001$ and $\eta = 0.4$ when used to track the principal component. The NOOja algorithm is initialized with $\alpha = 0.005$ and $\eta = 0.4$ and the forgetting factor of the OPASTd algorithm is set to 0.998. In order to analyze the performance of subspace tracking techniques, the above system is used with a 20 dB SNR. Figures 4.9 and 4.10 show the SINR and BER for the direct-MOE-IQRD w. max/min[1] algorithm with both the aforementioned subspace tracking algorithms and the optimum channel. The figures show the superiority of the Lagrange method when used to track the minor component and the NOOja algorithm when used to track the principal component.

1 The word "direct" is dropped in the figure for simplicity of notation.

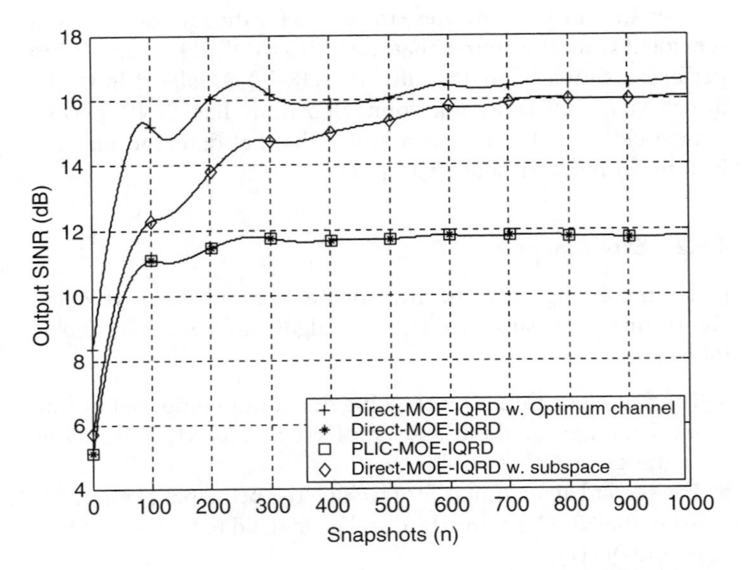

Figure 4.7 Output SINR for IQRD-based detectors versus snapshots.

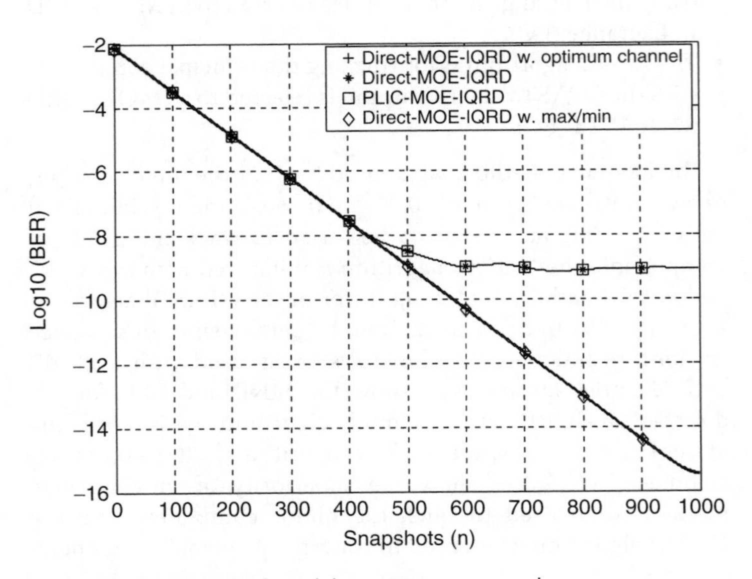

Figure 4.8 BER for IQRD-based detectors versus snapshots.

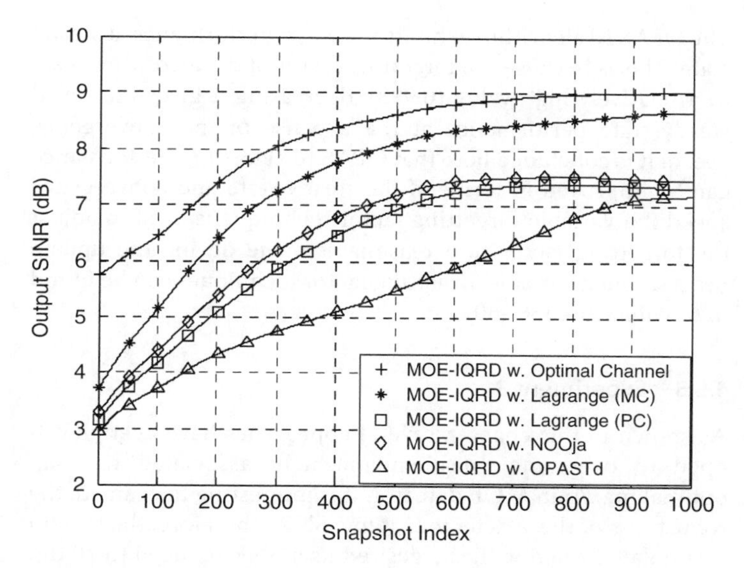

Figure 4.9 Output SINR of optimal MOE-IQRD with different subspace tracking algorithms.

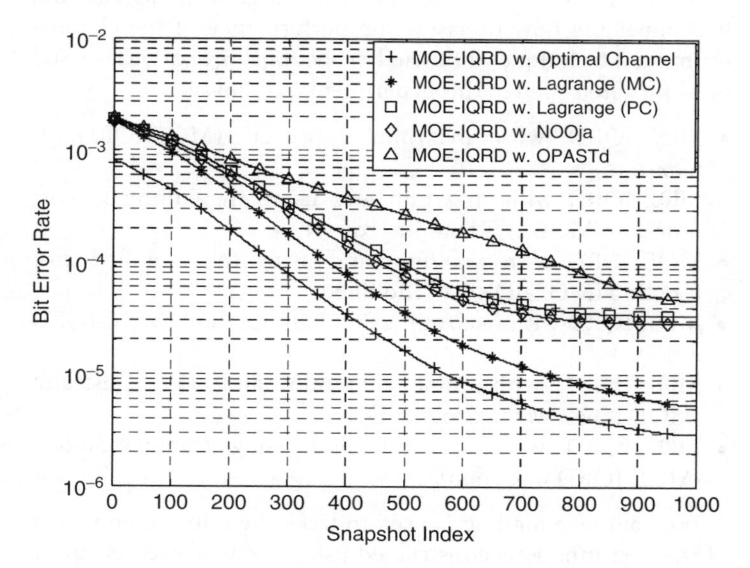

Figure 4.10 BER of optimal MOE-IQRD with different subspace tracking algorithms.

The OPASTd algorithm exhibits the worst performance at steady state. This is because the forgetting factor of the algorithm is set to a relatively high value (0.998); decreasing it gives improved steady-state performance at the expense of the convergence speed. It is concluded here that OPASTd algorithm performance can be improved in terms of the steady state and convergence speed if a variable forgetting factor technique is used to adjust the forgetting factor to a reasonable value during the update process. Such variable forgetting factor techniques can be found in the literature [58–60].

4.8.3 Experiment 3

As shown in this chapter, several approaches have been developed to overcome noise enhancement associated with an optimal max/min MOE detector. It is interesting to evaluate the robustness of these detectors at low SNR. Therefore, the system is simulated again with the desired user SNR reduced to 10 dB. In addition, the system experiences an acute near–far effect in a loaded multipath environment with large delay spread. So, it is appealing now to assess the performance of the channel estimation techniques discussed in Section 4.5. Six IQRD-based detectors are tested in this experiment as follows:

- MOE-IQRD with max/min approach (MOE-IQRD w. max/min)
- MOE-IQRD with max/min and using the improved cost function (MOE-IQRD w. Improved)
- MOE-IQRD with max/min and using the modified cost function (MOE-IQRD w. Modified)
- MOE-IQRD with max/min using POR method (MOE-IQRD w. POR)
- MOE-IQRD with joint max/min and QI constraint (MOE-IQRD w. max/min & QC)
- MOE-IQRD with max/min and using Capon method (MOE-IQRD w. Capon).

The Lagrange method is used to track the minor component of the cost functions constructed using the first five detectors, while the generalized subspace tracking algorithm of Rao *et al.* [55] is exploited to track the channel vector of the Capon

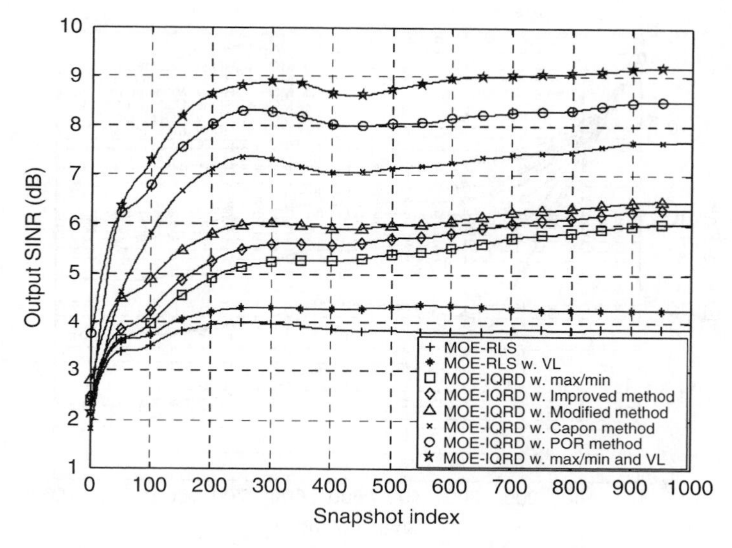

Figure 4.11 SINR versus iterations bits for various MOE detectors at 20 dB SNR.

method. Figures 4.11 and 4.12 show, respectively, the SINR and BER for these six detectors. In addition, MOE-RLS and MOE-RLS w. VL technique are shown in the same figures [56, 57, 61–65].

It is apparent from the figures that the robust approach of Tian *et al.* [57] (i.e. MOE-RLS w. VL) offers little improvement over the MOE-RLS detector with an equal-gain channel. The performance of the MOE-IQRD w. max/min detector is degraded compared with the previous experiment due to the power decrease of the user of interest. A little improvement over the MOE-IQRD w. max/min detector is introduced by the MOE-IQRD w. improved cost function detector. The MOE-IQRD w. modified cost function detector introduces a further improvement over the MOE-IQRD w. improved cost function detector. The constant τ of the modified cost function in (4.27) is adjusted so that $\tau\sigma_w^2 = \sigma_w^2 + 0.5$. If the constant τ is adjusted so that $0 \le \tau < 1$ [9], the MOE-IQRD w. modified cost function detector performs very similarly to the MOE-IQRD w. max/min detector [10]. This is due to the imperfect estimation of the noise power due to covariance matrix estimation errors.

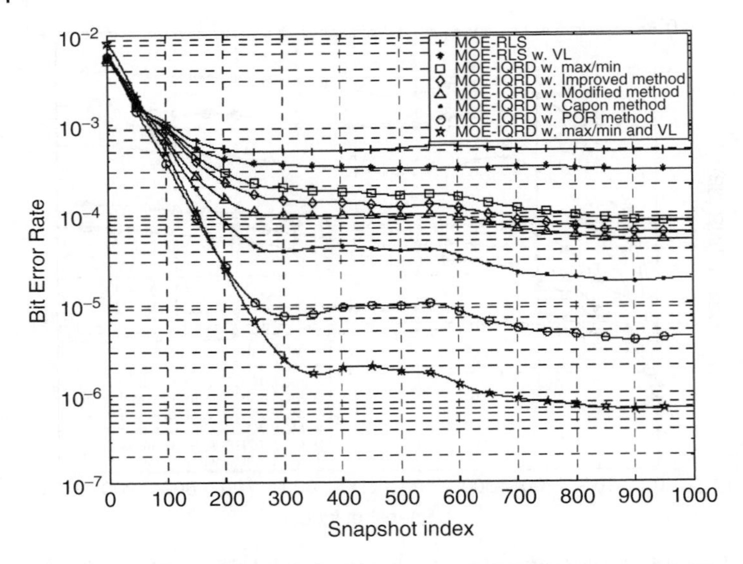

Figure 4.12 BER versus iterations bits for various MOE detectors at 20 dB SNR.

It is important to highlight that this selection is based on system parameters, noise power, and the simulation environment. However, we should be careful to avoid violation of data covariance matrix positiveness. The Capon method comes in the next position, bringing significant improvement. The POR receiver is better still. The MOE-IQRD w. max/min & VL detector exhibit the best performance in terms of SINR and BER [23].

4.8.4 Experiment 4

It is important now to analyze the effect of the QI constraint value on the performance of MOE-IQRD w. max/min & QC. Figures 4.13 and 4.14 show the SINR and BER, respectively, for the MOE-IQRD w. max/min & QC with three different QI constrained values. It is apparent from the figures that the performance of the detector is improved with a decrease of the QI constraint value. This is because the noise contribution decreases monotonically with a decrease of the detector norm. However, the performance improvement is converged at a low constrained value (0.001). No further improvement is seen beyond this value.

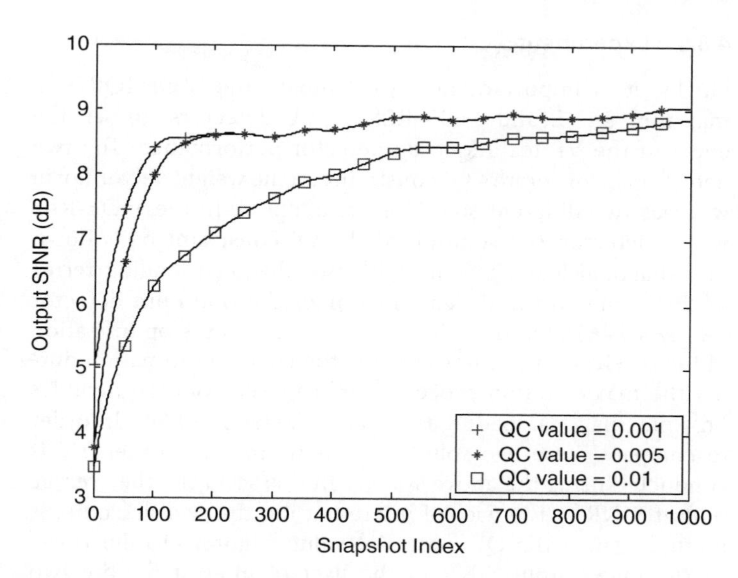

Figure 4.13 Output SINR of MOE-IQRD w. max/min & QC for different constrained values.

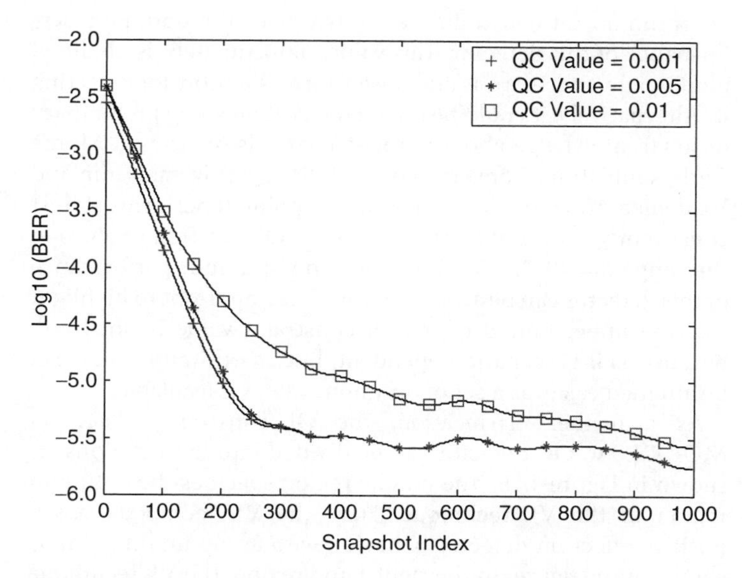

Figure 4.14 BER of MOE-IQRD w. max/min & QC for different constrained values.

4.8.5 Experiment 5

Finally, it is important now to compare the MOE-IQRD w. max/min & VL and MOE-RLS w. VL detectors, to see the effect of the VL technique on detector performance. The two detectors incorporate a QI constraint on the weight vector norm whereas two different solutions are adopted. In the MOE-RLS w. VL detector, the solution of the QI constraint boils down to a diagonal loading form, with two diagonal loading terms in the numerator and denominator of the optimum detector [56, 57, 61–65]. On the other hand, because of the optimization of the receiver output power using the channel estimation during the maximization process, the proposed solution includes only one diagonal loading term, as in Equation (4.54). In order to compare these two solutions, a new simulation scenario is conducted using the above system. In this scenario, the average output SINR of the user of interest using the two detectors is plotted against the QI constraint value. Figure 4.15 illustrates the average output SINR of the user of interest for the two detectors. By inspecting this figure, an important conclusion can be drawn, namely that the MOE-RLS w. VL detector has an optimum QI constraint value that gives the optimum performance of this detector. This value, unfortunately, is obtained empirically and there is no closed-form solution for selecting it. Alternatively, it could be selected based on some preliminary (coarse) knowledge about wireless channels or using a Monte Carlo simulation. Conversely, the MOE-IQRD w. max/min and VL detector does not have an optimum point. It performs almost consistently, with rigid optimality, over QI constraint values in the range $\rho = (0, 0.005)$. Therefore, the optimum performance of this detector can be attained easily. It is important to highlight that the upper bound of the QI constraint value is not fixed, because as it is scenario dependent, but an optimum value can be attained easily as a set of optimum values is available.

As a further clarification, the QI constraint effect on MOE-RLS w. QC detector can be divided into three regions, as shown in Figure 4.15. The middle region specifies the optimum region of the VL technique. Here the VL technique has a positive effect on detector behavior, with an optimum point as shown in the figure. In the right-hand region, the VL technique no longer affects the algorithm behavior and hence the detector

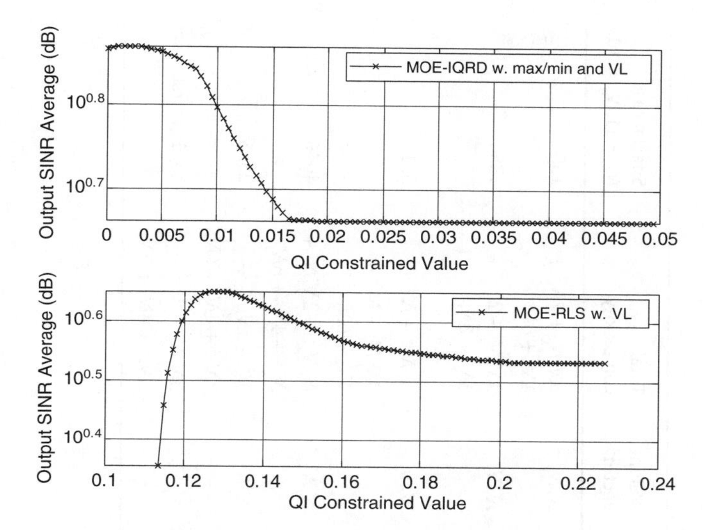

Figure 4.15 Comparison between MOE-IQRD w. max/min & VL and MOE-RLS w. VL.

performance is reduced to that of the MOE-RLS detector without QI constraint. In the left-hand region, the VL technique has an detrimental effect on the detector and its performance is dramatically affected.

The MOE-IQRD w. max/min & QC detector performance can also be divided into three regions. The first region is a steady-state region with rigid optimality. In the second region, the performance of the detector is monotonically decreasing with the increase of QI constraint value, until the effect of the QI constraint is negligible; then the detector boils down to a MOE-IQRD w. max/min detector.

4.9 Complexity Analysis

In order to assess the MOE-IQRD detectors simulated in the previous section, we need to compare their computational complexities. Table 4.1 compares the six detectors analyzed in the third experiment. It has been shown [66] that the recursive updating of the triangular matrix (i.e. Kalman gain) of

Table 4.1 Multiplication complexity comparison of MOE-IQRD detectors.

Detector	Kalman gain	Intermediate matrix update	Channel estimation and VL technique	Weight vector	Total complexity	Special case $N_f = 31, N_g = 10$
MOE-IQRD w. max/min	$6N_f$	$2N_fN_g + N_g^2$	$N_g^2 + 4N_g$ (CH)	N_fN_g	$3N_fN_g + 2N_g^2 + 4N_g + 6N_f$	1356
MOE-IQRD w. improved cost function	$6N_f$	$2N_fN_g + N_g^2$	$N_g^2 + 4N_g$ (CH)	N_fN_g	$3N_fN_g + 2N_g^2 + 4N_g + 6N_f$	1356
MOE-IQRD w. modified cost function	$6N_f$	$N_fN_g^2 + N_f^2 + 2N_fN_g$	$N_g^2 + 4N_g$ (CH)	N_fN_g	$N_fN_g^2 + N_f^2 + 3N_fN_g + N_g^2 + 4N_g + 6N_f$	5317
MOE-IQRD w. POR method	$6N_f$	$N_fN_g^2 + N_fN_g$	$N_g^2 + 4N_g$ (CH)	N_fN_g	$N_fN_g^2 + 2N_fN_g + N_g^2 + 4N_g + 6N_f$	4046
MOE-IQRD w. Capon method	$6N_f$	$2N_g^2 + 2N_g + 2N_fN_g$	$N_g^3 + 2N_g^2 + 2N_g$ (CH)	N_fN_g	$N_g^3 + 3N_fN_g + 4N_g^2 + 4N_g + 6N_f$	2556
MOE-IQRD w. max/min & VL	$6N_f$	$2N_fN_g + N_g^2$	$N_g^2 + 4N_g$ (CH) $2N_f^2 + 3N_f$ (VL)	N_fN_g	$3N_fN_g + 2N_g^2 + 2N_f^2 + 4N_g + 9N_f$	3371 (worst case)

IQRD-RLS algorithm requires $O(N_f)$ operations [12]. Therefore, the Cholesky factor $\boldsymbol{j}(n)$ requires $6N_f$ multiplications. The final detector will require $N_f N_g$ multiplications due to the matrix $\boldsymbol{\Delta}(n)$ being multiplied by the channel vector. Other complexities are summarized in Table 4.1. The last column provides the multiplication complexities for a special case exercised in the simulation experiments ($N_f = 31, N_g = 10$). It is evident from this column that the first and the second detectors have the same complexity and perform in almost the same way. The third detector involves very high complexity while not providing any improvement over the first and second detectors. The proposed robust detector, which is the last detector in the table, requires less computational complexity than the POR method, while it offers the best performance in terms of SINR and BER. The Capon method requires less computational complexity, while exhibiting worse performance than the POR method. In addition, it is noteworthy that the last detector will not perform the VL technique through all recursive steps. Therefore, the VL technique complexity is multiplied by a factor to produce the average complexity required per iteration. Therefore, the computational complexity in the table represents the worst-case scenario.

To further analyze the complexity analysis, we plotted the complexity analysis in terms of the number of multiplications per iteration versus the detector length while the channel length is fixed at 10. We also plotted the complexity analysis versus the channel length while the detector length is fixed at 62; that is., two bits. Figures 4.16 and 4.17 demonstrate these two scenarios respectively.

Figures 4.15 and 4.16, along with performance results in Experiment 3, show the following key points:

- The MOE-IQRD w. max/min & VL performs the best in terms of complexity and performance at lower detector length.
- At higher detector length, the MOE-IQRD w. Capon method is the best in terms of complexity.
- MOE-IQRD w. max/min & VL is the best detector in terms of complexity at higher channel size, followed by Capon method.
- MOE-IQRD w. POR requires high computational complexity, especially at high channel length.

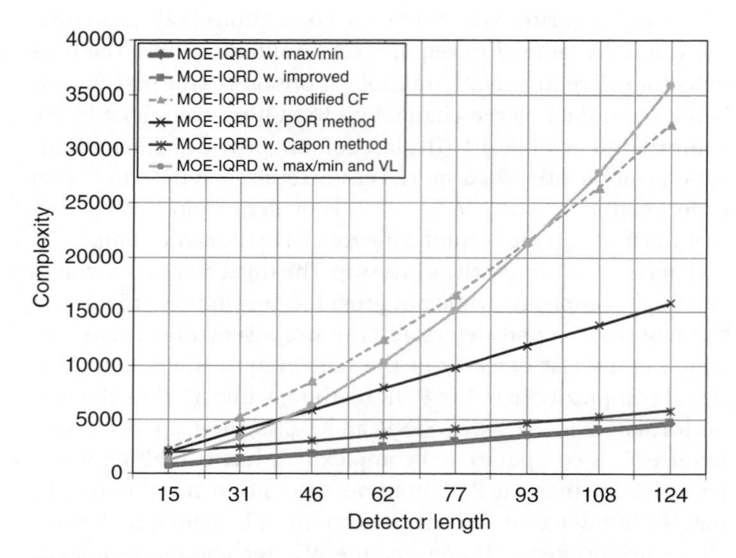

Figure 4.16 Complexity analysis versus detector length at fixed channel length.

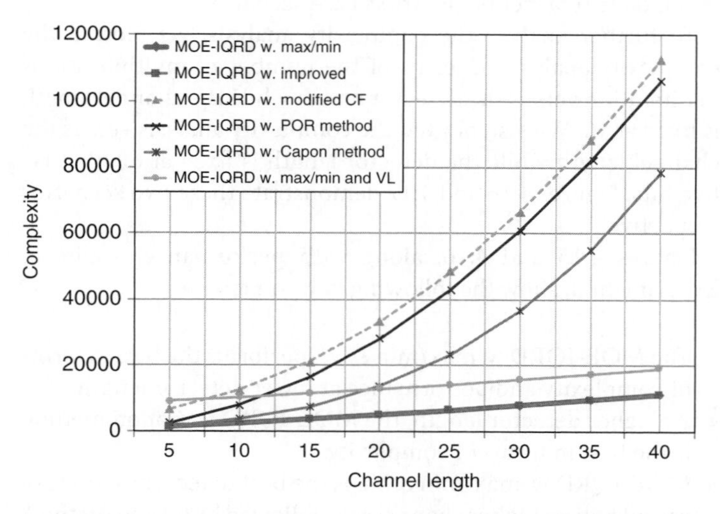

Figure 4.17 Complexity analysis versus channel length at fixed detector length.

- MOE-IQRD w. modified CF requires the highest computational load in the majority of cases, while its performance is similar to that of the MOE-IQRD w. max/min.
- The selection of the optimal detector is limited for max/min with VL, POR, and Capon, and the choice depends on the required computational load and the acceptable level of performance.

Additional information on QRD algorithms for application to MIMO-OFDM systems and beamforming can be found in the literature [67–75]. The author of this chapter used inverse QRD as the core of the presented algorithms. However, the same algorithms can be used with QRD decomposition and can be extended to other wireless communications systems such as MIMO-OFDM and beamforming.

Appendix 4.A Summary of Inverse QR Algorithm with Inverse Updating

Starting from the update equation for the direct QRD method:

$$\begin{bmatrix} \boldsymbol{R}(n) \\ 0^T \end{bmatrix} = \boldsymbol{T}(n) \begin{bmatrix} \sqrt{\lambda}\boldsymbol{R}(n-1) \\ \boldsymbol{x}^H(n) \end{bmatrix} \tag{4A.1}$$

where $\boldsymbol{T}(n)$ is a $(N_f + 1) \times (N_f + 1)$ orthogonal matrix; when multiplied, the last row of the right-hand side of (4A.1) should be converted to zeros to be similar to the left-hand side. It can be found using the Givens rotations method. We can solve our linearly constrained algorithm using the above equation, but we will use the back substitution method, which impedes the parallelization of the algorithm. Therefore, we will summarize the inverse QRD method, which allows us to develop a linearly constrained MVDR receiver without back substitutions.

Multiplying the matrices on each side of (4A.1) by the respective transpose of each side yields

$$\begin{aligned} &\begin{bmatrix} \boldsymbol{R}(n) & 0^T \end{bmatrix} \begin{bmatrix} \boldsymbol{R}(n) \\ 0^T \end{bmatrix} \\ &= \begin{bmatrix} \sqrt{\lambda}\boldsymbol{R}(n-1) & \boldsymbol{x}^H(n) \end{bmatrix} \boldsymbol{T}^H(n)\boldsymbol{T}(n) \begin{bmatrix} \sqrt{\lambda}\boldsymbol{R}(n-1) \\ \boldsymbol{x}^H(n) \end{bmatrix} \end{aligned} \tag{4A.2}$$

Since $\boldsymbol{T}(n)$ is orthogonal matrix, then $\boldsymbol{T}^H(n)\boldsymbol{T}(n) = \boldsymbol{I}$ and (4A.2) may be reduced to:

$$\boldsymbol{R}^H(n)\boldsymbol{R}(n) = \lambda \boldsymbol{R}^H(n-1)\boldsymbol{R}(n-1) + \boldsymbol{x}(n)\boldsymbol{x}^H(n) \qquad (4A.3)$$

By applying matrix inversion lemma, we get

$$\boldsymbol{R}^{-1}(n)\boldsymbol{R}^{-H}(n) = \frac{\boldsymbol{R}^{-1}(n-1)\boldsymbol{R}^{-H}(n-1)}{\lambda} - \frac{\boldsymbol{R}^{-1}(n-1)\boldsymbol{a}(n)\boldsymbol{a}^H(n)\boldsymbol{R}^{-H}(n-1)}{\lambda b^2(n)} \qquad (4A.4)$$

where

$$a(n) = \frac{R^{-H}(n-1)x(n)}{\sqrt{\lambda}} \qquad (4A.5)$$

The matrix $\boldsymbol{R}^{-H}(n-1)$ is lower triangular matrix. Equation (4A.4) can be further simplified by defining the following vector:

$$\boldsymbol{u}(n) = \frac{\boldsymbol{R}^{-1}(n-1)\boldsymbol{a}(n)}{b(n)\lambda} \qquad (4A.6)$$

Then Equation (4A.4) can be rewritten as follows:

$$\boldsymbol{R}^{-1}(n)\boldsymbol{R}^{-H}(n) + \boldsymbol{u}(n)\boldsymbol{u}^H(n) = \frac{\boldsymbol{R}^{-1}(n-1)\boldsymbol{R}^{-H}(n-1)}{\lambda} \qquad (4A.7)$$

Equation (4A.7) implies the existence of an $(N_f + 1) \times (N_f + 1)$ orthogonal matrix $\boldsymbol{P}(n)$ satisfying:

$$\begin{bmatrix} \boldsymbol{R}^{-H}(n) \\ \boldsymbol{u}^H(t) \end{bmatrix} = \boldsymbol{P}(n) \begin{bmatrix} \frac{\boldsymbol{R}^{-H}(n-1)}{\sqrt{\lambda}} \\ 0^T \end{bmatrix} \qquad (4A.8)$$

It is a straightforward to show that the orthogonal matrix $\boldsymbol{P}(n)$ is a rotation matrix, which successively annihilates the elements of vector $\boldsymbol{u}(n)$, from top to bottom, into $b(n)$ as shown in the following equation.

$$\boldsymbol{P}(n) \begin{bmatrix} \boldsymbol{u}(n) \\ 1 \end{bmatrix} = \begin{bmatrix} 0 \\ b(n) \end{bmatrix} \qquad (4A.9)$$

Appendix 4.B QR Decomposition Algorithms

Assume an real or complex matrix **A**. To decompose the matrix **A** into an upper triangular matrix **R**, the Gram–Schmidt algorithm projects the ith column vector a_i of matrix **A** onto the subspace spanned by $a_1, \ldots, a_{i-1}$ to obtain the element $r_{j,i}$ of **R** for $1 <= j < i$. Then the unit column vector of the unitary matrix **Q** is derived by normalizing the new orthogonal basis that is formed by removing these linearly dependent components from a_i. The pseudo codes of the Gram–Schmidt algorithm for processing both real and complex matrices are given in Table 4B.1. For an $N \times 1$ vector **b**, to perform $z = Q^H b$, additional matrix

Table 4B.1 Gram–Schmidt algorithm.

Real matrix	**Add.**	**Mul.**	**Div.**	**Sqrt.**
For $i = 1:N$				
$\mathbf{u}_i = \mathbf{a}_i$				
For $j = 1:i-1$				
$r_{j,i} = \mathbf{q}_j^T \mathbf{a}_i$	$N-1$	N		
$\mathbf{u}_i = \mathbf{u}_i - r_{j,i}\mathbf{q}_j$	N	N		
End				
$r_{i,i} = \|\mathbf{u}_i\|$	$N-1$	N		1
$\mathbf{q}_i = \mathbf{u}_i / r_{i,i}$			N	
$z_i = \mathbf{q}_i^T \mathbf{b}$	$N-1$	N		
End				
Complex matrix	Add.	Mul.	Div.	Sqrt.
For $i = 1:N$				
$\mathbf{u}_i = \mathbf{a}_i$				
for $j = 1:i-1$				
$r_{j,i} = \mathbf{q}_j^H \mathbf{a}_i$	$4N-2$	$4N$		
$\mathbf{u}_i = \mathbf{u}_i - r_{j,i}\mathbf{q}_j$	$4N$	$4N$		
End				
$r_{i,i} = \|\mathbf{u}_i\|$	$2N-1$	$2N$		1
$\mathbf{q}_i = \mathbf{u}_i / r_{i,i}$			$2N$	
$z_i = \mathbf{q}_i^H \mathbf{b}$	$4N-2$	$4N$		
End				

multiplication is needed. The complexity corresponding to each step is evaluated by its arithmetic operations and is also given in Table 4B.1. Although the modified Gram–Schmidt algorithm can be used for numerical stability considerations, the arithmetic complexity is basically the same.

The Householder transformation reflects a vector cross a hyperplane, called the Householder plane, to obtain a mirror vector αe_1, where e_1 is a unit vector with 1s in the first entry to the algorithm and α is related with the norm of the original vector, as shown in Table 4B.2. The vector v denotes the difference between the vector to be processed and the mirror vector. The subscript $(i : N)$ indicates the ith entry to the Nth entry in the vector or matrix. Unlike the Gram–Schmidt algorithm, it is not necessary to calculate the matrix $\mathbf{Q}$ explicitly in order to obtain $z = \mathbf{Q}^H b$. The complexity is evaluated in terms of $\gamma = N - i + 1$.

Table 4B.2 Householder transformation.

Real matrix	**Add.**	**Mul.**	**Div.**	**Sqrt.**
$\mathbf{R} = \mathbf{A}, \mathbf{z} = \mathbf{b}$				
For i $= 1{:}N - 1$				
$\alpha = \text{sign}(\mathbf{R}_{i,i})\|\mathbf{R}_{i:N,i}\|$	$\gamma - 1$	γ		1
$\mathbf{v} = \mathbf{R}_{i:N,i} - \alpha\mathbf{e}_1$	1			
$\mathbf{w} = \mathbf{v}\mathbf{v}^T/(\mathbf{v}^T\mathbf{v})$	$\gamma - 1$	$\gamma^2 + \gamma$	γ^2	
$\mathbf{R}_{i:N,i:N} = \mathbf{R}_{i:N,i:N}$	γ^3	γ^3		
$-2\mathbf{w}\mathbf{R}_{i:N,i:N}$				
$\mathbf{z}_{i:N} = \mathbf{z}_{i:N} - 2\mathbf{w}\mathbf{z}_{i:N}$	γ^2	γ^2		
End				
Complex matrix	Add.	Mul.	Div.	Sqrt.
$\mathbf{R} = \mathbf{A}, \mathbf{z} = \mathbf{b}$				
For i $= 1{:}N - 1$				
$\alpha = \|\mathbf{R}_{i:N,i}\|$	$2\gamma - 1$	2γ		1
$\mathbf{v} = \mathbf{R}_{i:N,i} - \alpha\mathbf{e}_1$	1			
$\mathbf{w} = \mathbf{v}\mathbf{v}^H/\mathbf{v}^H\mathbf{R}_{i:N,i}$	$5\gamma^2+$	$10\gamma^2+$	$2\gamma^2$	
	$4\gamma - 2$	4γ		
$\mathbf{R}_{i:N,i:N} = \mathbf{R}_{i:N,i:N}$	$4\gamma^3$	$4\gamma^3$		
$-\mathbf{w}\mathbf{R}_{i:N,i:N}$				
$\mathbf{z}_{i:N} = \mathbf{z}_{i:N} - \mathbf{w}\mathbf{z}_{i:N}$	$4\gamma^2$	$4\gamma^2$		
End				

As explained in IQRD algorithm, the Givens rotation tries to eliminate the undesired elements one by one to get an upper triangular matrix. The operations of the Givens rotation for deriving the upper triangular matrix **R** are summarized in Table 4B.3. Obviously, the same strategy can be employed to get vector $z = \boldsymbol{Q}^H b$ [15]. Similar to the IQRD algorithm, Givens rotation can also be applied by the famous coordinate rotation digital computer (CORDIC) algorithm. The architecture of the complex Givens rotation can be implemented as shown in Figure 4B.1.

Appendix 4.C Subspace Tracking Algorithms

Box 4C.1 Summary of PAST subspace tracking algorithm

For $n = 1,2,\cdots$, **do**

- $y(n) = \boldsymbol{v}^H(i)\boldsymbol{x}(n)$
- $\lambda(i) = \beta\lambda(i-1) + |y(i)|^2$
- $\boldsymbol{v}(i) = \boldsymbol{v}(i) + [\boldsymbol{x}(i) - \boldsymbol{v}(i-1)y(i)]\dfrac{y^*(i)}{\lambda(i)}$
- Return $\boldsymbol{v}(i)$ and $\lambda(i)$ principal eigenvector and corresponding maximum eigenvalue

End

Box 4C.2 Summary of OPAST subspace tracking algorithm

For $n = 1,2,\cdots$, **do**

- $y(n) = \boldsymbol{v}^H(i)\boldsymbol{x}(n)$
- $\boldsymbol{q}(i) = \dfrac{1}{\alpha}\boldsymbol{Z}(i-1)\boldsymbol{y}(i)$
- $\gamma(i) = \dfrac{1}{1 - \boldsymbol{y}^H(i)\boldsymbol{q}(i)}$
- $\boldsymbol{p}(i) = \gamma(i)(\boldsymbol{x}(i) - \boldsymbol{W}(i-1)\boldsymbol{y}(i))$
- $\boldsymbol{Z}(i) = \dfrac{1}{\alpha}\boldsymbol{Z}(i-1) - \gamma(i)\boldsymbol{q}(i)\boldsymbol{q}^H(i)$
- $\tau(i) = \dfrac{1}{\|\boldsymbol{q}(i)\|^2}\left(\dfrac{1}{\sqrt{1 + \|\boldsymbol{p}(i)\|^2\|\boldsymbol{q}(i)\|^2}} - 1\right)$

(Continued)

Table 4B.3 Givens rotations.

Real matrix	**Add.**	**Mul.**	**Div.**	**Sqrt.**	**CORDIC operations**
$\mathbf{R} = \mathbf{A}, \mathbf{z} = \mathbf{b}$					
For $i = 1 : N - 1$					
For $j = i + 1 : N$					
$\alpha = \Vert[\mathbf{R}_{i,i}\ \mathbf{R}_{j,i}]\Vert$	1	2		1	
$c = \mathbf{R}_{i,i}/\alpha$			1		
$s = \mathbf{R}_{j,i}/\alpha$			1		
$\mathbf{R}_{i,i:N} = c\mathbf{R}_{i,i:N} + s\mathbf{R}_{j,i:N}$	$(N - i + 1)$	$2(N - i + 1)$			$N - i + 1$
$\mathbf{R}_{j,i:N} = -s\mathbf{R}_{i,i:N} + c\mathbf{R}_{j,i:N}$	$(N - i + 1)$	$2(N - i + 1)$			
$z_i = cz_i + sz_j$	1	2			1
$z_j = -sz_i + cz_j$	1	2			
End					
End					
Complex matrix	Add.	Mul.	Div.	Sqrt.	CORDIC operations
$\mathbf{R} = \mathbf{A}, \mathbf{z} = \mathbf{b}$					
for $i = 1 : N - 1$					
for $j = i + 1 : N$					
$\alpha = \Vert[\mathbf{R}_{i,i}\ \mathbf{R}_{j,i}]\Vert$	3	4		1	
$c = \mathbf{R}_{i,i}^H/\alpha$			2		
$s = \mathbf{R}_{j,i}^H/\alpha$			2		
$\mathbf{R}_{i,i:N} = c\mathbf{R}_{i,i:N} + s\mathbf{R}_{j,i:N}$	$6(N - i + 1)$	$8(N - i + 1)$			$4(N - i) + 3$ or
$\mathbf{R}_{j,i:N} = -s^H\mathbf{R}_{i,i:N} + c^H\mathbf{R}_{j,i:N}$	$6(N - i + 1)$	$8(N - i + 1)$			$3(N - i) + 2$ [a]
$z_i = cz_i + sz_j$	6	8			4 or 3[a]
$z_j = -s^H z_i + c^H z_j$	6	8			
End					
End					
$\mathbf{R}_{N,N} = \mathbf{R}_{N,N}\mathbf{R}_{N,N}^H/\vert\mathbf{R}_{N,N}\vert$	1	2		1	1
$z_N = z_N\mathbf{R}_{N,N}^H/\vert\mathbf{R}_{N,N}\vert$	2	4	2		1

a) If $R_{i,i}$ is real.

Box 4C.2 (Continued)

- $\overline{\boldsymbol{p}}(i) = \tau(i)\boldsymbol{W}(i-1)\boldsymbol{q}(i) + (1 + \tau(i)\Vert\boldsymbol{q}(i)\Vert^2)\boldsymbol{p}(i)$
- $\boldsymbol{W}(i) = \boldsymbol{W}(i-1) + \overline{\boldsymbol{p}}(i)\boldsymbol{q}^H(i)$
- Return eigenvectors matrix $\boldsymbol{W}(i)$ and eigenvalues $\boldsymbol{Z}(i)$

End

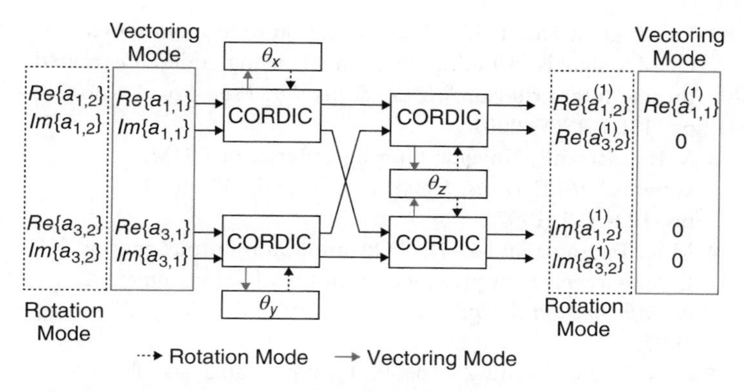

Figure 4B.1 Architecture of complex Givens rotation.

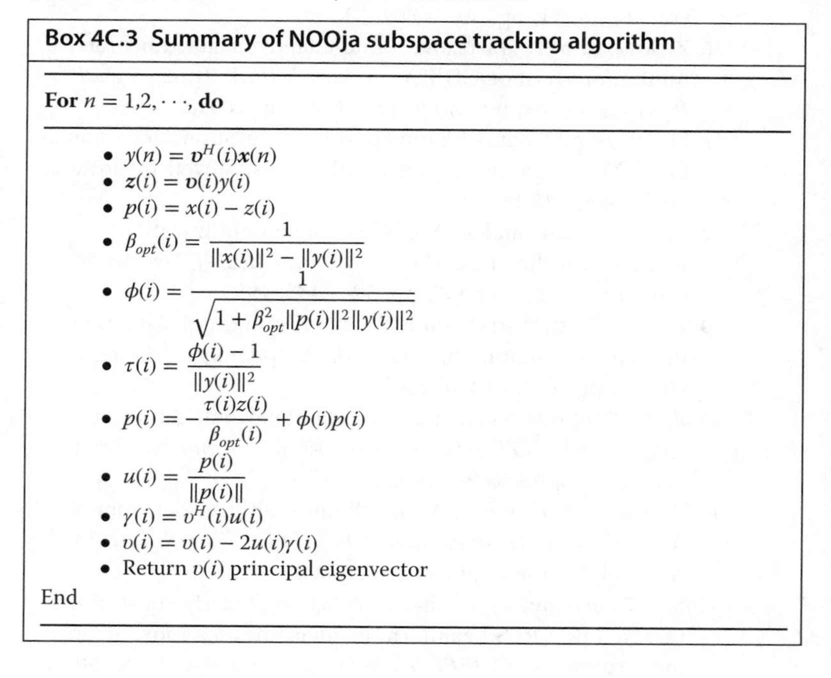

Box 4C.3 Summary of NOOja subspace tracking algorithm

For $n = 1,2,\cdots,$ **do**

- $y(n) = \boldsymbol{v}^H(i)\boldsymbol{x}(n)$
- $z(i) = \boldsymbol{v}(i)y(i)$
- $p(i) = x(i) - z(i)$
- $\beta_{opt}(i) = \dfrac{1}{\|x(i)\|^2 - \|y(i)\|^2}$
- $\phi(i) = \dfrac{1}{\sqrt{1 + \beta^2_{opt}\|p(i)\|^2\|y(i)\|^2}}$
- $\tau(i) = \dfrac{\phi(i) - 1}{\|y(i)\|^2}$
- $p(i) = -\dfrac{\tau(i)z(i)}{\beta_{opt}(i)} + \phi(i)p(i)$
- $u(i) = \dfrac{p(i)}{\|p(i)\|}$
- $\gamma(i) = v^H(i)u(i)$
- $v(i) = v(i) - 2u(i)\gamma(i)$
- Return $v(i)$ principal eigenvector

End

References

1 J.K. Tugnait and T. Li. "Blind detection of asynchronous CDMA signals in multipath channels using code-constrained inverse filter criterion," *IEEE Trans. Signal Process.*, vol. 49, no. 7, pp. 1300–1309, 2001.

2 J.K. Tugnait and T. Li, "Blind detection of asynchronous CDMA signals in multipath channels using code-constrained inverse filter criterion," *IEEE Trans. Sig. Proc.* vol. 49, no. 7, pp. 1300–1309, 2001.
3 M.K. Tsatsanis, "Inverse filtering criteria for CDMA systems," *IEEE Trans. Signal Process.*, vol. 45, no. 1, pp. 102–112, 1997.
4 M.K. Tsatsanis and Z. Xu, "On minimum output energy CDMA receives in presence of multipath." In: *Conference on Information Sciences and Systems (CISS'97)*, pp. 377–381, 1997.
5 M.K. Tsatsanis and Z. Xu, "Performance analysis of minimum variance CDMA receivers," *IEEE Trans. Signal Process.*, vol. 46, no. 11, pp. 3014–3022, 1998.
6 Z. Xu and M.K. Tsatsanis "Blind adaptive algorithms for minimum variance CDMA receivers", *IEEE Trans. Signal Process.*, vol. 49, no. 1, pp. 180–194, Jan. 2001.
7 Z. Xu, Blind Channel Estimation and Multiuser Detection for CDMA Communications, PhD thesis, Steven Institute of Technology, 1999.
8 M.K. Tsatsanis and Z. Xu, "Constrained optimization methods for direct blind equalization," *IEEE J. Sel. Areas Commun.*, vol. 17, no. 3, pp. 424–433, 1999.
9 Z. Xu, "Further study on MOE-based multiuser detection in unknown multipath," *EURASIP J. Appl. Signal Process.* vol. 12, pp. 1377–1386, 2002.
10 Z. Xu "Improved constraint for multipath mitigation in constrained MOE multiuser detection," *J. Commun. Netw.*, vol. 3, no. 3, pp. 189–198, 2001.
11 Z. Xu, L. Ping, and X. Wang "Blind multiuser detection: from MOE to subspace methods," *IEEE Trans. Signal Process.* vol. 52, no. 2, pp. 510–524, 2004.
12 S.-J. Chern and C.-Y. Chang, "Adaptive linearly constrained inverse QRD-RLS beamforming algorithm for moving jammers suppression", *IEEE Trans. Antennas Propag.*, vol. 50, no. 8, pp. 1138–1150, 2002.
13 S.T. Alexander and A.L. Ghirnikar, "A method for recursive least squares filtering based upon an inverse QR decomposition", *IEEE Trans. Signal Process.*, vol. 41, no. 1, pp. 20–30, 1993.

14 H.V. Poor and X. Wang "Code-aided interference suppression for DS/CDMA communications-Part II: Parallel blind adaptive implementations," *IEEE Trans. Commun.*, vol. 45, no. 9, pp. 1112–1122, 1997.

15 L. Gao and K. Parhi, "Hierarchical pipelining and folding of QRD-RLS adaptive filtering and Its application to digital beamforming" *IEEE Trans. Circuits Syst.*, vol. 47, no. 12, pp. 1503–1519, 2000.

16 J.M. Cioffi, "High speed systolic implementation of fast QR adaptive filters." In: *International Conference on Acoustics, Speech, and Signal Processing*, 11–14 April 1988, pp. 1584–1587.

17 M. Moonen and I.K. Proudler, "Generating 'Fast QR' algorithms using signal flow graph techniques." In: *1996 Conference Record of the Thirtieth Asilomar Conference on Signals, Systems and Computers*, 3–6 November 1996, vol. 1, pp. 410–414.

18 A. Rontogiannis and S. Theodoridis, "Multichannel fast QRD-RLS adaptive filtering: new technique and algorithms," *IEEE Trans. Signal Process.*, vol. 46, no. 11, November 1998.

19 A. Rontogiannis and S. Theodoridis, "New fast QR decomposition least squares adaptive algorithms," *IEEE Trans. Signal Process.*, vol. 46, no. 8, Aug. 1998.

20 L. Gao and K. Parhi, "Hierarchical pipelining and folding of QRD-RLS adaptive filtering and its application to digital beamforming." *IEEE Trans. Circuits Syst.*, vol. 47, no. 12, pp. 1503–1519, 2000.

21 M. Moonen and I.K. Roudler, "MVDR beamforming and generalized sidelobe cancellation based on inverse updating with residual extraction," *IEEE Trans. Circuits Syst.*, vol. 47, no. 4, pp. 352–358, 2000.

22 J. Ma, K. Parhi, and E.F. Deprettere, "Annihilation-reordering look-ahead pipelined CORDIC-based RLS adaptive filters and their application to adaptive beamforming," *IEEE Trans. Signal Process.*, vol. 48, no. 8, pp. 2414–2431, 2000.

23 A. Elnashar, S. Elnoubi, and H. Elmikati "Performance analysis of blind adaptive MOE multiuser receivers using inverse QRD-RLS algorithm," *IEEE Trans. Circuits Syst. I*, vol. 55, no. 1, pp. 398–411, 2008.

24 S.-J. Chern, C.-Y. Chang, and H.-C. Liu, "Multiuser wavelet based MC-CDMA receiver with linearly constrained constant modulus IQRD-RLS algorithm." In: *IEEE International Symposium on Circuits and Systems*, 26–29 May 2002, vol. 1, pp. 193–196.

25 S.-J. Chern and C.-Y. Chang, "Adaptive MC-CDMA receiver with constrained constant modulus IQRD-RLS algorithm for MAI suppression," *Signal Process.*, vol. 83, pp. 2209–2226, 2003.

26 Z. Fu and E.M. Dowling, "Conjugate gradient projection subspace tracking," *IEEE Trans. Signal Process.*, vol. 45, no. 6, pp. 1664–1668, 1997.

27 Y. Hua, Y. Xiang, T. Chen, K. Abed-Meriam, and Y. Miao, "Natural power method for fast subspace tracking," *Neural Networks for Signal Processing IX, 1999. Proceedings of the 1999 IEEE Signal Processing Society Workshop*, 23–25 August 1999, pp. 176–185.

28 B. Yang, "An extension of the PASTd algorithm to both rank and subspace tracking," *IEEE Signal Process. Letters*, vol. 2, no. 9, pp. 179–182, 1995.

29 B. Yang, "Projection approximation subspace tracking," *IEEE Trans. Signal Process.*, vol. 43, pp. 95–107, 1995.

30 K. Abed-Meraim, A. Chkeif, and Y. Hua, "Fast orthogonal PAST algorithm" *IEEE Signal Process. Lett.*, vol. 7, no. 3, pp. 60–62, 2000.

31 E. Oja, "A simplified neuron model as a principal component analyzer," *J. Math. Biol.*, vol. 15, pp. 267–273, 1982.

32 E. Oja, "Principal components, minor components, and linear networks," *Neural Networks*, vol. 5, pp. 927–935, 1992.

33 K. Abed-Meraim, S. Attallah, A. Chkeif, and Y. Hua, "Orthogonal Oja algorithm," *IEEE Signal Process. Lett.*, vol. 7 no. 5, pp. 116–119, 2000.

34 S. Attallah and K. Abed-Meraim, "Fast algorithms for subspace tracking" *IEEE Signal Process. Lett.*, vol. 8, no. 7, pp. 203–206, 2001.

35 M.Z.A. Bhotto and A. Antoniou, "Robust recursive least-squares adaptive-filtering algorithm for impulsive-noise environments," *IEEE Signal Process. Lett.*, vol 18, no. 3, pp. 185–188, 2011.

36 Y. Zou, S.C. Chan, and T.S. Ng, "A recursive least M-estimate (RLM) adaptive filter for robust filtering in impulsive noise," *IEEE Signal Process. Lett.*, vol. 7, no. 11, pp. 324–326, 2000.
37 S.C. Chan and Y. Zou, "A recursive least M-estimate algorithm for robust adaptive filtering in impulsive noise: Fast algorithm and convergene analysis," *IEEE Trans. Signal Process.*, vol. 52, no. 4, pp. 975–991, 2004.
38 L.R. Vega, H. Rey, J. Benesty, and S. Tressens, "A fast robust recursive least-squares algorithm," *IEEE Trans. Signal Process.*, vol. 57, no. 3, pp. 1209–1215, 2009.
39 K.-Y. Fan, P.-Y. Tsai, "An RLS tracking and iterative detection engine for mobile MIMO-OFDM systems," *IEEE Trans. Circuits Syst. I*, vol. 62, pp. 185–194, 2015.
40 A. Ibrahim, T.F. Al-Somani, and F. Gebali "New systolic array architecture for finite field inversion," *Can. J. Electr. Computer Eng.*, vol. 40, no. 1, pp. 23–30, 2017.
41 Z.-Y. Huang and P.-Y. Tsai, "Efficient implementation of QR decomposition for gigabit MIMO-OFDM systems," *IEEE Trans. Circuits Syst. I*, vol. 58, no. 10, pp. 2531–2542, 2001.
42 D. Chen and M. Sima, "Fixed-point CORDIC-based QR decomposition by Givens rotations on FPGA." In: 2011 *International Conference on Reconfigurable Computing and FPGAs*, pp. 327–332.
43 C.-W. Chen, H.-W. Tsao, and P.-Y. Tsai, "MIMO precoder design with a compensated QR-decomposition combination for CoMP downlink scenarios," *IEEE Trans. Vehic. Techn.*, vol. 65, pp. 7982–7992, 2016.
44 M.I.A. Mohamed, K. Mohammed, and B. Daneshrad, "Energy efficient programmable MIMO decoder accelerator chip in 65-nm CMOS," *IEEE Trans. Very Large Scale Integr. (VLSI) Syst.*, vol. 22, pp. 1481–1490, 2014.
45 R. Gangarajaiah, P. Nilsson, O. Edfors, and L. Liu, "Low complexity adaptive channel estimation and QR decomposition for an LTE-A downlink." In: *2014 IEEE 25th Annual International Symposium on Personal Indoor and Mobile Radio Communication (PIMRC)*, pp. 459–463, 2014.
46 A. Awasthi, R. Guttal, N. Al-Dhahir, and P.T. Balsara, "Complex QR decomposition using fast plane rotations for MIMO

applications," *IEEE Commun. Lett.*, vol. 18, pp. 1743–1746, 2014.

47 J.G. McWhirter "Recursive least-square minimization using a systolic array." In: *Proceedings of SPIE Real Time Signal Processing* VI, August 1983, pp. 105–112.

48 M.K. Tsatsanis and Z. Xu, "On minimum output energy CDMA receives in presence of multipath." In: *Conference on Information Sciences and Systems* 1997, pp. 377–381.

49 S.F. Hsieh, K.L.R. Liu, and K. Yao, "A unified square-root-free approach for QRD-based recursive least squares estimation" *IEEE Trans. Signal Process.*, vol. 41, no. 3, pp. 1405–1409, 1993.

50 S. Haykin, *Adaptive Filter Theory*, 4th edn, Prentice-Hall, 2002.

51 P. Strobach, "Square-root QR inverse iteration for tracking the minor subspace," *IEEE Trans. Signal Process.*, vol. 48, no. 11, pp. 2994–2999, 2000.

52 S. Choi and D. Shim "A novel adaptive beamforming algorithm for a smart antenna in a CDMA mobile communication environment" *IEEE Trans. Vehic. Tech.*, vol. 49, no. 5, pp. 1793–1806, 2000.

53 M.W. Ganz, R.L. Moses, and S.L. Wilson, "Convergence of the SMI and diagonally loaded SMI algorithm with weak interference," *IEEE Trans. Antennas Propag.*, vol. 38, no. 3, pp. 394–399, 1990.

54 S. Chi, J. Choi, H.J. Im, and B. Choi, "A novel adaptive beamforming algorithm for antenna array CDMA systems with strong interferes," *IEEE Trans. Vehic. Tech.* vol. 51, no. 5, pp. 808–816, 2002.

55 Y.N. Rao, J.C. Principe, and T.F. Wong, "Fast RLS-like algorithm for generalized eigendecomposition and its application" *J. VLSI Signal Process.* vol. 37, pp. 333–344, 2004.

56 Z. Tian, "*Blind Multi-user Detection with Adaptive Space-time Processing for DS-CDMA Wireless Communications*," PhD dissertation, George Mason University, 2000.

57 Z. Tian, K.L. Bell, and H.L. Van Trees, "Robust constrained linear receivers for CDMA wireless systems," *IEEE Trans. Signal Process.*, vol. 49, no. 7, pp. 1510–1522, 2001.

58 S. Song, J.-S. Lim, S.J. Baek, and K.-M. Sung, "Variable forgetting factor linear least squares algorithm," *IEEE Trans. Vehic. Tech.*, vol. 51, no. 3, pp. 613–616, 2002.
59 C.F. So and S.H. Leung, "Variable forgetting factor RLS algorithm based on dynamic equation of gradient of mean square error," *Electronics Lett.*, vol. 37, no. 3, pp. 202–203, 2001.
60 K. Nishikawa and H. Kiya, "Fast improvement techniques for improving throughput of RLS adaptive filter," *IEICE Trans. Fundamentals*, vol.E83-A, no. 8. 2000.
61 Z. Tian, K.L. Bell, and H.L. Van Trees, "A quadratically constrained decision feedback equalizer for DS-CDMA communication systems." In: *Proceedings of 2nd IEEE Workshop on Signal Processing and Advanced Wireless Communication*, 1999.
62 Z. Tian, K.L. Bell, and H.L. Van Trees, "A recursive least squares implementation for LCMP beamforming under quadratic constraint" *IEEE Trans. Signal Process.*, vol. 49, no. 6, pp. 1138–1145, 2001.
63 Z. Tian, K.L. Bell, and H.L. Van Trees, "A RLS implementation for adaptive beam-forming under quadratic constraint," *9th* IEEE Workshop on Statistical Signal and Array Processing, Portland, Oregon, September 1998.
64 Z. Tian, K.L. Bell, and H.L. Van Trees, "Quadratically constrained RLS filtering for adaptive beamforming and DS-CDMA multi-user detection," *Adaptive Sensor Array Processing Workshop*, MIT Lincoln Lab, Lexington, MA, 1999.
65 Z. Tian, K.L. Bell, and H.L. Van Trees, "Robust RLS implementation of the minimum output energy detector with multiple constraints," Technical paper, Center of Excellence in C3I, George Mason University, USA, 2000.
66 M. Moonen, "Systolic MVDR beamforming with inverse updating," *Proc. Inst. Elect. Eng.*, vol. 140, pp. 175–178, 1993.
67 Y.T. Hwang and W.D. Chen, "A low complexity complex QR factorization design for signal detection in MIMO OFDM systems." In: *Proceedings of IEEE International Symposium on Circuits and Systems (ISCAS)*, 2008, pp. 932–935.
68 C.K. Singh, S.H. Prasad, and P.T. Balsara, "VLSI architecture for matrix inversion using modified Gram–Schmidt

based QR decomposition." In: *Proceedings of International Conference on VLSI Design*, 2007, pp. 836–841.

69 R.C.-H. Chang, C.H. Lin, K.H. Lin, C.L. Huang, and F.C. Chen, "Iterative decomposition architecture using the modified Gram–Schmidt algorithm for MIMO systems," *IEEE Trans. Circuits Syst. I*, vol. 57, no. 5, pp. 1095–1102, 2010.

70 P. Salmela, A. Burian, H. Sorokin, and J. Takala, "Complex-valued QR decomposition implementation for MIMO receivers." In: *International Conference on Acoustics, Speech, and Signal Processing*, 2008, pp. 1433–1436.

71 D. Patel, M. Shabany, and P.G. Gulak, "A low-complexity high-speed QR decomposition implementation for MIMO receivers." In: *Proceedings of IEEE International Symposium on Circuits and Systems (ISCAS)*, 2009, pp. 1409–1412.

72 P.-L. Chiu, L.-Z. Huang, L.-W. Chai, and Y.-H. Huang, "Interpolation- based QR decomposition and channel estimation processor for MIMO-OFDM system," *IEEE Trans. Circuits Syst. I*, vol. 58, no. 5, pp. 1129–1141, 2011.

73 D. Cescato and H. Bölcskei, "Algorithms for interpolation-based QR decomposition in MIMO-OFDM systems," *IEEE Trans. Signal Process.*, vol. 59, no. 4, pp. 1719–1733, 2011.

74 R.C.H. Chang, C.H. Lin, K.H. Lin, C.L. Huang, and F.C. Chen, "Iterative Q R decomposition architecture using the modified Gram–Schmidt algorithm for MIMO systems," *IEEE Trans. Circuits Syst. I*, vol. 57, no. 5, pp. 1095–1102, 2010.

75 M. Sun, T. Juan, K. Lin, and T. Hsu, "Adaptive frequency-domain channel estimator in 4×4 MIMO-OFDM modems," *IEEE Trans. Very Large Scale Integr. (VLSI) Syst.*, vol. 17, no. 11, pp. 1616–1625, 2009.

5

Quadratically Constrained Simplified Robust Adaptive Detection

5.1 Introduction

Among different detection techniques, linear receivers are of great significance due to their ease of practical implementation. A linear receiver can be designed by minimizing some inverse filtering criterion [1, 2]. Appropriate constraints are imposed to avoid the trivial all-zero solution. The detector's output variance is minimized subject to appropriate constraints, which depend on the multipath structure of the signal of interest [3] or the steering vector direction in beamforming applications. Constrained optimization methods have received considerable attention as a means to derive blind multiuser receivers with low complexity [4–6] and robust minimum variance distortionless response (MVDR) beamformers. Minimum output energy (MOE) detection has been proposed as a blind adaptive technique for multiuser detection in direct-sequence code-division multiple-access (DS/CDMA) systems [5] as described in Chapter 4. The MOE detector is a scaled version of the MMSE detector. Since scaling does not affect the output SINR, MOE has the same optimal performance as MMSE if the channel is estimated properly. An extension to space-time processing using MOE has been conducted [7]. In addition, a rake receiver using blind adaptive MOE detection has been used for DS/CDMA over multipath channels [8].

Most of the proposed adaptive approaches in the literature [4, 9–11] are based on the well-known RLS algorithm. In the conventional recursive least squares (RLS) algorithm [12], the calculation of the Kalman gain requires matrix inversion

Simplified Robust Adaptive Detection and Beamforming for Wireless Communications,
First Edition. Ayman Elnashar.

Companion website: www.wiley.com/go/elnashar49

of the autocovariance matrix of the received signal. If the data matrix is ill-conditioned, the conventional RLS algorithm will rapidly become impossible to use [13–15]. An iterative algorithm for direct calculation of the MVDR beamformer without explicit matrix inversion/eigendecomposition/diagonalization has been developed [16]. The troublesome, data-dependent tuning of the real-valued least mean squares (LMS) learning gain parameter, the RLS initialization, or the SMI diagonal loading parameter are replaced by an integer choice from among the first few members of the estimator sequence. However, this iterative technique is based on dependent update steps and it may therefore lead to error accumulation. In addition, the behavior of this algorithm with data records of limited size is not fully analyzed. Moreover, least-square detectors require more robustness against pointing errors, round-off errors, perturbations in detector parameters and signature or steering vector mismatch errors [17, 18].

A robust low-complexity blind detector to overcome these shortcomings is presented in this chapter. It is based on a recursive fast steepest descent (RSD) adaptive algorithm rather than the RLS algorithm, and a quadratic inequality (QI) constraint on the weight vector norm [13, 14]. The QI constraint is employed to manage the residual signal mismatch and other random perturbation errors. In addition, the QI constraint will constrain the noise constituent in the output signal-to-interference-plus-noise ratio (SINR) to be constant and hence overcome noise enhancement at low SNR. Quadratic constraints have been used in adaptive beamforming for a variety of purposes, such as improving robustness against mismatch and modeling errors, controlling mainlobe response, and enhancing interference cancellation capability [19]. The quadratic constraint will be analyzed along with beamforming algorithms in Chapter 7 [20].

Analogous to the recursive conjugate gradient (RCG) algorithm [21–23], a fast RSD algorithm will be developed in this chapter [14]. A low-computational complexity recursive update equation for the gradient vector is derived. Furthermore, a variable step-size approach is introduced for the step-size update of the RSD algorithm based on an optimum step-size calculation [14]. The RSD algorithm is exploited to update the adaptive weight vector of the partition linear interference canceller (PLIC) structure to suppress MAI. The same technique will be extended to MVDR beamforming in Chapter 7. From this similarity, the reader can easily extend the algorithms developed in this book to other systems and even beyond the realm of wireless communications. We have therefore simplified the deployment of robust techniques such as the quadratic constraint, uncertainty constraint, worst-case constraint, and constrained optimization.

Cox *et al.* introduced a new robust beamforming algorithm that includes a QI constraint on the array gain against uncorrelated noise, while minimizing output power subject to multiple linear equality constraints [24]. They used a simple scaling projection (SP) technique to satisfy the QI constraint. In spite of the simplicity of the SP technique, it does not lead to the optimal solution. Fertig and McClellan developed a new form for norm-constrained beamforming [25]. This was based on a "dual" solution for the constrained minimization problem. The dual solution is unique in the sense that its update equations involve Lagrange multipliers rather than the adaptive filter weights. However, the dual-form algorithm they introduced gives an adaptive update whose dimensions equal the number of constraints rather than the array dimension, but its update equation is based on the SP technique.

Qian and van Veen proposed adding quadratic constraints to a linearly constrained minimum variance (LCMV) adaptive beamformer to prevent signal cancellation in coherent interference environments [26]. Since the true interference scenario is unknown in practice, they use preliminary estimates of the interference direction of arrival (DOA). But one of the key assumptions all high resolution DOA estimation techniques require is that the number of signal wavefronts including the interference signals must be less than the number of elements in the array; that is, the degree of freedom. Unfortunately, this is not a very realistic scenario for a practical wireless communication system.

Tian *et al.* proposed a technique for implementing the QI constraint into the RLS algorithm to improve robustness against pointing errors and random perturbations in sensor parameters [27]. They derived a closed-form solution for the amount of diagonal loading. This approach provides a robust LCMV beamformer based on the RLS algorithm, with a VL technique based on a Taylor series expansion approximation. They also extended their work to the robust MUD problem in a DS/CDMA system. They proposed a robust constrained optimization approach by using a fixed set of linear constraints that do not need to be optimized, plus a QI constraint on the weight vector norm [28]. This approach provides a robust DS/CDMA receiver based on the RLS algorithm and a QI constraint on the weight vector. Regrettably, this VL technique relies on an approximation that is only applicable for small loading levels. Therefore, for large loading levels it fails to reach the optimum loading.

The QI constraint robust approach embraced in this chapter can be interpreted as a general robust approach, where the constraint is imposed on the weight vector rather than the signature waveform. In contrast, recently proposed approaches adopt a worst-case performance optimization approach to solve an uncertainty constraint imposed on the signature waveform or steering vector directly [29–34]. However, all uncertainty constraint approaches (QI, ellipsoidal constraint, worst-case optimization) can be boiled down to the diagonal loading technique. The challenge facing these approaches is how to obtain the optimal diagonal loading term with low complexity and how to integrate it into a recursive update algorithm [14, 20]. In spite of the promising results produced by these algorithms [29–34], their recursive implementation seems to be cumbersome or impossible in practice. The following detailed comments corroborate this concern.

The approach proposed by Cui *et al.* for robust MUD explicitly models an arbitrary uncertainty in the desired user signature and uses worst-case performance optimization to improve the robustness of the MOE receiver [30]. This approach is based on convex optimization using second-order cone programming (SOCP). Although several efficient convex optimization software tools are currently accessible, such as the SeDuMe software [35], the SOCP-based method does not provide any closed-form solution, and does not have simple online implementations [36]. Moreover, the computational burden of this software seems to be cumbersome, which limits the practical implementation of the technique. In addition, the approach is regarded as a batch algorithm rather than a recursive scheme: the weight vector of the detector is individually computed in each step and cannot be updated recursively [20, 36]. The algorithm will be analyzed in Chapter 7 in the context of beamforming. The SeDuMi software MATLAB toolbox for optimization over symmetric cones, developed by J.F. Sturm [35], will be used for the beamforming calculation. SeDuMi is an excellent piece of software for optimization over symmetric cones. Currently SeDuMi is hosted and maintained by CORAL Lab at the Department of Industrial and Systems Engineering at Lehigh University [37]. Current versions can be found in the Downloads section of: http://sedumi.ie.lehigh.edu/. This SW package is provided with the Matlab package provided with this book.

Another approach involves the optimization of a lower-bound of the worst-case performance and uses the covariance matrix of the desired user signature rather than the signature itself [37, 38]. The algorithm boils down to a simple closed-form solution, somewhat analogous to the original MOE detector, with positive diagonal loading to the received covariance matrix and negative diagonal loading to the signature waveform covariance matrix. Although a closed-form solution is given to this detector, no closed-form solutions are developed for the two diagonal loading terms. Alternatively, they can be selected based on some preliminary knowledge about wireless channels or using Monte Carlo simulations. This algorithm has been shown to outperform MOE multiuser receivers with fixed diagonal loading, but its performance may be affected by the use of the lower bound on the worst-case performance instead of the worst-case performance itself [39]. In addition, the recursive implementation of this detector requires the inverse computation of the diagonally loaded received data covariance matrix and subspace tracking of matrix multiplication (the inverse of the positive diagonally loaded received data covariance matrix with the negative diagonally loaded signature waveform covariance matrix). Furthermore, improper selection of the negative diagonal loading term may violate the positive definiteness of the desired user signature covariance matrix.

The approach proposed in [39, 40] is based on robustness against possible uncertainties in the mean-square error (MSE) cost function. It achieves this by explicit modeling of uncertainties in both the desired user signature and the data covariance matrix. The approach is equivalent to the diagonal loading-based multiuser receiver with the optimal choice of the diagonal loading factor that was obtained based on the known level of uncertainty in the desired signal signature. The diagonal loading term contains the norm-bound constants of signature vector error and covariance matrix error in addition to the receiver vector norm $||f||$ [40]. The detector norm is the only parameter that can be estimated optimally. Unfortunately, the computation of the detector norm $||f||$ requires eigenvalue decomposition of the data covariance matrix (all the eigenvectors and eigenvalues are required). In addition, this approach is applicable only if the maximum norm of such an error does not exceed the norm of the presumed desired user signature itself, which cannot be always guaranteed [39], especially when there are large delay mismatches.

Generally, these approaches [36–40] seem to be applicable only to frequency non-selective fading channels, where the uncertainty constraint is imposed on the original signature waveform. Extension to the multipath case seems to be complicated. This is because, in frequency-selective fading channels with resolvable multipath components, the constraint vector form includes a matrix with replicas from the original signature waveform and hence multiple constraints are necessary. Furthermore, these approaches are robust against signature mismatches only; in the presence of any other mismatches (such as uncertainty in the autocovariance matrix or improper initialization), a joint robustness approach is obligatory, but this suffers from high computational complexity as well [41]. In Chapter 7, multiple worst-case constraints will be proposed for robust MVDR beamforming [42]. The same approach can be extended to frequency-selective channels, where multiple worst-case constraints can be imposed on the resolvable multipath components. However, the problem of selection of the optimum constraint value is not solved [42]. This is an interesting topic for future research work.

The shortcomings of the diagonal loading technique are tackled in this chapter (further details are available in the literature [13, 14, 20]). An alternative approach to robust adaptive detection is presented. This is based on the RSD adaptive algorithm, with an accurate technique for precisely computing the diagonal loading level without approximation or eigendecomposition. We combined the QI constraint with the RSD algorithm to produce a robust recursive implementation with $O(N^2)$ complexity. A new optimal VL technique is developed and integrated into the RSD adaptive algorithm. In addition, the diagonal loading term is optimally computed, with $O(N)$ complexity, using a simple quadratic equation. The geometrical interpretations of the SP and VL techniques along with RLS and RSD [14] algorithms are illustrated and clarified. The performance of the robust RSD with VL detector is compared with traditional detectors and is shown to be more accurate and more robust against signal mismatch and random perturbations. Finally, the presented approach can be reformulated to handle an uncertainty constraint imposed on a signature waveform or on a steering vector, such as the ellipsoidal constraint proposed by Li *et al.* [43]. RSD with VL can use any of these approaches [38, 39, 44] to produce a simple recursive implementation.

5.2 Robust Receiver Design

5.2.1 Quadratic Inequality Constraint

By applying the QI constraint to the adaptive weight portion ($\boldsymbol{f}_a$) of the PLIC structure given in (3.39), the optimal robust weight vector can be obtained from the solution of the following constrained optimization problem:

$$\boldsymbol{f}_a = \min_{\boldsymbol{f}_a} (\boldsymbol{f}_c - \boldsymbol{B}\boldsymbol{f}_a)^H \boldsymbol{R}_{xx} (\boldsymbol{f}_c - \boldsymbol{B}\boldsymbol{f}_a) \; Subject\ to\, \boldsymbol{f}_a^H \boldsymbol{f}_a \leq \beta^2 \tag{5.1}$$

The constrained value is:

$$\beta^2 = \delta^2 - \boldsymbol{f}_c^H \boldsymbol{f}_c \tag{5.2}$$

and

$$\delta^2 = t.(\|\boldsymbol{f}_c\|^2) \tag{5.3}$$

t is set to a suitable value.

The method of Lagrange multipliers is now used to solve this constrained minimization problem by forming the following real-valued Lagrangian function:

$$\boldsymbol{\Psi}_{\boldsymbol{f}_a}(n) = (\boldsymbol{f}_c - \boldsymbol{B}\boldsymbol{f}_a)^H \boldsymbol{R}_{xx}(\boldsymbol{f}_c - \boldsymbol{B}\boldsymbol{f}_a) + \frac{1}{2}\lambda_0 t(\boldsymbol{f}_a^H \boldsymbol{f}_a - \beta^2) \tag{5.4}$$

The Lagrange multiplier λ_0 is a real scalar, which is determined from the QI constrained value. It must be non-negative to ensure that the diagonally loaded autocovariance data matrix is positive definite. The problem is converted from a QI *constrained* minimization to an *unconstrained* minimization problem. The optimal solution is obtained by first taking the gradient of $\boldsymbol{\Psi}_{\boldsymbol{f}_a}(n)$ with respect to the real and imaginary components of $\boldsymbol{f}_a^H$ and then setting the resulting quantities to zero. This yields:

$$\boldsymbol{f}_{a(opt)} = (\boldsymbol{R}_B + \lambda_0 \boldsymbol{I})^{-1} \boldsymbol{p}_B \tag{5.5}$$

where

$$\boldsymbol{R}_B = \boldsymbol{B}^H \boldsymbol{R}_{xx} \boldsymbol{B} \tag{5.6}$$

$$\boldsymbol{p}_B = \boldsymbol{P}_B \boldsymbol{f}_c \tag{5.7}$$

$$\boldsymbol{P}_B = \boldsymbol{B}^H \boldsymbol{R}_{xx} \tag{5.8}$$

The adaptive implementation of the detector given in (5.5) incorporates three difficulties. Firstly, matrix inversion is required for the diagonally loaded blocked data matrix (lower branch of PLIC structure). Secondly, the value of the diagonally loaded term (λ_0) cannot be easily determined where there is no closed-form expression for the optimal loading level. Thirdly, there is no convenient way to incorporate the diagonal loading term into R_B or its inverse directly. In the next three subsections, three different approaches are introduced to implement the robust quadratically constrained detector.

5.2.1.1 SP Approach

Cox *et al.* described an iteration algorithm for solving this problem [24]. This is called the scaled projection algorithm, but it does not lead to the optimal solution. The SP technique simply scales the unconstrained updated vector to satisfy the

QI constraint. That is:

$$\tilde{\boldsymbol{f}}_a(n) \Rightarrow \beta \frac{\tilde{\boldsymbol{f}}_a(n)}{\|\tilde{\boldsymbol{f}}_a(n)\|} \tag{5.9}$$

5.2.1.2 Tian Approach

Recently, Tian *et al.* developed a variable loading technique based on a Taylor series approximation for the term $(\boldsymbol{R}_B + \lambda_0 \boldsymbol{I})^{-1}$ [28], and they adaptively incorporated it into the RLS algorithm. The robust MUD of [28] can be summarized as follows:

$$\begin{aligned}\boldsymbol{f}_a(n) &= (I + \lambda_0 \boldsymbol{R}_B^{-1}(n))^{-1} \boldsymbol{R}_B^{-1}(n) \boldsymbol{p}_B(n) \\ &= (I + \lambda_0 \boldsymbol{R}_B^{-1}(n))^{-1} \bar{\boldsymbol{f}}_a(n)\end{aligned} \tag{5.10}$$

$$\boldsymbol{f}_a(n) \approx \bar{\boldsymbol{f}}_a(n) - \gamma \hat{\boldsymbol{f}}_a(n) \quad \rightarrow \text{based on Taylor series approximation)} \tag{5.11}$$

$$\bar{\boldsymbol{f}}_a(n) = \boldsymbol{R}_B^{-1}(n) \boldsymbol{p}_B(n) \quad \rightarrow \text{the updated detector without QI} \tag{5.12}$$

$$\hat{\boldsymbol{f}}_a(n) = \boldsymbol{R}_B^{-1}(n) \bar{\boldsymbol{f}}_a(n) \quad \rightarrow \text{new correction vector} \tag{5.13}$$

The diagonal loading term is changed to γ to distinguish it from the optimal one λ_0. We solve for γ by plugging (5.11) into $\boldsymbol{f}_a^H(n)\boldsymbol{f}_a(n) \leq \beta^2$. This means that γ can be calculated as follows:

$$\gamma = \frac{-\tilde{b} \pm \mathrm{Re}\left\{\sqrt{\tilde{b}^2 - 4\tilde{a}\tilde{c}}\right\}}{2\tilde{a}} \tag{5.14}$$

where

$$\tilde{a} = \hat{\boldsymbol{f}}_a^H(n)\hat{\boldsymbol{f}}_a(n) \tag{5.15a}$$

$$\tilde{b} = -2\mathrm{Re}\left\{\hat{\boldsymbol{f}}_a^H(n)\bar{\boldsymbol{f}}_a(n)\right\} \tag{5.15b}$$

$$\tilde{c} = \bar{\boldsymbol{f}}_a^H(n)\bar{\boldsymbol{f}}_a(n) - \beta^2 \tag{5.15c}$$

This detector is referred to as MOE-RLS w. QC and can be summarized as follows. The matrix $\boldsymbol{R}_B^{-1}$ is updated using the RLS algorithm and the detector $\bar{\boldsymbol{f}}_a(n)$ is updated using (5.12) or using the update equation of the RLS algorithm [28]. If the QI constraint is not satisfied, the approximate loading level is calculated using (5.14) and the correction vector $\hat{f}_a(n)$ is calculated using (5.13) to update the detector using (5.11). This VL technique relies on the approximation given in (5.11), which is only valid for small loading levels. Therefore, for large loading levels the technique fails to reach the optimum loading. More specifically, if $\tilde{b}^2 - 4\tilde{a}\tilde{c} < 0$ then the roots in (5.14) form a complex conjugate pair. Neither of the roots provides an acceptable solution, which indicates that the norm constraint cannot be met with any real γ using the Taylor approximation in (5.11) [45–50]. When the QI constraint cannot be achieved with a real solution, the real part of the complex conjugate roots is the best loading level in the sense that it brings the weight vector norm as close as possible to the quadratic constraint. Therefore, the real operator Re{·} is added to (5.14) to prevent complex solutions. In addition, the loading required at that step in the VL algorithm is an *incremental* loading level that supplements the loading incorporated at all previous steps. This is usually a small amount and the optimal incremental loading level to satisfy the quadratic constraint can usually be achieved. However, in dynamic scenarios abrupt mismatches may lead to performance degradation due to the requirement for a large diagonal loading level, which cannot be achieved in this technique [14]. In addition to this, the approximate diagonal loading term γ requires $O(N_a^2 + 2N_a)$ multiplications due to the matrix-vector multiplication in (5.13).

Box 5.1 summarizes the MOE-RLS w. QC detector.

Box 5.1 Summary of the robust MOE-RSD w. QC.

Initialization

- $\boldsymbol{R}_B^{-1}(0) = \boldsymbol{B}^H(\zeta \boldsymbol{I}_{N_a})\boldsymbol{B}; \boldsymbol{f}_c = \boldsymbol{C}_1(\boldsymbol{C}_1{}^H\boldsymbol{C}_1)^{-1}\boldsymbol{g}$
- $\delta^2 = t.(\|\boldsymbol{f}_c\|^2), \beta^2 = \delta^2 - \|\boldsymbol{f}_c\|^2,$
- $\gamma = 0; \boldsymbol{f}_a(n) = \boldsymbol{0}_{N_a}; \boldsymbol{p}_B = \boldsymbol{P}_B\boldsymbol{f}_c$

For $n = 1,2,\cdots,$ **do**

- $\boldsymbol{z}(n) = \boldsymbol{B}^H\boldsymbol{x}(n); N_aN_f$
- $\boldsymbol{R}_B^{-1}(n) = \eta\boldsymbol{R}_B^{-1}(n-1) + \boldsymbol{z}^H(n)\boldsymbol{z}(n); N_a{}^2$

- $\bar{f}_a(n) = R_B^{-1}(n)p_B(n)$

- *if*$(\|f_a(n)\|^2 > \beta^2)$; (QI constraint not satisfied) N_a
  - $\hat{f}_a(n) = R_B^{-1}(n)\bar{f}_a(n)$, $\tilde{a} = \hat{f}_a^H(n)\hat{f}_a(n)$,
    $\tilde{b} = -2\text{Re}\left\{\hat{f}_a^H(n)\bar{f}_a(n)\right\}$, $\tilde{c} = \bar{f}_a^H(n)\bar{f}_a(n) - \beta^2$; $2N_a$
  - $\gamma = \dfrac{-\tilde{b} \pm \text{Re}\left\{\sqrt{\tilde{b}^2 - 4\tilde{a}\tilde{c}}\right\}}{2\tilde{a}}$
  - $f_a(n) \approx \bar{f}_a(n) - \gamma\hat{f}_a(n)$
- *else*; (QI constraint satisfied)
  - $f_a(n) = \bar{f}_a(n)$, $\gamma = 0$
- End if

- $f(n) = f_c - Bf_a(n)$; (only for evaluation)

End for

5.2.1.3 A Simplified VL Approach

A new optimal VL technique that is capable of precisely computing the optimal diagonal loading term λ_0 was developed by Elnashar *et al.* [14]. In addition, the simplified VL technique will be efficiently integrated into a fast RSD algorithm for recursively estimating the optimal robust detector. The value of the optimal diagonal loading level is estimated without approximation. The method of steepest descent is employed to iteratively update the weight vector that minimizes the Lagrangian functional (5.4). As a result,

$$f_a(n) = f_a(n-1) - \mu\nabla_{f_a}(n) \tag{5.16}$$

where μ is the step-size of the algorithm, $\nabla_{f_a}(n)$ is the conjugate derivative of $\Psi_{f_a}(n)$ with respect to the real and imaginary parts of $f_a^H(n)$ and is given by:

$$\begin{aligned}\nabla_{f_a}(n) &= -B^H R_{xx}(n)f_c + B^H R_{xx}(n)Bf_a(n-1) + \lambda_0 f_a(n-1)\\ &= -P_B f(n-1) + \lambda_0 f_a(n-1)\end{aligned} \tag{5.17}$$

Therefore, the adaptive implementation of $f_a(n)$ can be obtained by substituting from Equation 5.17 into Equation 5.16 as follows:

$$f_a(n) = f_a(n-1) - \mu(R_B(n)f_a(n-1) - p_B(n)) - \mu\lambda_0 f_a(n-1) \tag{5.18}$$

Notice that the constraint should be satisfied at each iteration step; that is, $\boldsymbol{f}_a^H(n)\boldsymbol{f}_a(n) \le \beta^2$. Assuming the constraint is satisfied on the previous step and using (5.18), we get:

$$\left(\widetilde{\boldsymbol{f}}_a(n) - \mu\lambda_0\boldsymbol{f}_a(n-1)\right)^H \left(\widetilde{\boldsymbol{f}}_a(n) - \mu\lambda_0\boldsymbol{f}_a(n-1)\right) \le \beta^2 \tag{5.19}$$

The detector $\widetilde{\boldsymbol{f}}_a(n)$ is the non-robust version of the proposed detector (referred to as MOE-RSD) and is given by:

$$\widetilde{\boldsymbol{f}}_a(n) = \boldsymbol{f}_a(n-1) - \mu[\boldsymbol{R}_B(n)\boldsymbol{f}_a(n-1) - \boldsymbol{p}_B(n)] \tag{5.20}$$

The value of λ_0 can be found by solving the following quadratic equation:

$$\mu^2\boldsymbol{f}_a{}^H(n-1)\boldsymbol{f}_a(n-1)\lambda_0^2 - 2\mu\mathrm{Re}\left\{\widetilde{\boldsymbol{f}}_a^H(n)\boldsymbol{f}_a(n-1)\right\}\lambda_0 + \widetilde{\boldsymbol{f}}_a^H(n)\widetilde{\boldsymbol{f}}_a(n) - \beta^2 = 0 \tag{5.21}$$

Then, from (5.21) the value for λ_0 that satisfies the inequality constraint is:

$$\lambda_0 = \frac{-b \pm \sqrt{b^2 - 4ac}}{2a} \tag{5.22}$$

where

$$a = \mu^2\|\boldsymbol{f}_a(n-1)\|^2 \tag{5.23a}$$

$$b = -2\mu\mathrm{Re}\left\{\widetilde{\boldsymbol{f}}_a^H(n)\boldsymbol{f}_a(n-1)\right\} \tag{5.23b}$$

$$c = \left\|\widetilde{\boldsymbol{f}}_a(n)\right\|^2 - \beta^2 \tag{5.23c}$$

If the QI constraint is not satisfied; that is, $\widetilde{\boldsymbol{f}}_a^H(n)\widetilde{\boldsymbol{f}}_a(n) > \beta^2$, then $C > 0$. Hence, the roots of the quadratic equation (5.21) are either two real positive values or a conjugate pair whose real parts are positive and hence the condition of λ_0 to be positive is satisfied. Furthermore, the complex roots can be avoided by properly setting the step-size μ. This cannot be guaranteed for the VL technique of Tian's approach, as presented in the previous section.

It is important now to derive conditions for the step-size which will guarantee that the roots of (5.21) will be real positive. The following two lemmas state the necessary and sufficient conditions under which Equation 5.21 has real positive solutions:

Lemma 5.1

$$(b^2 - 4ac \geq 0)$$

By substituting from (5.23) and (5.20) into $b^2 - 4ac \geq 0$, the following inequality is obtained:

$$\begin{aligned}&[(\boldsymbol{f}_a(n-1) - \mu\nabla_{\boldsymbol{f}_a}(n))\boldsymbol{f}_a(n-1)]^2 \\ &\quad \geq \beta^2[(\boldsymbol{f}_a(n-1) - \mu\nabla_{\boldsymbol{f}_a}(n))^H(\boldsymbol{f}_a(n-1) - \mu\nabla_{\boldsymbol{f}_a}(n)) - \beta^2]\end{aligned} \tag{5.24}$$

where

$$\nabla_{\boldsymbol{f}_a}(n) = \overline{\nabla}_{\boldsymbol{f}_a}(n) + \lambda_0^1\boldsymbol{f}_a(n-1) \tag{5.25}$$

$$\overline{\nabla}_{\boldsymbol{f}_a}(n) = \boldsymbol{R}_B(n)\boldsymbol{f}_a(n-1) - \boldsymbol{p}_B(n) \tag{5.26}$$

After some manipulations to (5.24) we can have the following inequality:

$$\mu \leq \frac{\beta\|\boldsymbol{f}_a(n-1)\|}{\sqrt{\|\boldsymbol{f}_a(n-1)\|^2\left\|\overline{\nabla}_{\boldsymbol{f}_a}(n)\right\|^2 - \overline{\nabla}_{\boldsymbol{f}}^H(n)\boldsymbol{f}_a(n-1)\boldsymbol{f}_a^H(n-1)\overline{\nabla}_{\boldsymbol{f}_a}(n)}} \tag{5.27}$$

Therefore, this upper bound on μ guarantees real positive roots for (5.21) and consequently, the optimal loading level can be obtained. The equality in (5.27) represents the one positive real root solution to (5.21); that is,

$$b^2 - 4ac = 0 \tag{5.28}$$

and hence,

$$\lambda_0 = \frac{\tilde{\boldsymbol{f}}_a^H(n)\boldsymbol{f}_a(n-1)}{\mu\beta^2} \tag{5.29}$$

Additionally, based on the well-known Cauchy–Schwarz inequality [51] and $\boldsymbol{f}_a^H(n-1)\boldsymbol{f}_a(n-1) \leq \beta^2$, it is easily verified that:

$$\|\boldsymbol{f}_a(n-1)\|^2\left\|\overline{\nabla}_{\boldsymbol{f}_a}(n)\right\|^2 \geq \overline{\nabla}_{\boldsymbol{f}_a}^H(n)\boldsymbol{f}_a(n-1)\left(\overline{\nabla}_{\boldsymbol{f}_a}^H(n)\boldsymbol{f}_a(n-1)\right)^H \tag{5.30}$$

Therefore, this inequality guarantees the existence of the step-size upper bound for any gradient vector $\overline{\nabla}_{\boldsymbol{f}_a}(n)$. The equality in (5.30) is obtained when the gradient vector $\overline{\nabla}_{f_a}(n)$ has the same direction as the detector $\boldsymbol{f}_a(n-1)$. As a consequence, the simplified VL technique is reduced to the SP technique. Therefore, the SP technique is a special case of the simplified robust VL technique.

Lemma 5.2

$(b \leq 0)$

In addition to (5.27), another constraint on the step-size is required to guarantee positive diagonal loading. Substituting (5.23) into $b \leq 0$, we get:

$$\mu \left[\boldsymbol{f}_a(n-1)\boldsymbol{R}_B(n)\boldsymbol{f}_a(n-1) - \boldsymbol{f}_a(n-1)\boldsymbol{P}_B(n)\boldsymbol{f}_c\right] \leq \|\boldsymbol{f}_a(n-1)\|^2 \tag{5.31}$$

The sign of the left-hand side of (5.31) is determined from the constrained value $\beta^2 = (t-1).(\|\boldsymbol{f}_c\|^2)$. In most practical cases, the constrained parameter t is selected within $(1, 2]$. Therefore, the left-hand side sign is negative and subsequently positive diagonal loading is always guaranteed.

5.2.2 Optimum Step-size Estimation

The steepest descent algorithm convergence speed is affected by the choice of step size. The step size can be set to a fixed value, but it is better to develop a closed-form for the optimum step-size and calculate it with each iteration update. The variable step-size approach will increase the convergence speed of the algorithm and will reduce the algorithm's sensitivity to step-size selection. Furthermore, properly setting the step size will guarantee real roots for (5.21) and hence the optimal loading level can be obtained.

Consider now the cost function of the Lagrange method, as given in (5.4):

$$\begin{aligned}\boldsymbol{\Psi}_{f_a}(n+1) &= (\boldsymbol{f}_c - \boldsymbol{B}\boldsymbol{f}_a(n+1))^H \boldsymbol{R}_{xx}(\boldsymbol{f}_c - \boldsymbol{B}\boldsymbol{f}_a(n+1)) \\ &\quad + \frac{1}{2}\lambda_0 t(\boldsymbol{f}_a^H(n+1)\boldsymbol{f}_a(n+1) - \beta^2)\end{aligned} \tag{5.32}$$

Using the update equation (5.20) in (5.32) and omitting the QC term, we obtain

$$\boldsymbol{\Psi}_{f_a}(n+1) = (\boldsymbol{f}(n) + \mu\boldsymbol{B}\nabla_{f_a}(n))^H \boldsymbol{R}_{xx}(\boldsymbol{f}(n) + \mu\boldsymbol{B}\nabla_{f_a}(n)) \tag{5.33}$$

Incorporating the variable step size $\mu(n)$ into the above equation instead of the fixed step-size, and after some manipulations, we obtain:

$$\begin{aligned}\boldsymbol{\Psi}_{f_a}(n+1) = \boldsymbol{\Psi}_{f_a}(n) + 2\mu(n)\nabla^H_{f_a}(n)\boldsymbol{P_B f}(n) \\ + \mu^2(n)\nabla^H_{f_a}(n)\boldsymbol{R}_B\nabla_{f_a}(n)\end{aligned} \tag{5.34}$$

This means that $\boldsymbol{\Psi}_{f_a}(n+1)$ is a quadratic function of $\mu(n)$, and hence it has a global minimum. Therefore, we can obtain the optimum step-size by differentiating (5.34) with respect to $\mu(n)$ and assuming that $\boldsymbol{\Psi}_{f_a}(n)$ is independent of $\mu(n)$. Therefore,

$$\frac{\partial \boldsymbol{\Psi}_{f_a}(n+1)}{\partial \mu(n)} = 2\nabla^H_{f_a}(n)\boldsymbol{P_B f}(n) + 2\mu(n)\nabla^H_{f_a}(n)\boldsymbol{R}_B\nabla_{f_a}(n) \tag{5.35}$$

By setting (5.35) to zero, the optimum step-size can be obtained as follows:

$$\mu_{opt}(n) = \frac{\alpha\left\|\nabla_{f_a}(n)\right\|^2}{\nabla^H_{f_a}(n)\boldsymbol{R_B}\nabla_{f_a}(n) + \sigma} \tag{5.36}$$

where α, σ are two positive constants, the former added to improve the numerical stability of the algorithm and the latter to avoid dividing by zero. The optimum step size derived in (5.36) is somewhat similar to the step size used in the recursive conjugate gradient (RCG) algorithm [52, 53]. More interestingly, this conjugate gradient algorithm can also be combined with the simplified VL approach to produce another robust detector based on the RCG algorithm. This approach does not appear to offer any additional improvement because the QI constraint compensates for the convergence speed difference, as will be demonstrated in Section 5.4. The previous statement is not theoretically validated and it can be further analyzed. This algorithm is summarized in Appendix 5A for the sake of comprehensiveness.

5.2.3 Low-complexity Recursive Implementation based on PLIC

Summarizing the simplified algorithm, the matrices $\boldsymbol{R_B}(n)$ and $\boldsymbol{P_B}(n)$ of the PLIC structure presented in Chapter 3, are updated using an exponentially decaying data window as follows:

$$\boldsymbol{R}_B(n) = \eta \boldsymbol{R}_B(n-1) + \boldsymbol{z}(n)\boldsymbol{z}^H(n) \tag{5.37}$$

$$\boldsymbol{P}_B(n) = \eta \boldsymbol{P}(n-1) + \boldsymbol{z}(n)\boldsymbol{x}^H(n) \tag{5.38}$$

where

$$\boldsymbol{z}(n) = \boldsymbol{B}^H \boldsymbol{x}(n) \tag{5.39}$$

and η is the usual forgetting factor with $0 \ll \eta \leq 1$. The unconstrained detector $\widetilde{\boldsymbol{f}}_a(n)$ is updated using (5.20). If $\widetilde{\boldsymbol{f}}_a(n)$ does not satisfy the QI constraint the optimal loading level is calculated using (5.22) and the quadratically constrained vector $\boldsymbol{f}_a(n)$ is calculated according to:

$$\boldsymbol{f}_a(n) = \widetilde{\boldsymbol{f}}_a(n) - \mu\lambda_0 \boldsymbol{f}_a(n-1) \tag{5.40}$$

From Equation 5.40, the simplified VL technique depends only on the previous update of the weight vector. Consequently, the total number of multiplications required to calculate the diagonal loading term in (5.40) is about $O(2N_a)$. Compared with the previously derived equivalent in (5.10)–(5.13), the new VL technique has substantially lower complexity: the number of multiplications required to calculate the approximate diagonal loading term γ in (5.14) is about $O(N_a^2 + 2N_a)$ due to the matrix vector multiplication in (5.12).

In spite of the simplicity of the VL technique, recursive implementation requires high computational loads because of the need to store and update the two matrices $\boldsymbol{R}_B$ and $\boldsymbol{P}_B$. To overcome this drawback, the gradient vector can be updated recursively by substituting from (5.37), (5.38), and (5.18) into (5.17). After some manipulations, the following update equation is obtained:

$$\overline{\nabla}_{f_a}(n) = \eta \overline{\nabla}_{f_a}(n-1) - y(n)\boldsymbol{z}(n) - \mu(n-1)\boldsymbol{R}_B(n)\nabla_{f_a}(n-1) \tag{5.41}$$

Therefore, only matrix $\boldsymbol{R}_B$ will be updated. The second gradient vector on the right-hand side ($\nabla_{f_a}(n-1)$) includes the diagonal loading effects from (5.25). The adaptive algorithm based on the RSD method and the VL technique (referred to as MOE-RSD w. QC) with the required amount of multiplications at each step is summarized in Box 5.2.

Box 5.2 Summary of the robust MOE-RSD w. QC.

Initialization

- $\boldsymbol{R_B}(0) = \boldsymbol{B}^H(\zeta \boldsymbol{I}_{N_a})\boldsymbol{B}\,; \boldsymbol{f}_c = \boldsymbol{C}_1(\boldsymbol{C}_1{}^H\boldsymbol{C}_1)^{-1}\boldsymbol{g}$
- $\delta^2 = t.(\|\boldsymbol{f}_c\|^2),\ \beta^2 = \delta^2 - \|\boldsymbol{f}_c\|^2,\ \alpha = \sigma = 0.1$
- $\lambda_0^1(n) = 0; \boldsymbol{f}_a(n) = \boldsymbol{0}_{N_a}; \overline{\nabla}_{f_a}(0) = \boldsymbol{B}^H(\zeta \boldsymbol{I}_{N_a})\boldsymbol{f}_c$

For $n = 1,2,\cdots$, **do**

- $\boldsymbol{z}(n) = \boldsymbol{B}^H\boldsymbol{x}(n);\ N_aN_f$
- $y(n) = \boldsymbol{x}^H(n)\boldsymbol{f}_c + \boldsymbol{z}^H(n)\boldsymbol{f}_a(n-1);\ N_f + N_a$
- $\boldsymbol{R}_B(n) = \eta \boldsymbol{R}_B(n-1) + \boldsymbol{z}^H(n)\boldsymbol{z}(n);\ N_a{}^2$
- $\nabla_{f_a}(n-1) = \overline{\nabla}_{f_a}(n-1) + \lambda_0^1(n-1)\boldsymbol{f}_a(n-1)$
- $\mu_{opt}(n) = \dfrac{\alpha\|\nabla_{\boldsymbol{f}_a}(n-1)\|^2}{\nabla_{\boldsymbol{f}_a}^H(n-1)\boldsymbol{R_B}(n)\nabla_{\boldsymbol{f}_a}(n-1)+\sigma};\ N_a^2 + N_a$
- $\overline{\nabla}_{f_a}(n) = \eta\overline{\nabla}_{f_a}(n-1) - y(n)\boldsymbol{z}(n) - \mu_{opt}(n)\boldsymbol{R}_B(n)\nabla_{f_a}(n-1)$

  - *if* $\left(\mu_{opt}(n) > \dfrac{\beta\|\boldsymbol{f}_a(n-1)\|}{\sqrt{\|\boldsymbol{f}_a(n-1)\|^2\left\|\overline{\nabla}_{\boldsymbol{f}_a}(n)\right\|^2 - \overline{\nabla}_{\boldsymbol{f}}^H(n)\boldsymbol{f}_a(n-1)\boldsymbol{f}_a^H(n-1)\overline{\nabla}_{\boldsymbol{f}_a}(n)}}\right);\ N_a$
  - $\mu_{opt}(n) = \dfrac{\beta\|\boldsymbol{f}_a(n-1)\|}{\sqrt{\|\boldsymbol{f}_a(n-1)\|^2\left\|\overline{\nabla}_{\boldsymbol{f}_a}(n)\right\|^2 - \overline{\nabla}_{\boldsymbol{f}}^H(n)\boldsymbol{f}_a(n-1)\boldsymbol{f}_a^H(n-1)\overline{\nabla}_{\boldsymbol{f}_a}(n)}}$
  - End

- $\widetilde{\boldsymbol{f}}_a(n) = \boldsymbol{f}_a(n-1) - \mu_{opt}(n)\overline{\nabla}_{f_a}(n)$

  - *if* $(\|\widetilde{\boldsymbol{f}}_a(n)\|^2 > \beta^2)$; (QI constraint not satisfied) N_a
    - $a = \mu(n)^2\|\boldsymbol{f}_a(n)\|^2,\ b = -2\mu\widetilde{\boldsymbol{f}}_a^H(n)\boldsymbol{f}_a(n-1),\ c = \|\widetilde{\boldsymbol{f}}_a(n)\|^2 - \beta^2;\ 2N_a$
    - $\lambda_0^1(n) = \dfrac{-b-(\sqrt{b^2-4ac})}{2a}$
    - $\boldsymbol{f}_a(n) = \widetilde{\boldsymbol{f}}_a(n) - \mu_{opt}(n)\lambda_0^1(n)\boldsymbol{f}_a(n-1)$
  - *else*; (QI constraint satisfied)
    - $\boldsymbol{f}_a(n) = \widetilde{\boldsymbol{f}}_a(n),\ \lambda_0^1(n) = 0$
  - End if

- $\boldsymbol{f}(n) = \boldsymbol{f}_c - \boldsymbol{B}\boldsymbol{f}_a(n)$; (only for evaluation)

End for

We add an *if-else* condition statement to ensure that $C > 0$. If this condition is achieved, the VL subroutine will be executed and a new update procedure for the detector is performed based on (5.22) and (5.40). Otherwise, the algorithm will continue without executing the VL subroutine. The smaller root of the quadratic equation (5.21) is selected to guarantee more algorithm stability.

5.2.4 Convergence Analysis

To analyze the convergence of the presented algorithm, we will derive conditions for the step size that will guarantee that the detector $\boldsymbol{f}_a(n)$ asymptotically converges to the optimum value given in (5.5). This can be done by evaluating the difference between the expected value of the updated detector $\boldsymbol{f}_a(n)$ and its optimal value [54].

$$\varepsilon_{f_a} = E\{\boldsymbol{f}_a(n)\} - \boldsymbol{f}_{a(opt)} \tag{5.42}$$

By inserting (5.5) and (5.18) into (5.42) we get:

$$\begin{aligned}\varepsilon_{f_a}(n) = {} & (I - \mu(n)(\boldsymbol{R}_B(n) + \lambda_0(n)\boldsymbol{I}))E\{\boldsymbol{f}_a(n-1)\} \\ & + \mu(n)\boldsymbol{p}_B - \boldsymbol{f}_{a(opt)}\end{aligned} \tag{5.43}$$

After some manipulations, we can reformulate this as follows:

$$\begin{aligned}\varepsilon_{f_a}(n) = {} & (I - \mu(n)(\boldsymbol{R}_B(n) + \lambda_0(n)\boldsymbol{I}))\varepsilon_{f_a}(n-1) \\ & + \mu(n)\boldsymbol{p}_B - \mu(n)(\boldsymbol{R}_B(n) + \lambda_0(n)\boldsymbol{I})\boldsymbol{f}_{a(opt)}\end{aligned} \tag{5.44}$$

It can be seen that the last two terms are cancelled and hence we obtain a recursive formula for ε_{f_a} as follows:

$$\varepsilon_{f_a}(n) = (\boldsymbol{I} - \mu(n)(\boldsymbol{R}_B(n) + \lambda_0(n)\boldsymbol{I}))\varepsilon_{f_a}(n-1) \tag{5.45}$$

Therefore, the convergence of the algorithm depends on the eigenvalue spread of the diagonally loaded blocked data matrix. If the eigenvalues of $(\boldsymbol{R}_B(n) + \lambda_0(n)\boldsymbol{I})$ are denoted by ρ_i, then the updated detector converges to its optimal value if [54, 55]:

$$0 < \mu < \frac{1}{\rho_{\max}} \tag{5.46}$$

The eigenvalues of the diagonally loaded matrix are individually increased by the addition of the VL term λ_0, which is equivalent to increasing the matrix diagonal by this VL level.

The eigenvectors remain unchanged by the diagonal loading [35]. In addition to (5.46), the inequality in (5.27) must be considered to guarantee positive diagonal loading. As a result, the presented robust detector is guaranteed to converge with optimum diagonal loading computation if:

$$0 < \mu(n) < \min\left\{\frac{1}{\rho_{\max}}, \frac{\beta\|\boldsymbol{f}_a(n-1)\|}{\sqrt{\begin{array}{l}\|\boldsymbol{f}_a(n-1)\|^2\left\|\overline{\nabla}_{\boldsymbol{f}_a}(n)\right\|^2\\ \quad -\overline{\nabla}_{\boldsymbol{f}}^H(n)\boldsymbol{f}_a(n-1)\boldsymbol{f}_a^H(n-1)\overline{\nabla}_{\boldsymbol{f}_a}(n)\end{array}}}\right\} \tag{5.47}$$

5.3 Geometric Approach

To elucidate the variable loading techniques, let us consider a simple 2-D case, as proposed by Song *et al.* [56]. Figures 5.1 and 5.2 give a geometrical representation of Tian's VL technique

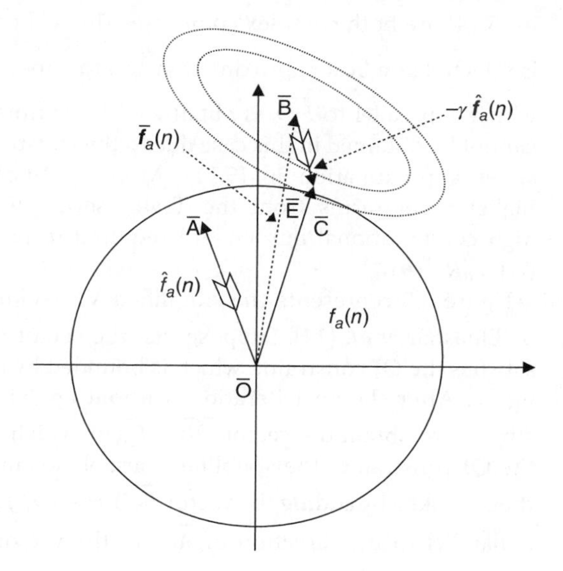

Figure 5.1 MOE-RLS w. QC (real roots).

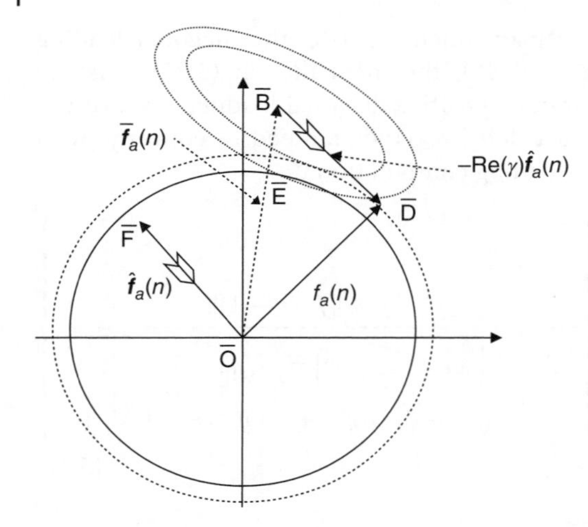

Figure 5.2 MOE-RLS w. QC (imaginary roots).

[28, 45–50]. The figures represent two cases, the first with a small loading level and the second is with a relatively large one. In the first figure, the roots in (5.14) are both positive and the smaller root is selected. Hence, the QI constraint can be satisfied, and the updated vector $\overrightarrow{\overline{O}\,\overline{C}}$, which is parallel to $\overrightarrow{\overline{A}\,\overline{O}}$, is obtained. Figure 5.2 represents the second case, where the roots in (5.14) are both complex conjugate. The real part of the roots is selected as a best approximation and the updated vector $\overrightarrow{\overline{O}\,\overline{D}}$, which is parallel to $\overrightarrow{\overline{F}\,\overline{O}}$, is obtained. The optimal loading level cannot be obtained in this case due to the violation of the Taylor series approximation in (5.11). More specifically, additional higher-order terms from the Taylor series, with a resultant high computational burden, are required to best approximate $(I + \lambda_0 \boldsymbol{R}_B^{-1}(n))^{-1}$.

Figure 5.3 represents the simplified VL technique proposed by Elnashar *et al.* [14]. Suppose that the vector $\overrightarrow{OA} = \boldsymbol{f}_a(n-1)$ satisfies the QI constraint, which is bounded by the circle in the figure. After the next iteration, without applying the QI constraint, we obtain the vector $\overrightarrow{OB} = \widetilde{\boldsymbol{f}}_a(n)$, which may not satisfy the QI constraint. The simplified variable loading technique is then invoked by adding the vector $\overrightarrow{BC1} = -\mu\lambda_0^1 \boldsymbol{f}_a(n-1)$, which is parallel to the direction of $\overrightarrow{AO}$, to the vector $\overrightarrow{OB}$. We then

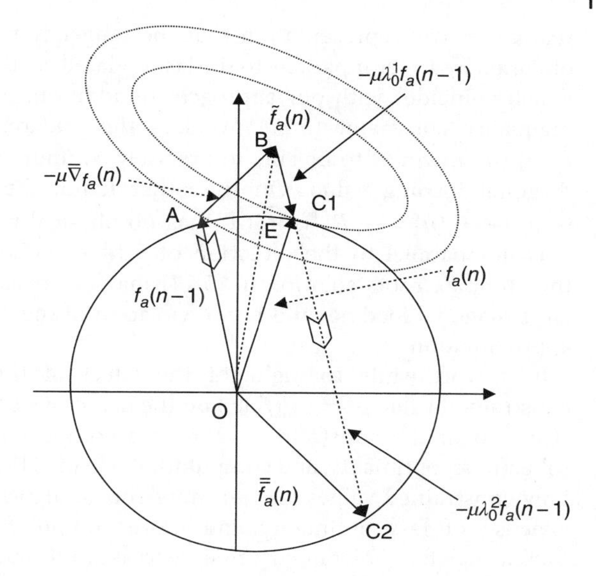

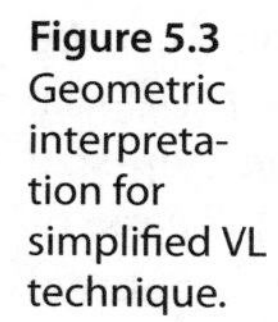
Figure 5.3 Geometric interpretation for simplified VL technique.

obtain the constrained vector $\overrightarrow{OC1}$ that satisfies the QI constraint. The VL value λ_0^1 represents the smaller root of (5.21) and λ_0^2 represents the larger root. If the larger root is selected, the point $C2$ is obtained by adding the vector $\overrightarrow{BC2} = -\mu\lambda_0^2 \boldsymbol{f}_a(n-1)$ to the vector $\overrightarrow{OB}$, which is parallel to the direction of $\overrightarrow{AO}$ as well, and hence we obtain a new constrained vector $\overrightarrow{OC2} = \bar{\bar{\boldsymbol{f}}}_a(n)$. We would choose the smaller root (λ_0^1) to be the VL value for algorithm stability.

The vectors $\overline{\overline{OE}}$ and $\overrightarrow{OE}$ in the two VL techniques represent, respectively, the SP updated vectors [24] where the SP technique simply scales the unconstrained updated vector $(i.e.\, \tilde{\boldsymbol{f}}_a(n) \Rightarrow \beta\tilde{\boldsymbol{f}}_a(n)/\|\tilde{\boldsymbol{f}}_a(n)\|)$, to satisfy the QI constraint.

An important observation is that the constrained region is a closed sphere, centered at the origin O. The concentric ellipses are equivalued contours representing the unconstrained cost function in (5.1) and their center is the minimum point without constraints. Hence, the tangency point C with the QI constrained boundary surface is the optimum loading point, in the sense of satisfying the QI constraint and the unconstrained cost function. Also, it would appear from the curves and with real geometric representation that the tangency point can be obtained by a vector parallel to the last updated vector $\boldsymbol{f}_a(n-1)$, which coincides with our approach. In addition, the step-size inequality constraint (5.27) oversees the update process of $\widetilde{\boldsymbol{f}}_a(n)$ (*in advance*) to be able to precisely compute the optimal diagonal loading value. On the other hand, the correction term $(-\gamma\widehat{\boldsymbol{f}}_a(n) = -\gamma \boldsymbol{R}_B^{-1}(n)\overline{\boldsymbol{f}}_a(n))$ has no physical meaning and it is not parallel to the direction of $\boldsymbol{f}_a(n-1)$. Consequently, the complex roots solution in (5.14) may occur as a result of large diagonal loading, and hence violation of the Taylor series approximation.

It is worthwhile to highlight the effect of the quadratic constrained value $\rho^2 = t.(\|\boldsymbol{f}_c\|^2)$ on the detector's behavior. The selection of the constrained value is a compromise between robustness, optimality, and computational load of the algorithm. Low constrained values lead to more robust algorithms at the expense of less optimality and more computational loads. Increasing the constrained value decreases the number of VL subroutine executions and hence decreases the computational loads, until a certain limit where the algorithm robustness is affected. The essence of any QC technique is to boost the robustness of the algorithm without affecting its optimality. Therefore, the constrained value can be set within a certain range of values. Note that the reasonable range for the constrained value is not static; it is a dynamic range that depends on the system parameters, the required robustness, and the algorithm optimality.

5.4 Simulation Results

In this section, the performance of the simplified robust detector will be investigated and compared with the traditional blind MOE detector updated using the RLS algorithm (referred to as MOE-RLS) and a robust MOE-RLS w. QC detector [28]. Five synchronous users using Gold codes in a multipath Rayleigh fading channel with five multipath components are simulated. The channel length (maximum delay spread) is assumed to be ten delayed components and multipath delays are randomly distributed. The detector is synchronized to the user of interest.

Users are assumed to have equal power except the required user, which is set to be 10 dB less than the other users. The SNR is 30 dB. Gold codes with 31 chips are used and the detector length is assumed to be one bit length (31 samples). Two performance measures are adopted here: the output SINR versus bit iteration and bit error rate (BER) versus bit iteration.

In order to verify the robustness of the detectors, five simulation scenarios are analyzed. The first scenario is referred to as the ideal scenario, where all the algorithm parameters are properly selected. The initialization of this scenario is as follows; $\eta = 1$, $\delta^2 = 1.35(\|\boldsymbol{f}_c\|^2), \boldsymbol{f}_{\boldsymbol{a}}(n) = \mathbf{0}_{N_a}$, and $\zeta = 1$. The selection of the constrained value is based on a compromise between robustness and performance optimality. Figures 5.4 and 5.5 illustrate the SINR and BER for the three detectors in the ideal scenario. The performance of the three detectors in terms of convergence speed is almost the same, and the ideal scenario decreases the necessity for robustness (QI constraint). Therefore the VL subroutine will be only infrequently executed. Note, however, that the figure shows the superiority of the simplified robust detector in terms of steady-state performance.

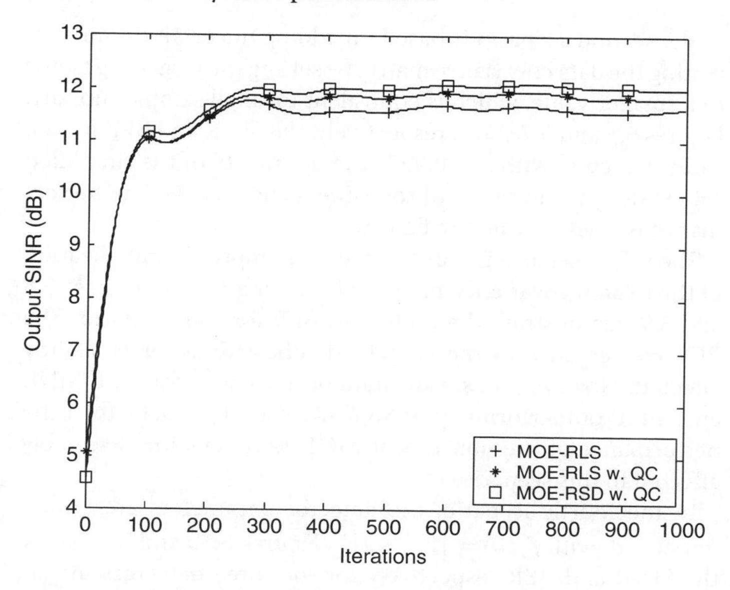

Figure 5.4 SINR versus iterations for the first (ideal) initializations scenario.

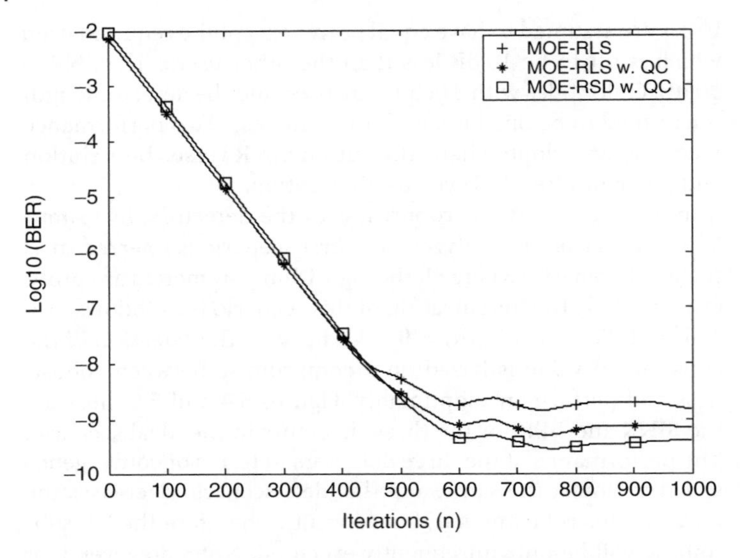

Figure 5.5 BER versus iterations for the first (ideal) initializations scenario.

The second scenario is based on adding uncertainties in estimating the data covariance matrix by setting the forgetting factor to a smaller value, which is equivalent to small sample support. Figures 5.6 and 5.7 show, respectively, the SINR and BER for the three detectors with $\eta = 0.997$. The optimality of the simplified robust detector in terms of robustness and steady-state performance is evident from the figures.

The third scenario is constructed with improper initialization of the related covariance matrices by setting $\zeta = 10$. Figures 5.8 and 5.9 demonstrate the SINR and BER for this scenario. The BER convergence of the simplified robust detector is slightly affected. However, its steady-state performance for both SINR and BER outperforms the MOE-RLS w. QC detector. The performance of the non-robust MOE-RLS detector is severely affected in this scenario.

In the fourth scenario, the detector parameters are poorly initialized, with $\boldsymbol{f}_a(0) = \left[1 \ \cdots \ 1\right]^T$. Figures 5.10 and 5.11 show the SINR and BER respectively for the three detectors in this scenario.

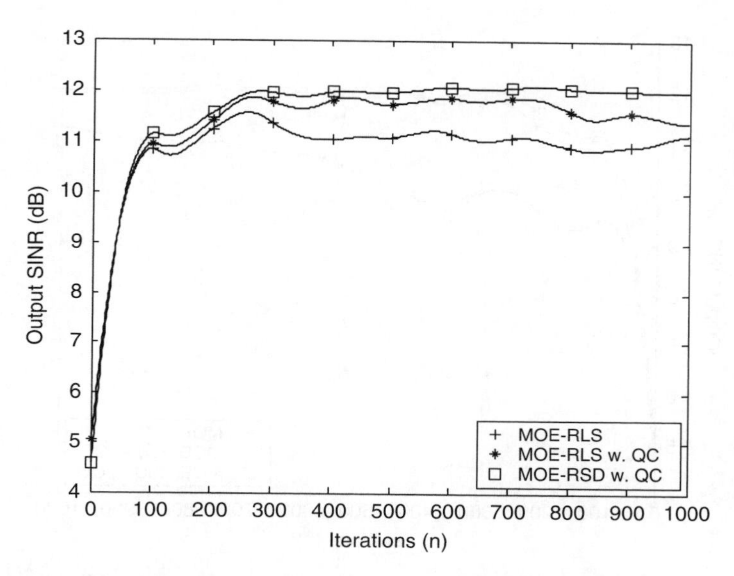

Figure 5.6 SINR versus iterations for the second initialization scenario.

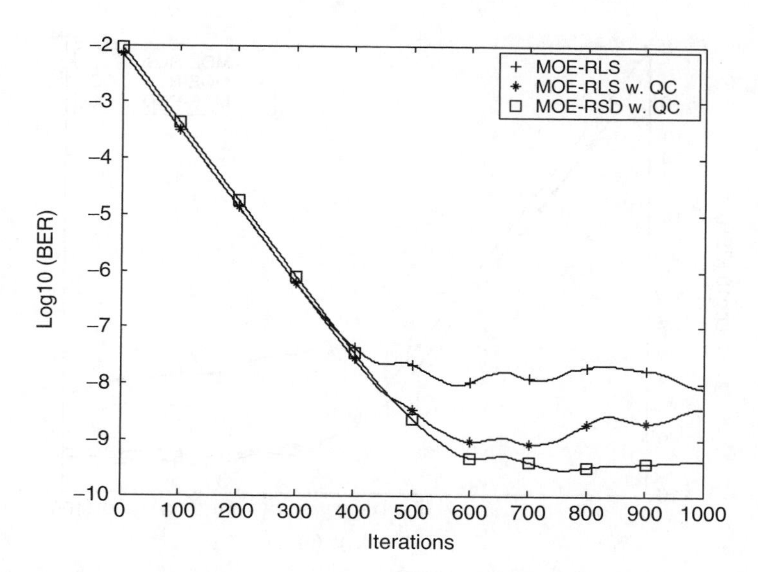

Figure 5.7 BER versus iterations for the second initialization scenario.

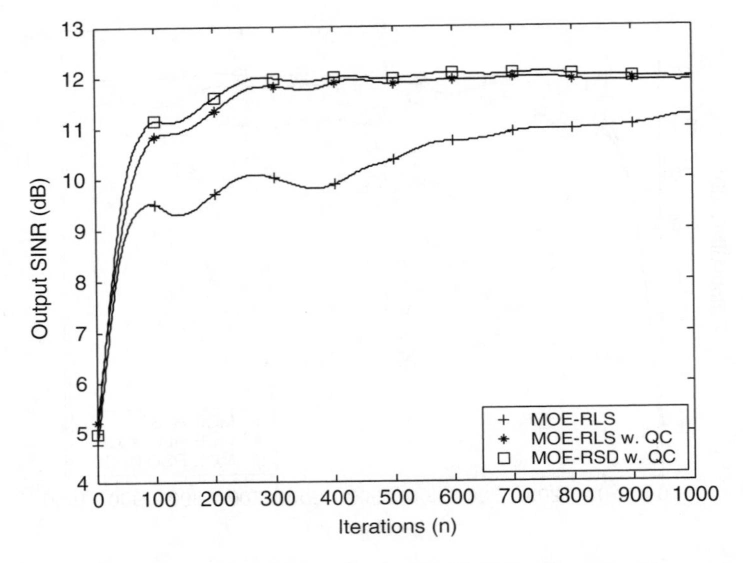

Figure 5.8 SINR versus iterations for the third initialization scenario.

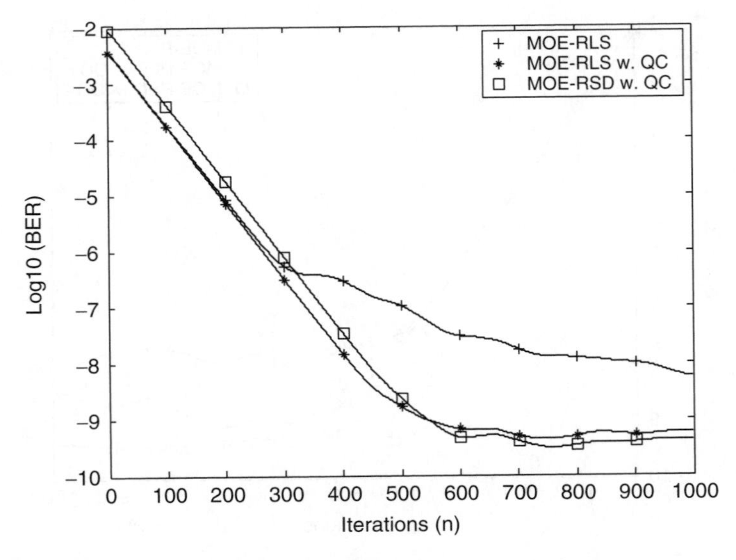

Figure 5.9 BER versus iterations for the third initialization scenario.

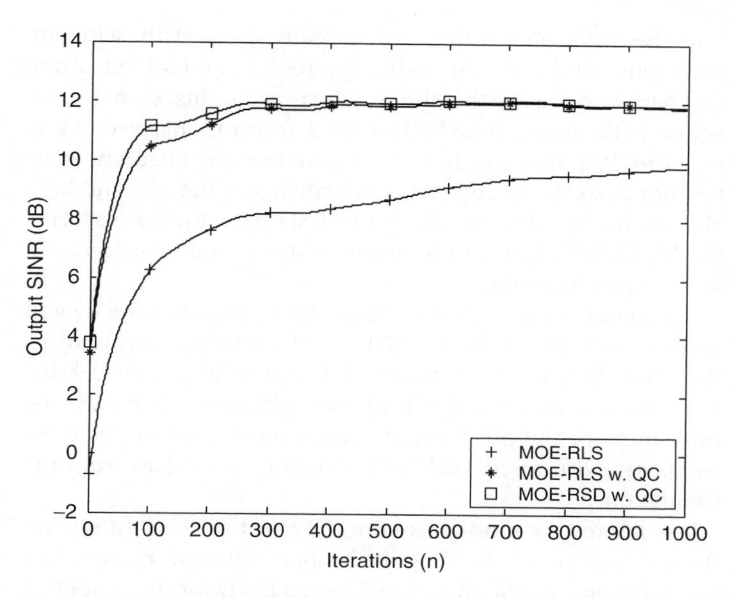

Figure 5.10 SINR versus iterations for the fourth initialization scenario.

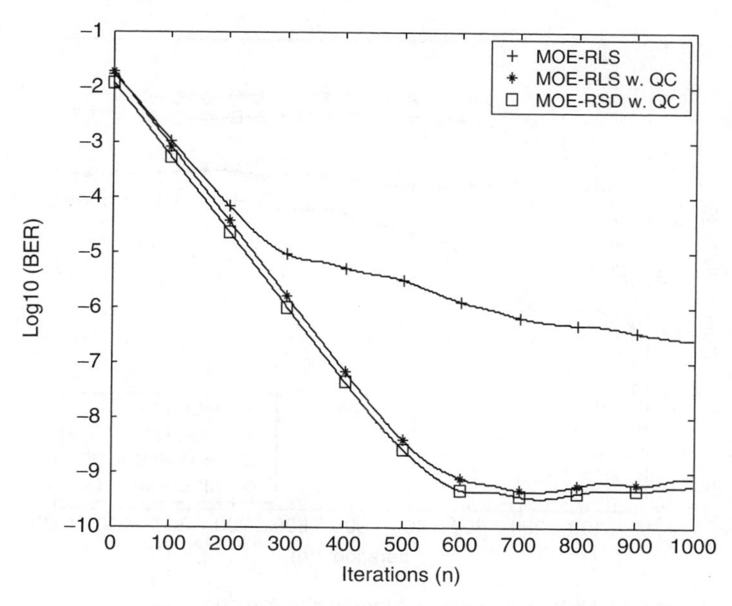

Figure 5.11 BER versus iterations for the fourth initialization scenario.

In the last scenario the SNR is reduced to 20 dB, with the parameters of the ideal scenario. Figures 5.12 and 5.13 illustrate the performance of the three detectors in this scenario. In addition, the non-robust MOE-RSD detector is analyzed in this scenario. It is apparent from the figure that the quadratic constraint boosts the steady-state performance of the RLS and RSD algorithms. In addition, the MOE-RSD detector outperforms the MOE-RLS algorithm in terms of steady-state performance and convergence rate.

In general, we can observe from the figures that the overall performance of the MOE-RSD w. QC detector outperforms the MOE-RLS w. QC detector and represents a considerable improvement over the MOE-RLS algorithm. Additionally, the robustness of the quadratically constrained class of detectors compared with traditional MOE detectors is evident from the simulations.

It is interesting now to explore the effect of the step size on the convergence of the simplified robust detector. Figures 5.14 and 5.15 show the SINR and BER respectively for the simplified robust detector, with different step sizes and with the optimum

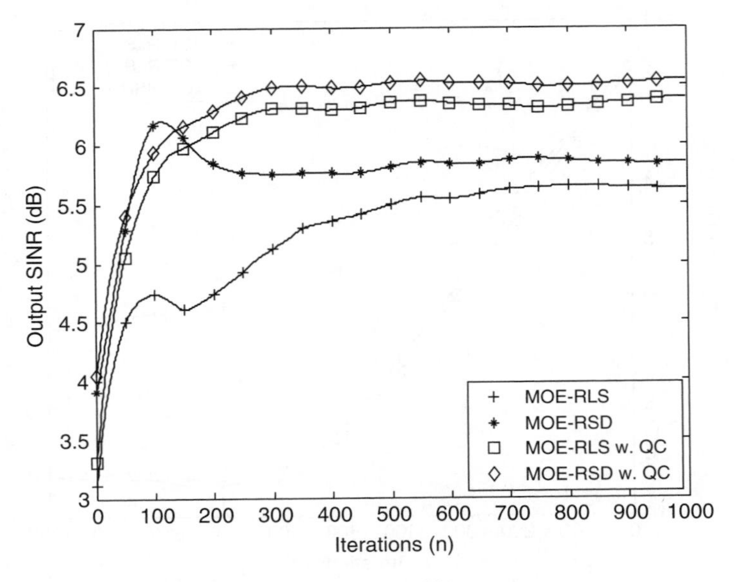

Figure 5.12 SINR versus iterations for the fifth scenario.

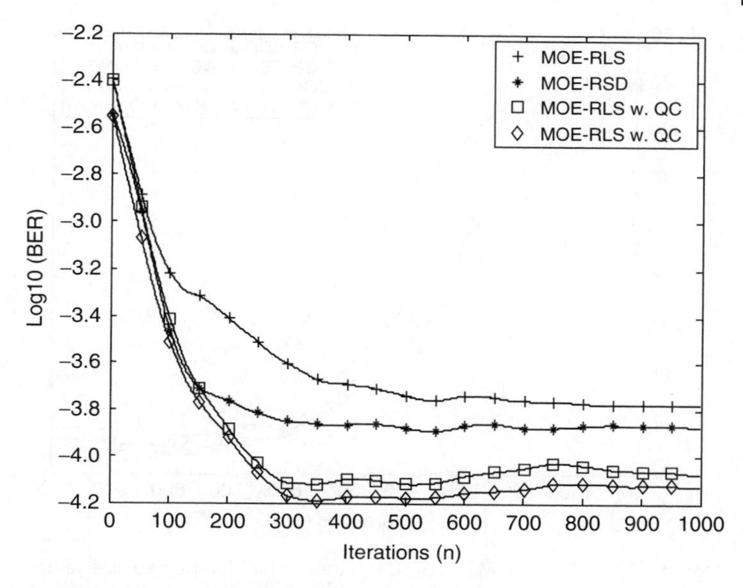

Figure 5.13 BER versus iterations for the fifth scenario.

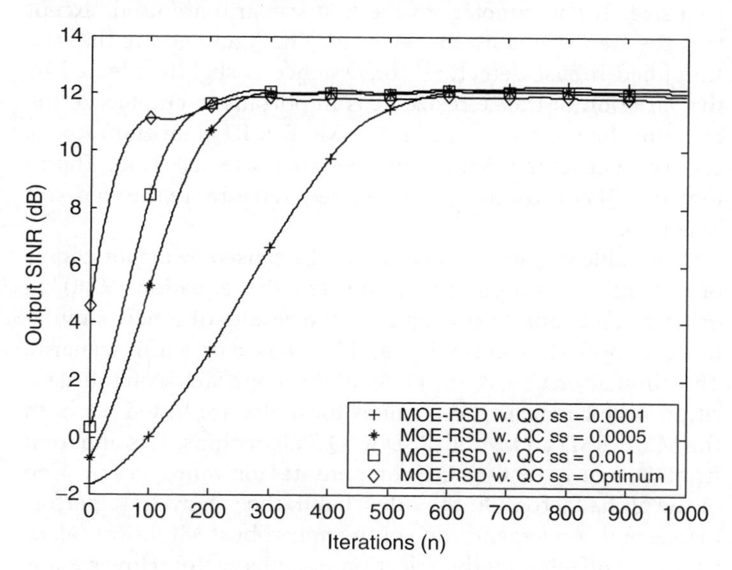

Figure 5.14 SINR for the MOE-RSD Detectors with different step-size values.

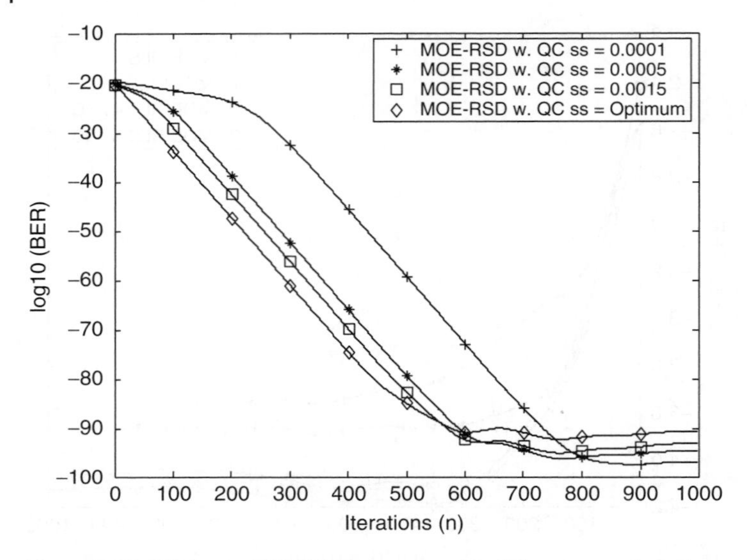

Figure 5.15 BER for the MOE-RSD Detectors with different step-size values.

step size. The parameters of the first scenario are used, except that the step size is no longer fixed. The figure shows that the simplified robust detector's convergence is slightly affected by the variation of the step size. It is important to emphasize the effect of the QI constraint on the MOE-RSD algorithm; it acts as a compensator for the improper step-size selection. Therefore, the QI constraint algorithm is less sensitive to the step-size selection.

Unsuitable step-size selection may be caused by a poor choice of the parameter alpha (α) in the step-size equation (5.36). In order to corroborate this finding, the results of a new simulation are shown in Figures 5.16 and 5.17 based on a fifth scenario. This simulation reveals the effect of the alpha factor on the algorithm performance. Three values for alpha are tested for both the MOE-RSD and MOE-RSD w. QC algorithms. It is apparent from the figures that the QC compensates for improper selection of the alpha factor. On the other hand, the steady-state performance and convergence rate of the non-robust MOE-RSD algorithm are affected by the selection of alpha factor. However, by comparing Figures 5.14 and 5.15 against Figures 5.16 and 5.17, the worst performance of the MOE-RSD algorithm (at $\alpha = 0.9$) is analogous to that of the MOE-RSL detector.

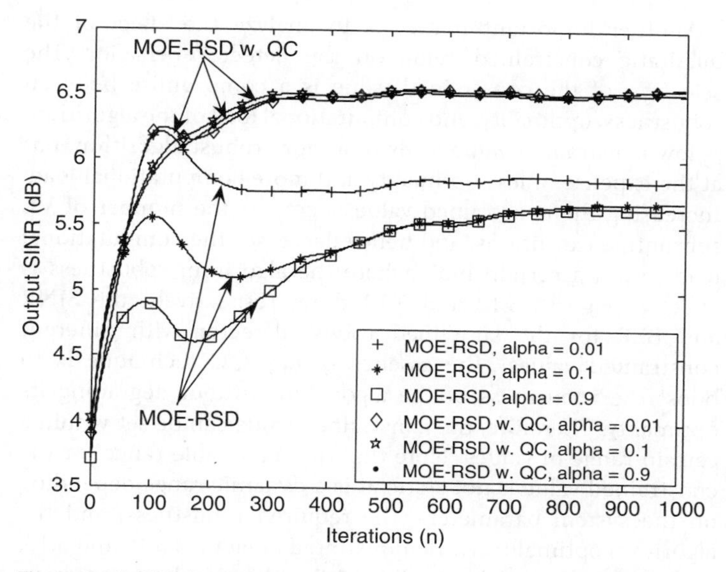

Figure 5.16 SINR for the MOE-RSD Detectors with different alpha values.

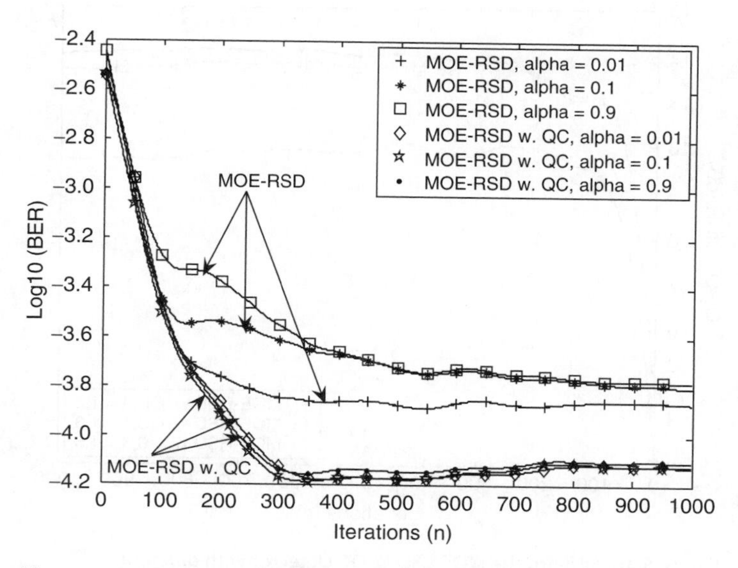

Figure 5.17 BER for the MOE-RSD Detectors with different alpha values.

Another important matter is to analyze the effect of the quadratic constrained value on the detector behavior. The selection of the constrained value is a compromise between robustness, optimality, and computational load of the algorithm. A low constrained value leads to a more robust algorithm, but at the expense of less optimality and more computational load. Increasing the constrained value decreases the number of VL subroutine executions, and hence decreases the computational loads, until a certain limit where the algorithm robustness is affected. Figures 5.18 and 5.19 show, respectively, the SINR and BER for the simplified robust detector with different constrained values. The essence of any QC technique is to boost the robustness of the algorithm without degrading its optimality. Therefore, the constrained value can be set within a certain range of values. Note that the reasonable range for the constrained value is not static; it is a dynamic range depending on the system parameters, the required robustness, and the algorithm optimality. As demonstrated in Figures 5.18 and 5.19, the optimum constrained value is 1.3 in this simulation exercise.

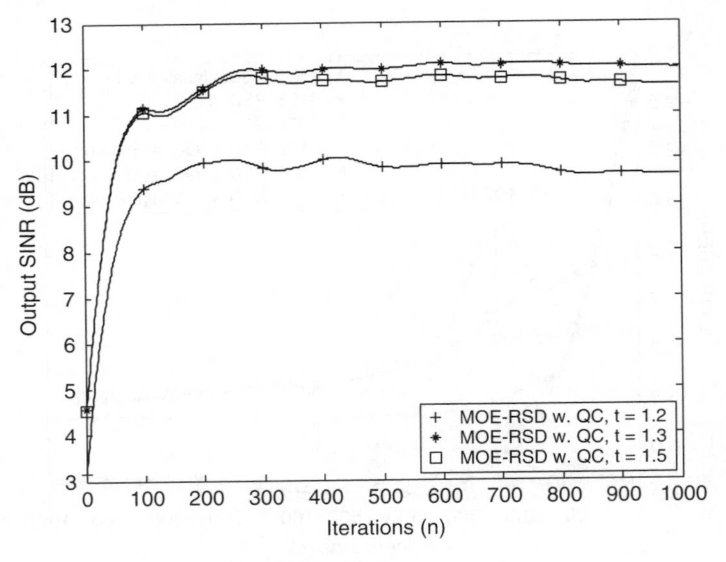

Figure 5.18 SINR for the MOE-RSD w. QC Detector with different constrained values.

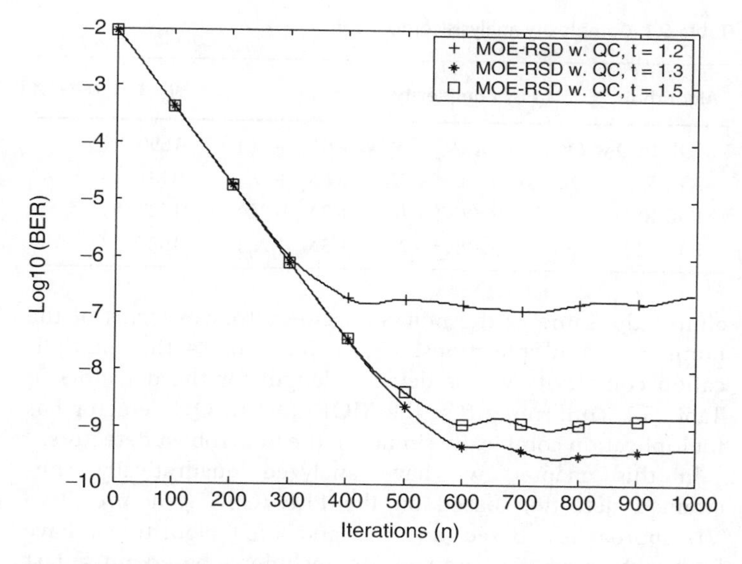

Figure 5.19 BER for the MOE-RSD w. QC detector with different constrained values.

5.5 Complexity Analysis

As shown in Table 5.1, the total number of multiplications required at each snapshot of the MOE-RSD detector is about $O(2N_a^2 + N_aN_f + 6N_a + N_f)$, including the matrix update, the VL technique, and the vector update, The multiplication expense for the MOE-RLS w. QC detector [18, 28] is about $O(3N_a^2 + N_aN_f + 6N_a + N_f)$. The non-robust version of the RSD detector (referred to as MOE-RSD) can be obtained by removing the VL subroutine from Box 5.1. The computational complexity of this detector is about $O(2N_a^2 + N_aN_f + 3N_a + N_f)$. Analogous to MOE-RSD detector, the MOE-RSL detector can be obtained by removing the VL routine from the MOE-RLS w. QC detector. This detector requires $O(2N_a^2 + N_aN_f + 3N_a + N_f)$ multiplications, which is similar to MOE-RSD detector. Table 5.1 summarizes the number of multiplications required in the simulated scenario. The robust MOE-RSD w. QC detector requires fewer multiplications than the MOE-RLS w. QC detector. Furthermore, the MOE-RSD w. QC algorithm is

Table 5.1 Complexity analysis comparison.

Algorithm	Complexity	$N_a = 21, N_f = 31$
MOE-RSD w. QC	$O(2N_a^2 + N_aN_f + 6N_a + N_f)$	1690
MOE-RLS w. QC	$O(3N_a^2 + N_aN_f + 6N_a + N_f)$	2131
MOE-RSD	$O(2N_a^2 + N_aN_f + 3N_a + N_f)$	1627
MOE-RLS	$O(2N_a^2 + N_aN_f + 3N_a + N_f)$	1627

almost the same as the non-robust detectors in terms of the number of multiplications. Figure 5.20 shows the multiplication complexity versus detector length for the detectors in Table 5.1, confirming that the MOE-RSD w. QC detector has multiplication complexity similar to the non-robust detectors.

In this chapter, we have analyzed quadratically constrained detection based on the PLIC/GSC structure. Two VL approaches, based on RLS and RSD algorithms, have been presented. An optimal VL technique based on a fast low-complexity RSD algorithm is used to satisfy the quadratic constraint. The diagonal loading term is computed using a simple quadratic equation with low complexity. Other attractive merits of the RSD-based approach are the simplicity and the low

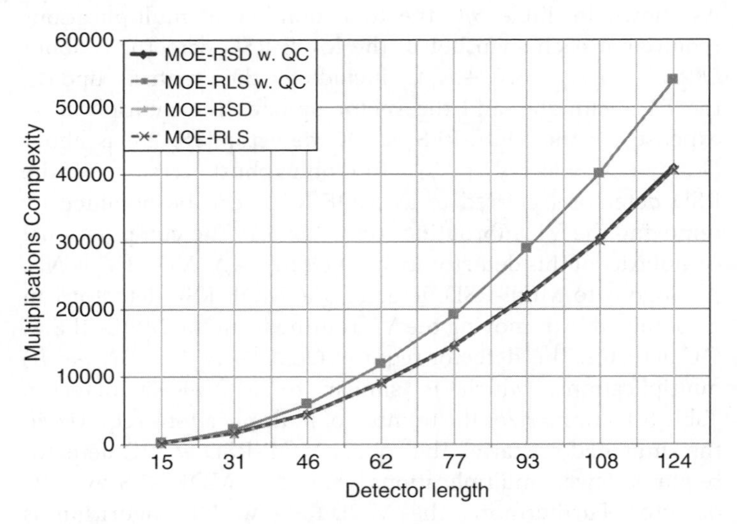

Figure 5.20 Multiplication complexity versus detector length.

computational complexity. Moreover, the adaptive procedure does not require matrix inversion and the algorithm is not sensitive to the step-size, because the optimum step-size is adaptively computed. Future work may include combining the RSD-VL algorithm with the channel estimation techniques in MIMO-OFDM, or addressing other mismatch constraints, or using worst-case constraint beamforming applications to produce double-constraint robust detectors. Details of possible algorithms that may be realized using the RSD-VL approach can be found in the literature [56–88].

Appendix 5.A Robust Recursive Conjugate Gradient (RCG) Algorithm

By combining the RCG algorithm [21–23, 52, 53], used to estimate the optimal weight vector, and the simplifed VL technique in [14], the following recursive equations can be obtained:

$$\boldsymbol{R}_B(n) = \boldsymbol{R}_B(n-1) + \boldsymbol{z}^H(n)\boldsymbol{z}(n) \tag{5A.1}$$

$$\mu(n) = \frac{\alpha \boldsymbol{p}^T(n)\boldsymbol{g}(n-1)}{\boldsymbol{p}^T(n)\boldsymbol{R}_B(n)\boldsymbol{p}(n) + \sigma} \tag{5A.2}$$

$$\bar{\boldsymbol{f}}_a(n) = \boldsymbol{f}_a(n-1) - \mu(n)\boldsymbol{p}(n) \tag{5A.3}$$

$$\bar{g}(n) = -\boldsymbol{B}^H\boldsymbol{R}_{xx}(n)\boldsymbol{f}_c + \boldsymbol{B}^H\boldsymbol{R}_{xx}(n)\boldsymbol{B}\boldsymbol{f}_a(n-1) \tag{5A.4}$$

$$\bar{g}(n) = \eta g(n-1) - y(n)\boldsymbol{z}(n) - \mu(n)\boldsymbol{R}_B(n)\boldsymbol{p}(n) \tag{5A.5}$$

$$\boldsymbol{g}(n) = \bar{\boldsymbol{g}}(n) + \lambda_0(n)\boldsymbol{f}_a(n) \tag{5A.6}$$

$$\omega(n) = \frac{(\boldsymbol{g}(n) - \boldsymbol{g}(n-1))^T\boldsymbol{g}(n)}{\boldsymbol{g}(n-1)^T\boldsymbol{g}(n-1)} \tag{5A.7}$$

$$\boldsymbol{p}(n+1) = \boldsymbol{g}(n) + \omega(n)\boldsymbol{p}(n) \tag{5A.8}$$

$$\boldsymbol{f}_a(n) = \bar{\boldsymbol{f}}_a(n) - \mu(n)\lambda_0(n)\boldsymbol{f}_a(n-1) \tag{5A.9}$$

where $\boldsymbol{z}(n) = \boldsymbol{B}^H\boldsymbol{x}(n)$ is the blocked received signal, $\mu(n)$ is the step-size of the algorithm, and $\boldsymbol{g}(n)$ is the conjugate derivative of $\boldsymbol{\Psi}_{f_a}(n)$ with respect to the real and imaginary parts of $\boldsymbol{f}_a^H(n)$. In addition, $\bar{\boldsymbol{g}}(n)$ is the conjugate gradient without the QI term, and the adaptive implementation of $\bar{\boldsymbol{g}}(n)$ can be found by substituting (5A.1), (5A.3), and (5A.8) into (5A.4). The complete algorithm is summarized in Table 5A.1.

Table 5A.I Summary of the robust RCG algorithm.

Initialization

- $\boldsymbol{R_B}(0) = \boldsymbol{B}^H(\zeta \boldsymbol{I}_{N_a})\boldsymbol{B}; \boldsymbol{f}_c = \boldsymbol{C}_1(\boldsymbol{C}_1{}^H\boldsymbol{C}_1)^{-1}\boldsymbol{g}$
- $\delta^2 = t.(\|\boldsymbol{f}_c\|^2),\ \beta^2 = \delta^2 - \|\boldsymbol{f}_c\|^2,\ \alpha = \sigma = 0.1$
- $\lambda_0^1(n) = 0; \boldsymbol{f}_a(n) = \boldsymbol{0}_{N_a}; \boldsymbol{p}(1) = \boldsymbol{g}(0) = \boldsymbol{B}^H(\zeta \boldsymbol{I}_{N_a})\boldsymbol{f}_c$

For $n = 1,2,\cdots$, **do**

- $z(n) = \boldsymbol{B}^H\boldsymbol{x}(n);\ N_aN_f$
- $y(n) = \boldsymbol{x}^H(n)\boldsymbol{f}_c + \boldsymbol{z}^H(n)\boldsymbol{f}_a(n-1);\ N_f + N_a$
- $\boldsymbol{R}_B(n) = \eta \boldsymbol{R}_B(n-1) + \boldsymbol{z}^H(n)\boldsymbol{z}(n);\ N_a{}^2$
- $\mu(n) = \dfrac{\alpha \boldsymbol{p}^H(n)\boldsymbol{g}(n-1)}{\boldsymbol{p}^H(n)\boldsymbol{R}_B(n)\boldsymbol{p}(n) + \sigma} N_a^2 + 2N_a$

  - $if\left(\mu_{opt}(n) > \sqrt{\dfrac{\beta^2}{\beta^2 \boldsymbol{p}^H(n)\boldsymbol{p}(n) - \boldsymbol{p}^H(n)\boldsymbol{f}_a(n-1)\boldsymbol{f}_a^H(n-1)\boldsymbol{p}(n)}}\right)$
    - $\mu_{opt}(n) = \sqrt{\dfrac{\beta^2}{\beta^2 \boldsymbol{p}^H(n)\boldsymbol{p}(n) - \boldsymbol{p}^H(n)\boldsymbol{f}_a(n-1)\boldsymbol{f}_a^H(n-1)\boldsymbol{p}(n)}};\ N_a$
  - end
- $\overline{\boldsymbol{f}}_a(n) = \overline{\boldsymbol{f}}_a(n-1) - \mu(n)\boldsymbol{p}(n)$

  - $if(\|\overline{\boldsymbol{f}}_a(n)\|^2 > \beta^2)$; (QI constraint not satisfied)
    - $a = \mu^2\beta^2,\ b = -2\mu\overline{\boldsymbol{f}}_a^H(n)\boldsymbol{f}_a(n-1),\ c = \overline{\boldsymbol{f}}_a^H(n)\overline{\boldsymbol{f}}_a(n) - \beta^2;$ $2N_a$
    - $\lambda_0^1(n) = \dfrac{-b - \left(\sqrt{b^2 - 4ac}\right)}{2a}$
    - $\boldsymbol{f}_a(n) = \overline{\boldsymbol{f}}_a(n) - \mu(n)\lambda_0^1(n)\boldsymbol{f}_a(n-1)$
  - $else$; (QI constraint satisfied)
    - $\boldsymbol{f}_a(n) = \overline{\boldsymbol{f}}_a(n)$
    - $\lambda_0^1(n) = 0$

- $g(n) = \eta g(n-1) - y(n)\boldsymbol{z}(n) - \mu(n-1)\boldsymbol{R}_B(n)\boldsymbol{p}(n) + \lambda_0^1(n)\boldsymbol{f}_a(n)$
- $\omega(n) = \dfrac{(\boldsymbol{g}(n) - \boldsymbol{g}(n-1))^H\boldsymbol{g}(n)}{\boldsymbol{g}(n-1)^H\boldsymbol{g}(n-1)};\ 2N_a$
- $\boldsymbol{p}(n+1) = \boldsymbol{g}(n) + \omega(n)\boldsymbol{p}(n)$
- $\boldsymbol{f}(n) = \boldsymbol{f}_c - \boldsymbol{B}\boldsymbol{f}_a(n)$; (only for evaluation)

End

References

1 M.K. Tsatsanis and Z. Xu, "On minimum output energy CDMA receives in presence of multipath." In: *Conference on Information Sciences and Systems (CISS'97)*, pp. 377–381, March 1997.

2 J.K. Tugnait and T. Li, "Blind detection of asynchronous CDMA signals in multipath channels using code-constrained inverse filter criterion," *IEEE Trans. Sig. Proc.* vol. 49, no. 7, pp. 1300–1309, 2001.

3 M.K. Tsatsanis, "Inverse filtering criteria for CDMA systems," *IEEE Trans. Signal Process.*, vol. 45, no. 1, pp. 102–112, 1997.

4 M. Hoing and M.K. Tsatsanis, "Adaptive techniques for multiuser CDMA receivers," *IEEE Signal Process. Mag.*, vol. 17, no. 3, pp. 49–61, 2000.

5 M. Honig, U. Madhow, and S. Verdu, "Blind adaptive multiuser detection," *IEEE Trans. Info. Theory*, vol. 41, no. 4, pp. 944–960, 1995.

6 M.K. Tsatsanis and Z. Xu, "Performance analysis of minimum variance CDMA receivers," *IEEE Trans. Signal Process.*, vol. 46, no. 11, pp. 3014–3022, 1998.

7 Z. Tian, K.L. Bell, and H.L. Van Trees, "Constrained space-time MOE detection for CDMA wireless systems." In: *34th Asilomar Conference on Signals, Systems and Computers*, vol. 1, pp. 526–531, 2000.

8 J.F. Weng and T. Le-Ngoc, "Rake receiver using blind adaptive minimum output energy detection for DS/CDMA over multipath fading channels," *IEE Proc. Commun.*, vol. 148, no. 6, pp. 385–392, 2001.

9 Z. Xu "Further study on MOE-based multiuser detection in unknown multipath" *EURASIP J. Appl. Signal Process.*, vol. 12, pp. 1377–1386, 2002.

10 Z. Xu "Improved constraint for multipath mitigation in constrained MOE multiuser detection" *J. Commun. Netw.*, vol. 3, no. 3, pp. 249–256, 2001.

11 Z. Xu and M.K. Tsatsanis "Blind adaptive algorithms for minimum variance CDMA receivers", *IEEE Trans. Signal Process.*, vol. 49, no. 1, pp. 180–194, 2001.

12 J.M. Cioffi and T. Kailath, "Fast recursive least squares transversal filters for adaptive filtering," *IEEE Trans. Acoust. Speech Signal Process*, vol. 32, no. 2, pp. 304–337, 1984.
13 A. Elnashar, S. Elnoubi, and H. Elmikati "Performance analysis of blind adaptive moe multiuser receivers using inverse QRD-RLS algorithm," *IEEE Trans. Circuits Syst. I*, vol. 55, no. 1, pp. 398–411, 2008.
14 A. Elnashar, S. Elnoubi, and H. Elmikati, "Low-complexity robust adaptive generalized sidelobe canceller detector for DS/CDMA systems," *Int. J. Adapt. Cont. Sig. Process.*, vol. 23, no. 3, pp. 293–310, 2008.
15 S.-J. Chern and C.-Y. Chang, "Adaptive linearly constrained inverse QRD-RLS beamforming algorithm for moving jammers suppression", *IEEE Trans. Antennas Propag.*, vol. 50, no. 8, pp. 1138–1150, 2002.
16 D.A. Pados and G.N. Karystinos, "An iterative algorithm for computation of the MVDR filter," *IEEE Trans. Signal Process.* vol. 49, no. 2, pp. 290–200, 2001.
17 L.S. Resende, J.M. T. Romano, and M.G. Bellanger, "A fast least-squares algorithm for linearly constrained adaptive filtering," *IEEE Trans. Signal Process.* vol. 44, no. 5, pp. 1168–1174, 1996.
18 Z. Tian, Blind Multi-user Detection with Adaptive Space-time Processing for DS-CDMA Wireless Communications, PhD thesis, George Mason University, 2000.
19 F. Qian and B.D. van Veen, "Quadratically constrained adaptive beamforming for coherent signals and interference" *IEEE Trans. Signal Process.*, vol. 43, no. 8, pp. 1890–1900, 1995.
20 A. Elnashar, S. Elnoubi, and H. Elmikati "Further study on robust adaptive beamforming with optimum diagonal loading," *IEEE Trans. Antennas Propag.*, vol. 54, no. 12, pp. 3647–3658, 2006.
21 P.S. Chang and A.N. Willson, Jr., "Analysis of conjugate gradient algorithms for adaptive filtering," *IEEE Trans. Signal Process.*, vol. 48, no. 2, pp. 409–418, 2000.
22 G.K. Boray and M.D. Srinath, "Conjugate gradient techniques for adaptive filtering," *IEEE Trans. Circuits Syst. I*, vol. 39, no. 1, pp. 1–10, 1992.
23 P.S. Chang and A.N. Willson, Jr., "Adaptive filtering using modified conjugate gradient." In: *38th Midwest Symposium*

on Circuits and Systems, Rio de Janeiro, Brazil, pp. 243–246, 1995.

24 H. Cox, Zeskind R., and Qwen M., “Robust adaptive beamforming”, *IEEE Trans. Acoustics, Speech, Signal Process.*, vol. ASSP-35, no. 10, pp. 1365–1376, 1987.

25 L. Fertig and J. McClellan, “Dual forms for constrained adaptive filtering,” *IEEE Trans. Signal Process.*, vol. 42, no. 1, pp. 11–23, 1994.

26 Qian F. and Van Veen B.D. “Quadratically constrained adaptive beamforming for coherent signals and interference,” *IEEE Trans. Signal Process.*, vol. 43, no. 8, pp. 1890–1900, 1995.

27 Z. Tian, K.L. Bell, and H.L. Van Trees, “A recursive least squares implementation for lcmp beamforming under quadratic constraint” *IEEE Trans. Signal Process.*, vol. 49, no. 6, pp. 1138–1145, 2001.

28 Z. Tian, K.L. Bell, and H.L. Van Trees, “Robust constrained linear receivers for CDMA wireless systems,” *IEEE Trans. Signal Process.*, vol. 49, no. 7, pp. 1510–1522, 2001.

29 S. Cui, Z. Luo, and Z. Ding, “Robust blind multiuser detection against CDMA signature waveform mismatch.” In: *IEEE International Conference on Acoustics, Speech, and Signal Processing*, vol. 4, pp. 2297–2300, 2001.

30 S. Cui, Z.-Q. Luo, and Z. Ding “Robust blind multiuser detection against CDMA signature mismatch.” In: *IEEE International Conference on Acoustics, Speech, and Signal Processing*, 2001, vol. 4, pp. 2297–2300, 2001.

31 K. Zarifi, S. Shahbazpanahi, A.B. Gershman, and Z.-Q. Luo, “Robust blind multiuser detection based on worst-case MMSE performance optimization.” In: *IEEE International Conference on Acoustics, Speech, and Signal Processing*, May 2004, Montreal, Canada, pp. 897–900.

32 S. Shahbazpanahi and A.B. Gershman, “Robust multiuser CDMA receivers based on the worst-case performance optimization.” In: *IEEE Signal Processing Advances in Wireless Communications (SPAWC) Workshop*, June 2003, Rome, Italy, pp. 537–541.

33 K. Zarifi, S. Shahbazpanahi, A.B. Gershman, and Z-Q Luo, “Robust blind multiuser detection based on the worst-case

performance optimization of the MMSE receiver," *IEEE Trans. Signal Process.*, vol. 53, no. 1, pp. 295–305, 2005.

34 S. Shahbazpanahi and A.B. Gershman, "Robust blind multiuser detection for synchronous CDMA systems using worst-case performance optimization," *IEEE Trans. On Wireless Commun.*, vol. 3, no. 6, pp. 2232–2245, 2004.

35 J.F. Sturm, "Using SeDuMi 1.02, a MATLAB toolbox for optimization over symmetric cones," *Optim. Meth. Softw.*, vol. 11–12, pp. 625–653, 1999.

36 S. Shahbazpanahi, A.B. Gershman, Z.Q. Luo, and K.M. Wong, "Robust adaptive beamforming for general-rank signal model," *IEEE Trans. Signal Process.*, vol. 51, pp. 2257–2269, 2003.

37 K. Zarifi, S. Shahbazpanahi, A.B. Gershman, and Z-Q Luo, "Robust blind multiuser detection based on worst-case MMSE performance optimization." In: *IEEE International Conference on Acoustics, Speech, and Signal Processing*, May 2004, Montreal, Canada, pp. 897–900.

38 S. Shahbazpanahi and A.B. Gershman, "Robust blind multiuser detection for synchronous CDMA systems using worst-case performance optimization," *IEEE Trans. Wireless Commun.*, vol. 3, no.6, pp. 2232–2245, 2004.

39 K. Zarifi, S. Shahbazpanahi, A.B. Gershman, and Z-Q Luo, "Robust blind multiuser detection based on the worst-case performance optimization of the MMSE receiver," *IEEE Trans. Signal Process.*, vol. 53, no. 1, pp. 295–305, 2005.

40 S. Shahbazpanahi and A.B. Gershman, "Robust multiuser CDMA receivers based on the worst-case performance optimization," *IEEE Signal Processing Advances in Wireless Communications (SPAWC) Workshop*, June 2003, Rome, Italy, pp. 537–541.

41 S.A. Vorobyov, A.B. Gershman, Z.-Q. Luo, and N. Ma, "Adaptive beamforming with joint robustness against mismatched signal steering vector and interference nonstationarity," *IEEE Signal Process. Lett.*, vol. 11, no. 2, pp. 108–111, 2004.

42 A. Elnashar, "On efficient implementation of robust adaptive beamforming based on worst-case performance optimization" *IET Signal Process.*, vol. 2, no. 4, pp. 381–393, 2008.

43 J. Li, P. Stoica, and Z. Wang "On robust Capon beamforming and diagonal Loading," *IEEE Trans. Signal Process.*, vol. 51, no. 7, pp. 1702–1715, 2003.

44 S. Cui, Z. Luo, and Z. Ding, "Robust CDMA signal detection in the presence of user and interference signature mismatch." In: *IEEE Third Workshop on Signal Processing Advances in Wireless Communications,* pp. 221–224, 2001.

45 Z. Tian, Blind Multi-user Detection with Adaptive Space-time Processing for DS-CDMA Wireless Communications, PhD thesis, George Mason University, 2000.

46 Z. Tian, K.L. Bell, and H.L. Van Trees, "A quadratically constrained decision feedback equalizer for DS-CDMA communication systems." In: *Proceedings of 2nd IEEE Workshop on Signal Processing Advances in Wireless Communication,* May 1999.

47 Z. Tian, K.L. Bell, and H.L. Van Trees, "A recursive least squares implementation for LCMP beamforming under quadratic constraint," *IEEE Trans. Signal Process.*, vol. 49, no. 6, pp. 1138–1145, 2001.

48 Z. Tian, K.L. Bell, and H.L. Van Trees, "A RLS implementation for adaptive beam-forming under quadratic constraint," 9th *IEEE Workshop on Statistical Signals and Array Processing*, Portland, Oregon, September 1998.

49 Z. Tian, K.L. Bell, and H.L. Van Trees, "Quadratically constrained RLS Filtering for adaptive beamforming and DS-CDMA multi-user detection," *Adaptive Sensor Array Processing Workshop,* MIT Lincoln Lab, Lexington, MA, March 1999.

50 Z. Tian, K.L. Bell, and H.L. Van Trees, "Robust RLS implementation of the minimum output energy detector with multiple constraints," *Center of Excellence in C3I*, George Mason University, USA, July 2000.

51 E.W. Swokowski, M. Olinick, and D. Pence, *Calculus*, 6th edn, PWS, 1994, pp. 869–871.

52 J.A. Apoliario, Jr., M.L.R. de Campos, and C.P. Bernal, "The constrained conjugate gradient algorithm," *IEEE Signal Process. Lett.*, vol. 7, no. 12, pp. 351–354, 2000.

53 P. Sheng Chang and A.N. Willson, Jr., "Analysis conjugate gradient algorithms for adaptive filtering," *IEEE International*

Symposium on Circuits and Systems, Hong Kong, 9–12 June 1997, pp. 2292–2295.

54 L. Fertig and J. McClellan, "Dual forms for constrained adaptive filtering," *IEEE Trans. Signal Process.*, vol. 42, no. 1, pp. 11–23, 1994.

55 D.A. Pados and G.N. Karystinos, "An iterative algorithm for computation of the MVDR filter," *IEEE Trans. Signal Process.* vol. 49, no. 2, pp. 290–200, 2001.

56 Yoo S. Song, Hyuck M. Kwon, Byung J. Min, "Computationally efficient smart antennas for CDMA wireless communications" *IEEE Trans. Vehic. Techn.*, vol. 50, no. 6, pp. 1613–1628, 2001.

57 J. Li and P. Stoica (eds), *Robust Adaptive Beamforming*. Wiley, 2006.

58 S. Vorobyov, A.B. Gershman, and Z.-Q. Luo, "Robust adaptive beam- forming using worst-case performance optimization: A solution to the signal mismatch problem," *IEEE Trans. Signal Process.*, vol. 51, no. 2, pp. 313–324, 2003.

59 J. Li, P. Stoica, and Z. Wang, "On robust capon beamforming and diagonal loading," *IEEE Trans. Signal Process.*, vol. 51, no. 7, pp. 1702–1714, 2003.

60 S. Shahbazpanahi, A.B. Gershman, Z.-Q. Luo, and K.M. Wong, "Robust adaptive beamforming for general-rank signal models," *IEEE Trans. Signal Process.*, vol. 51, no. 9, pp. 2257–2269, 2003.

61 K.L. Bell, Y. Ephraim, and H.L. Van Trees, "A Bayesian approach to robust adaptive beamforming," *IEEE Trans. Signal Process.*, vol. 48, no. 2, pp. 386–398, 2000.

62 B.D. Van Veen, "Minimum variance beamforming with soft response constraints," *IEEE Trans. Signal Process.*, vol. 39, no. 9, pp. 1964–1972, 1991.

63 D.D. Feldman and L.J. Griffiths, "A projection approach for robust adaptive beamforming," *IEEE Trans. Signal Process.*, vol. 42, no. 4, pp. 867–876, 1994.

64 H.L. Van Trees, *Detection, Estimation, and Modulation Theory, Part IV, Optimum Array Processing*. Wiley, 2002.

65 X. Jiang, W.-J. Zeng, A. Yasotharan, H.C. So, and T. Kirubarajan, "Quadratically constrained minimum dispersion beamforming via gradient projection," *IEEE Trans. Signal Process.*, vol. 63, no. 1, pp. 192–205, 2015

66 S. Boyd and L. Vandenberghe, *Convex Optimization.* Cambridge University Press, 2006.
67 A.B. Gershman, N.D. Sidiropoulos, S. Shahbazpanahi, M. Bengtsson, and B. Ottersten "Convex optimization-based beamforming," *IEEE Signal Process. Mag.*, vol. 27, no. 4, pp. 62–75, 2010.
68 S.A. Vorobyov, "Principles of minimum variance robust adaptive beamforming design," *Signal Process.*, vol. 93, no. 12, pp. 3264–3277, 2013.
69 C.Y. Chen and P.P. Vaidyanathan, "Quadratically constrained beam- forming robust against direction-of-arrival mismatch," *IEEE Trans. Signal Process.*, vol. 55, no. 8, pp. 4139–4150, 2007.
70 J. Xu, G. Liao, S. Zhu, and L. Huang "Response vector constrained robust LCMV Beamforming based on semidefinite programming," *IEEE Trans. Signal Process.*, vol. 63, no. 21, pp. 5720–5732, 2015.
71 J. Li, P. Stoica, and Z. Wang, "Doubly constrained robust Capon beamformer," *IEEE Trans. Signal Process.*, vol. 52, no. 9, pp. 2407–2423, 2004.
72 A. Beck and Y.C. Eldar, "Doubly constrained robust Capon beamformer with ellipsoidal uncertainty sets," *IEEE Trans. Signal Process.*, vol. 55, no. 2, pp. 753–758, 2007.
73 S.A. Vorobyov, H. Chen, and A.B. Gershman, "On the relationship between robust minimum variance beamformers with probabilistic and worst-case distrortionless response constraints," *IEEE Trans. Signal Process.*, vol. 56, no. 11, pp. 5719–5724, 2008.
74 S.D. Somasundaram, "Linearly constrained robust Capon beamforming," *IEEE Trans. Signal Process.*, vol. 60, no. 11, pp. 5845–5856, 2012.
75 Z.L. Yu, W. Ser, M.H. Er, Z.H. Gu, and Y.Q. Li, "Robust adaptive beamformers based on worst-case optimization and constraints on magnitude response," *IEEE Trans. Signal Process.*, vol. 57, no. 7, pp. 2615–2628, 2009.
76 Z.L. Yu, M.H. Er, and W. Ser, "A novel adaptive beamformer based on semidefinite programming (SDP) with constraints on magnitude response," *IEEE Trans. Antennas Propag.*, vol. 56, no. 5, pp. 1297–1307, 2008.
77 Z.L. Yu, Z.H.Gu, J.J. Zhou, et al., "A robust adaptive beamformer based on worst-case semi-definite programming,"

IEEE Trans. Signal Process., vol. 58, no. 11, pp. 5914–5919, 2010.

78 D.J. Xu, R. He, and F. Shen, "Robust beamforming with magnitude response constraints and conjugate symmetric constraint," *IEEE Commun. Lett.*, vol. 17, no. 3, pp. 561–564, 2013.

79 Z.Q. Luo, W.K. Ma, A.M. So, Y.Y. Ye, and S.Z. Zhang, "Semidefinite relaxation of quadratic optimization problems," *IEEE Signal Process. Mag.*, vol. 27, no. 3, pp. 20–34, 2010.

80 Y. Huang and D.P. Palomar, "Rank-constrained separable semidefinite programming with applications to optimal beamforming," *IEEE Trans. Signal Process.*, vol. 58, no. 2, pp. 664–678, 2010.

81 A. Khabbazibasmenj, S.A. Vorobyov, and A. Hassanien, "Robust adaptive beamforming based on steering vector estimation with as little as possible prior information," *IEEE Trans. Signal Process.*, vol. 60, no. 6, pp. 2974–2987, 2012.

82 Y.C. Eldar, A. Nehorai, and P.S. Rosa, "A competitive mean-squared error approach to beamforming," *IEEE Trans. Signal Process.*, vol. 55, no. 11, pp. 5143–5154, 2007.

83 Y. Huang and D.P. Palomar, "Randomized algorithms for optimal solutions of double-sided QCQP with applications in signal processing," *IEEE Trans. Signal Process.*, vol. 62, no. 5, pp. 1093–1108, 2014.

84 K. Law, X. Wen, M. Vu, and M. Pesavento, "General rank multiuser downlink beamforming with shaping constraints using real-valued OSTBC," *IEEE Trans. Signal Process.*, vol. 63, no. 21, pp. 5758–5771, 2015.

85 M. Yukawa, Y. Sung, and G. Lee, "Dual-domain adaptive beamformer under linearly and quadratically constrained minimum variance," *IEEE Trans. Signal Process.*, vol. 61, no. 11, pp. 2874–2886, 2013.

86 Z.1. Yu., W. Ser, M.H. Err, Z.H. Gu, and Y.Q. Li, "Robust adaptive beamformers based on worst-case optimization and constraints on magnitude response," *IEEE Trans. Signal Process.*, vol. 57, no. 7, pp. 2615–2628, 2009.

87 D. Wei, C.K. Sestok, and A.V. Oppenheim, "Sparse filter design under a quadratic constraint: Low-complexity algorithms," *IEEE Trans. Signal Process.*, vol. 61, no. 4, pp. 857–870, 2013.

6

Robust Constant Modulus Algorithms

6.1 Introduction

Blind equalization of intersymbol interference (ISI) in communications channels and blind separation of multiple users are crucial signal processing techniques in particular communications systems design. Because of the similarity between ISI and multiple access interference (MAI), many researchers have attempted to use blind equalization techniques, such as the constant modulus algorithm (CMA) to suppress MAI [1]. In addition to blind equalization and blind beamforming [2], the CMA can be used also for code acquisition in the context of DS-CDMA systems [3, 4]. One of the earliest blind receiver designs is the Godared algorithm [5–7]. Godared [5] observed from simulations that blind receivers that minimize the CM cost function have similar mean-squared error (MSE) performance to non-blind Wiener receivers. Similar observations were also made by Treichler and Agee [6]. However, due to the complexity of the CM cost function, the performance of CM receivers has primarily been evaluated using numerical simulations or using Monte Carlo simulations [7]. Theoretical analysis is typically based on either the noiseless case or approximations of the cost function [8]. Zeng *et al.* presented a geometrical approach that relates the CM to Wiener (or minimum MSE) receivers [9]. The authors provided a generalization of this geometrical approach [10] using a subspace constraint imposed on the CM cost function during the optimization. The analysis their first paper [9] shows that, while in some cases blind CM receivers

Simplified Robust Adaptive Detection and Beamforming for Wireless Communications, First Edition. Ayman Elnashar.

Companion website: www.wiley.com/go/elnashar49

perform almost as well as non-blind Wiener receivers, it is also possible that the CM receiver may be considerably worse.

The existence of multiple CM minima, however, makes it difficult for CM-minimizing schemes to generate estimates of the desired source (as opposed to interference) in multiuser environments or to steer the beam towards the desired direction of arrival (DOA) in beamforming applications. Schniter and and Johnson showed three separate sufficient conditions under which gradient descent (GD) minimization of the CM cost will locally converge to an estimator of the desired source at particular delay [11]. It has been shown that the CMA can perform similarly to non-blind receivers if undesirable local minima can be avoided [9]. Perhaps the greatest challenge facing successful application of the CM criterion in arbitrary interference environments is the difficulty in determining CM-minimizing estimates of the desired source (as opposed to interference capture). The potential for "interference capture" is a direct consequence of the fact that the CM criterion exhibits multiple local minima in the estimator parameter space, each corresponding to a particular user at a particular delay or corresponding to side loops in beamforming.

Several multiuser modifications of the CM criterion have been proposed that aim to avoid capturing interference instead of the required source [12]. Some of these techniques add a cross-correlation term to the CM cost function to penalize correlation between the desired user and any estimated sources [13–17]. These approaches may be applicable only in uplink scenarios where the base stations have already recovered all users. In the downlink, it will require high complexity because the handset receiver will need to estimate all users in order to apply these techniques. A performance analysis of a blind channel identification scheme based on the cross-correlation of CM-minimizing blind symbol estimates is provided by Schniter *et al.* [18].

Schniter and Johnson proposed a near–far resistance initialization procedure for the CMA to uplink DS-CDMA based on pre-whitening of the received signal [4]. Unfortunately, pre-whitening is costly to implement in an adaptive manner. Other authors have used suitable initialization if the CMA to guarantee desired source convergence [11, 19, 20]. However, even with proper initialization, there are conditions that must be met to guarantee global convergence. These conditions are based on CM cost, kurtosis [21], and signal-to-interference-plus-noise ratio (SINR).

An alternative criterion is the higher-order statistics (HOS) of the received signals – say the cumulants, or kurtosis, or through use of the Shalvi-Weinstein (SW) algorithm [22] – since criteria based on second-order statistics like the CM cost function in most cases do not suffice for the complete separation of the sources [21, 23, 24]. The SW cost function evaluates the fourth-order cumulants under a power constraint. It has been shown that the minima of the CM criterion and the maxima of the SW criterion are equivalent in the sense that they differ only by a scaling factor [25]. Domains of attraction of SW receivers have been analyzed [26] and the existence and performance of SW estimators have been detailed [27]. Although global convergence can generally be guaranteed using HOS methods, they carry a high computational burden, which limits the practical implementations of these methods.

Lin proposed a minimum-disturbance technique, which leads to simultaneous equalization and phase recovery [28]. This study was limited to ISI equalization and does not extend to multiple-user cases [28–30]. A new multimodulus algorithm for blind equalization of complex communication channels has been derived by solving a constrained optimization problem with relaxation [29]. The authors analyzed the ISI optimization and phase-recovery capabilities of the proposed algorithm, and showed, through simulations, the superiority of its performance to Lin's algorithm [28]. A new multimodulus algorithm is presented by Abrar and Shah [30] for blind equalization of complex-valued communication channels. The proposed algorithm is obtained by solving a novel deterministic optimization criterion that constitutes the dispersion minimization of *a-priori* as well as *a-posteriori* quantities, leading to an update equation having a particular zero-memory continuous nonlinearity. Analyses of the automatic phase-recovery and interference cancellation capabilities of the proposed algorithm are provided by the authors [30].

Miguez and Castedo [31] proposed a linearly constrained CMA (LCCMA) to prevent the capture of an interference signal instead of the desired user. Two approximate closed-form solutions of the optimum LCCMA were presented [32]. In the noise-free case, the LCCMA proposed in these two papers [31, 32] can completely remove the MAI if and only if the desired user's relative amplitude is no less than the critical value $1/\sqrt{3}$ [33–37]. Unfortunately, this is not a very realistic scenario for a commercial system due to the near–far effect. Several enhancements to the LCCMA algorithm can be found [38–51].

The other most important drawback of traditional CMA, like the celebrated LMS algorithm, is that it involves a constant step size that controls the speed of convergence. The selection of the CMA step size is case sensitive and can affect the algorithm convergence [52]. Many researchers have attempted to increase the convergence speed of CMA [42, 48, 51–58]. Chern and Wang [59] used a recursive conjugate gradient algorithm presented by Chang and Willson [60], combining it with the LCCMA and showing that this gave faster convergence than the classical SD-based LCCMA.

The code-constrained constant modulus algorithm (CCM) implemented with a stochastic gradient (SG) technique such as LMS is a very effective and efficient blind approach for interference suppression when a communication channel is frequency selective [42]. Despite the fast convergence of recursive least squares (RLS) algorithms, however, it is preferable to implement adaptive receivers with stochastic gradient (SG) algorithms such as LMS, due to complexity and cost issues [61]. Moreover, in non-stationary wireless environments, users frequently enter and exit the system, making it very difficult for the receiver to compute a predetermined step size. This suggests the deployment of mechanisms to automatically adjust the step size in order to ensure good tracking of the interference and the channel. The performance of blind CCM adaptive receivers for DS-CDMA systems that employ stochastic gradient (SG) algorithms with variable step-size mechanisms has been investigated [42]. The authors proposed a novel low-complexity variable step-size mechanism for blind CCM CDMA receivers [42], and carried out convergence and tracking analyses of the proposed adaptation techniques for multipath channels in non-stationary environments [42].

Previous studies have demonstrated significant gains in performance through use of averaging methods (AV) or an adaptive gradient step-size (AGSS) mechanism, where one SG algorithm adapts the parameter vector and another SG recursion adapts the step-size [42, 61]. Other authors have borrowed the idea of averaging originally developed by Polyak [64] and applied it to CDMA receivers with the MV criterion [62, 63]. The AGSS algorithms [67] can be considered to be MV and CCM extensions of the approach in the papers by Mathews and Xie [65] and Kushner and Yang [66]. All these methods require an additional number of operations (additions and multiplications) proportional to the processing gain and to the number of multipath components [42].

An alternative approach to robust adaptive blind multiuser detection is presented in this chapter. This is based on the LCCMA and with a QI constraint on the weight vector norm [51]. The LCCMA is adapted with the fast-recursive steepest descent (RSD) algorithm presented in Chapter 5, which is based on the partition linear interference canceller (PLIC) structure with multiple constraints. The Lagrange multiplier method is used to solve the QI constraint problem. The optimum step size of the RSD algorithm derived in Chapter 5 is used with the LCCMA. The VL technique is used to precisely determine the required amount of diagonal loading. The geometrical interpretation of the VL technique is similar to the one presented in Chapter 5. It is shown from computer simulations that the robust LCCMA can handle the near–far effect without restrictions. In addition, simulation results show that the proposed approach exhibits superior performance to other CM-based detectors, in terms of BER and SINR.

The constrained CMA has been generalized to multipath channels using multiple constraints [68, 69]. A stochastic gradient algorithm was developed by Xu and Liu [68] and a computationally efficient RLS algorithm by de Lamare and Sampaio Neto [69]. They showed that the blind receiver design based on a constrained constant modulus (CCM) criterion increases robustness against signature mismatch and provides improved performance over constrained minimum variance (CMV) algorithms [68, 69]. In addition to the LCCMA and block Shanno CMA (BSCMA) algorithms [51], a class of subspace or low-rank blind multiuser algorithms that can speed up convergence, improve tracking, and which are also robust against signature mismatch have been developed [70]. These algorithms employ eigenvalue-based subspace techniques [70], Krylov subspace methods [71], and an iterative approach for the construction of the subspace [72].

Unfortunately, the robust LCCMA approach discussed above uses RLS-like algorithm with $O(N^2)$ complexity to estimate the blind multiuser receiver. Moreover, it will be demonstrated in Section 6.7 that the constant modulus LMS-like algorithm with $O(N)$ complexity does not offer any performance improvement by incorporating the quadratic inequality constraint [51].

A modified version of the block Shanno algorithm, with $O(N)$ complexity, has been shown to offer good convergence speeds at low cost [73, 74]. Additionally, a combined BS-CMA and cross-correlation approach (termed BS-CC-CMA) has been developed [75–77]. However, the BS-CMA [73, 74] is notorious for its sensitivity to step size and no clear vision of step size updating is given. In addition, the algorithm involves a gradient vector norm check step and in this case, if the norm starts to increase, the algorithm will stop the adaptation and start with the initial weight vector. This block of data will therefore not be an improvement from previous updates. Moreover, the BS-CMA is more sensitive to the number of iterations required in every block of data and no clear break point is determined to stop iteration inside the blocks.

A space–time blind adaptive processing algorithm has been proposed for the uplink base station receiver in a DS-CDMA cellular system [78]. The proposed algorithm, called the space-time block Shanno constant modulus algorithm (STB-SCMA), minimizes a CMA objective function in a bilinear function structure, with penalty terms for both spatial and temporal domains. The proposed algorithm rapidly converges to the stationary points in fewer iterations than block conjugate gradient CMA (CGCMA) and block gradient descent CMA (GDCMA). STBSCMA finds the critical points of the objective function. STBSCMA also recovers rapidly from an abrupt change of channel. The proposed algorithm structure requires relatively low computational complexity to improve the receiver performance for space–time processing in the DS-CDMA system [78, 79].

A class of adaptive beamforming algorithms has been proposed based on minimizing the BER cost function directly instead of the MSE [80]. Unfortunately, the popular least minimum BER (LMBER) stochastic beamforming algorithm suffers from low convergence speeds. Gradient Newton algorithms have been proposed to speed up the convergence rate and enhance performance but at the expense of complexity. Samir *et al.* formulated a block processing objective function for the minimum BER (MBER) [80], and developed a nonlinear optimization strategy, which produces the so-called block-Shanno MBER (BSMBER) algorithm. A complete discussion for the complexity calculations of the proposed algorithm is given. This algorithm will be presented in Chapter 8 for DS-CDMA system with antenna arrays and MIMO-OFDM systems.

In this chapter, we apply the quadratic inequality constraint to the weight vector norm of the LCCMA and BSCMA algorithms in order to enhance their performance. The weight norm constraint will control the gradient vector norm, and hence there is no need to check the gradient vector norm increase in BSCMA. Additionally, the iteration inside the block can continue without affecting algorithm stability, due to the weight vector norm constraint. We will investigate the effect of adding a quadratic inequality constraint on the LCCMA and BSCMA algorithms. The simplified VL technique presented in Chapter 5 is exploited to estimate the optimum diagonal loading value. The LCCMA and BSCMA algorithms are used to update the adaptive vector of the PLIC structure. The PLIC structure with multiple constraints is employed to identify the MAI and hence help prevent interference capture. Moreover, the different forms of BS-CMA, the block-conjugate gradient CMA, and the block gradient descent CMA are investigated as well. The resistance of BSCMA-based algorithms to the near–far effect is discussed and evaluated. Details on robust LCCMA and BSCMA algorithms [51, 80] and other relevant algorithms [81–86] can be found in the literature.

The rest of the chapter is organized as follows. The robust LCCMA detector design and its adaptive implementation of robust LCCMA are provided in Section 6.2 and Section 6.3 respectively. The BSCMA algorithm is outlined in Section 6.4

and the robust BSCMA with the QI constrained is introduced in Section 6.5. Block processing and adaptive implementation of robust BSCMA is provided in Section 6.6. Finally, computer simulations and performance comparison are presented in Section 6.7.

6.2 Robust LCCMA Formulation

To avoid the cancellation of the signal of interest scattered over different multipaths during the minimization of the dispersion of the receiver output (the CM cost function), we can generally impose a set of linear constraints (as we did with the MOE detector in Chapter 5) of the form $\boldsymbol{C}_1{}^H\boldsymbol{f} = \boldsymbol{g}$, where $\boldsymbol{g}$ is the constraint vector. Consequently, the so-called LCCMA detector can be obtained by solving the following constrained minimization problem:

$$\min_{\boldsymbol{f}} J_1(\boldsymbol{f}) \triangleq E\left\{(|\boldsymbol{f}^H\boldsymbol{x}|^2 - r)^2\right\} \ s.t. \quad \boldsymbol{C}_1{}^H\boldsymbol{f} = \boldsymbol{g} \tag{6.1}$$

An alternative formulation for the LCCMA can be obtained as follows [59]:

$$\min_{\boldsymbol{f}} J_2(\boldsymbol{f}) \triangleq E\left\{(\boldsymbol{f}^H\overline{\boldsymbol{x}} - r)^2\right\} \ s.t. \quad \boldsymbol{C}_1{}^H\boldsymbol{f} = \boldsymbol{g} \tag{6.2}$$

where $\overline{\boldsymbol{x}}(i) = \boldsymbol{x}(i)y(i)$. The cost function in (6.2) can be expressed as follows:

$$J_2(\boldsymbol{f}) \triangleq -\boldsymbol{f}^H E\left\{r\overline{\boldsymbol{x}}\right\} + \boldsymbol{f}^H E\left\{\overline{\boldsymbol{x}}\,\overline{\boldsymbol{x}}^H\right\}\boldsymbol{f} \tag{6.3}$$

Then

$$J_2(\boldsymbol{f}) \triangleq -\boldsymbol{f}^H\tilde{\boldsymbol{x}} + \boldsymbol{f}^H\overline{\boldsymbol{R}}(n)\boldsymbol{f} \tag{6.4}$$

where $\tilde{\boldsymbol{x}} = E\left\{r\overline{\boldsymbol{x}}\right\}$ and $\overline{\boldsymbol{R}}(n) = E\{\overline{\boldsymbol{x}}\,\overline{\boldsymbol{x}}^H\}$ denote a new cross-correlation vector and autocorrelation matrix respectively.

By adopting the PLIC structure with the cost functions of (6.1) and (6.4) and then applying the QI constraint on the adaptive weight portion ($\boldsymbol{f}_a$) of the PLIC structure, the optimal robust weight vector corresponding to (6.1) and (6.4) can be obtained, from the solution of the following constrained optimization problem:

$$\min_{\boldsymbol{f}_a} J_2(\boldsymbol{f}_a) \triangleq E\left\{(|(\boldsymbol{f}_c - \boldsymbol{B}\boldsymbol{f}_a)^H\boldsymbol{x}|^2 - r)^2\right\} \ s.t. \ \boldsymbol{f}_a^H\boldsymbol{f}_a \le \beta^2 \tag{6.5}$$

$$\min_{f_a} J_2(\boldsymbol{f}_a) \triangleq -(\boldsymbol{f}_c - \boldsymbol{B}\boldsymbol{f}_a)^H \tilde{\boldsymbol{x}} + (\boldsymbol{f}_c - \boldsymbol{B}\boldsymbol{f}_a)^H \overline{R}(n)(\boldsymbol{f}_c - \boldsymbol{B}\boldsymbol{f}_a) \; s.t. \; \boldsymbol{f}_a^H \boldsymbol{f}_a \leq \beta^2 \tag{6.6}$$

We will consider the constrained optimization problem given in (6.6); the other one (6.5) can be straightforwardly estimated. Consider now the Lagrange multiplier method to solve this constrained minimization problem. This involves forming the following real valued Lagrangian function:

$$J_2(\boldsymbol{f}_a) = -(\boldsymbol{f}_c - \boldsymbol{B}\boldsymbol{f}_a)^H \tilde{\boldsymbol{x}} + (\boldsymbol{f}_c - \boldsymbol{B}\boldsymbol{f}_a)^H \overline{\boldsymbol{R}}(n)(\boldsymbol{f}_c - \boldsymbol{B}\boldsymbol{f}_a) + \frac{1}{2}\lambda_0 t(\boldsymbol{f}_a^H \boldsymbol{f}_a - \beta^2) \tag{6.7}$$

The problem is then converted from a QI *constrained* minimization to an *unconstrained* minimization problem. The optimal solution is obtained by first taking the gradient of $J_2(\boldsymbol{f}_a)$ with respect to the real and imaginary components of $\boldsymbol{f}_a^H$, and then setting the resulting quantities to zero. This leads to:

$$\boldsymbol{f}_{a(opt)} = (\overline{\boldsymbol{R}}_B + \lambda_0 \boldsymbol{I})^{-1}(\overline{\boldsymbol{p}}_B - \boldsymbol{B}^H \tilde{\boldsymbol{x}}) \tag{6.8}$$

where

$$\overline{\boldsymbol{R}}_B = \boldsymbol{B}^H \overline{\boldsymbol{R}} \boldsymbol{B}, \; \overline{\boldsymbol{p}}_B = \overline{\boldsymbol{P}}_B \boldsymbol{f}_c \text{ and } \overline{\boldsymbol{P}}_B = \boldsymbol{B}^H \overline{\boldsymbol{R}}$$

Similar to the approach proposed in Chapter 5, we will develop a VL technique that is capable of precisely computing the optimal diagonal loading term λ_0. The RSD method is employed to iteratively update and estimate the weight vector that minimizes the Lagrangian functional (6.7). Therefore,

$$\boldsymbol{f}_a(n) = \boldsymbol{f}_a(n-1) - \mu \nabla_{f_a} \boldsymbol{\Psi}_{f_a}(n) \tag{6.9}$$

where μ is the step size of the algorithm and $\nabla_{f_a} \boldsymbol{\Psi}_{f_a}(n)$ is the conjugate derivative of $\boldsymbol{\Psi}_{f_a}(n)$ with respect to the real and imaginary parts of $\boldsymbol{f}_a^H(n)$ and is given by:

$$\boldsymbol{d}(n) = \boldsymbol{B}^H \tilde{\boldsymbol{x}} - \boldsymbol{B}^H \overline{\boldsymbol{R}}(n) \boldsymbol{f}_c + \boldsymbol{B}^H \overline{\boldsymbol{R}}(n) \boldsymbol{B} \boldsymbol{f}_a(n-1) + \lambda_0 \boldsymbol{f}_a(n-1) \tag{6.10}$$

Therefore, the adaptive implementation of $\boldsymbol{f}_a(n)$ can be obtained by substituting from (6.10) into (6.9) as follows:

$$\begin{aligned}\boldsymbol{f}_a(n) &= \boldsymbol{f}_a(n-1) - \mu(\boldsymbol{B}^H\tilde{\boldsymbol{x}} + \overline{\boldsymbol{R}}_B(n)\boldsymbol{f}_a(n-1) - \overline{\boldsymbol{p}}_B(n)) \\ &\quad - \mu\lambda_0\boldsymbol{f}_a(n-1) \end{aligned} \tag{6.11}$$

Notice that the constraint should be satisfied at each iteration step; that is, $\boldsymbol{f}_a^H(n)\boldsymbol{f}_a(n) \leq \beta^2$. Substituting from (6.9) and assuming the constraint is satisfied on the previous step. This yields:

$$(\overline{\boldsymbol{f}}_a(n) - \mu\lambda_0\boldsymbol{f}_a(n-1))^H(\overline{\boldsymbol{f}}_a(n) - \mu\lambda_0\boldsymbol{f}_a(n-1)) \leq \beta^2 \tag{6.12}$$

$$\overline{\boldsymbol{f}}_a(n) = \boldsymbol{f}_a(n-1) - \mu\overline{\boldsymbol{d}}(n) \tag{6.13}$$

$$\overline{\boldsymbol{d}}(n) = \boldsymbol{B}^H\tilde{\boldsymbol{x}} + \overline{\boldsymbol{R}}_B(n)\boldsymbol{f}_a(n-1) - \overline{\boldsymbol{p}}_B(n) \tag{6.14}$$

The value of λ_0 can be found by solving the following quadratic equation:

$$\begin{aligned}&\mu(n)^2\boldsymbol{f}_a^H(n-1)\boldsymbol{f}_a(n-1)\lambda_0^2 - 2\mu(n)\overline{\boldsymbol{f}}_a^H(n)\boldsymbol{f}_a(n-1)\lambda_0 \\ &\quad + \overline{\boldsymbol{f}}_a^H(n)\overline{\boldsymbol{f}}_a(n) - \beta^2 = 0\end{aligned} \tag{6.15}$$

Therefore, the optimal diagonal loading term λ_0 which satisfies the inequality constraint is:

$$\lambda_0^{1,2}(n) = \frac{b \pm \sqrt{b^2 - 4ac}}{2a} \tag{6.16}$$

where

$$\begin{aligned}&a = \mu(n)^2\|\boldsymbol{f}_a(n-1)\|^2,\ b = 2\mu(n)\overline{\boldsymbol{f}}_a^H(n)\boldsymbol{f}_a(n-1), \\ &c = \overline{\boldsymbol{f}}_a^H(n)\overline{\boldsymbol{f}}_a(n) - \beta^2\end{aligned} \tag{6.17}$$

The two lemmas provided in Chapter 5 are necessary in order to guarantee real and positive diagonal loading.

6.3 Low-complexity Recursive Implementation of LCCMA

The matrices $\overline{\boldsymbol{R}}_B(n)$ and $\overline{\boldsymbol{P}}_B(n)$ and the cross-correlation vector $\tilde{\boldsymbol{x}}(n)$ are updated using exponentially decaying data windows as follows:

$$\overline{R}_B(n) = \eta \overline{R}_B(n-1) + \overline{z}(n)\overline{z}^H(n) \quad (6.18)$$

$$\overline{P}_B(n) = \eta \overline{P}(n-1) + \overline{z}(n)\overline{x}^H(n) \quad (6.19)$$

$$\tilde{z}(n) = \eta \tilde{z}(n) + \overline{z}(n) \quad (6.20)$$

where $\overline{z}(n) = B^H \overline{x}(n)$, $\tilde{z}(n) = B^H \tilde{x}(n)$, and η is the usual forgetting factor with $0 \ll \eta \leq 1$.

The gradient vector can be updated recursively by substituting (6.13) and (6.18)–(6.20) into (6.14). After some manipulations, the following update equations are obtained:

$$\overline{d}(n) = \eta \overline{d}(n-1) - e(n)\tilde{z}(n) - \mu(n-1)\overline{R}_B(n)d(n-1) \quad (6.21)$$

$$d(n-1) = \overline{d}(n-1) + \lambda_0^1(n-1)f_a(n-1) \quad (6.22)$$

$$f_a(n) = \overline{f}_a(n) - \mu\lambda_0^1(n)f_a(n-1) \quad (6.23)$$

The optimum step size can be estimated, as in Chapter 5, as follows:

$$\mu_{opt}(n) = \frac{\gamma d^H(n)d(n)}{d^H(n)\overline{R}_B d(n) + \upsilon} \quad (6.24)$$

where γ, υ are two positive constants added to improve the numerical stability of the algorithm and to avoid dividing by zero, respectively.

The robust adaptive LCCMA algorithm based on the RSD method and the simplified VL technique with the required amount of computation at each step is summarized in Box 6.1. The total number of multiplications required at each snapshot for the detector is about $O(2N_a^2 + N_a N_f + 6N_a + N_f)$, including the matrix update, the VL technique, and the vector update.

Similar to the analysis provided in Section 5.2.4, the convergence of the algorithm depends on the eigenvalue spread of the matrix $(\overline{R}_B + \lambda_0 I)$. If the eigenvalues of $(\overline{R}_B + \lambda_0 I)$ are denoted by ρ_i, then the updated detector converges to its optimal value if

$$0 < \mu < \frac{1}{\rho_{\max}} \quad (6.25)$$

Box 6.1 Summary of the robust LCCMA.

Initialization:

- $\overline{\boldsymbol{R}}_B(0) = \boldsymbol{B}^H(\zeta \boldsymbol{I}_{N_a})\boldsymbol{B}\,; \boldsymbol{f}_c = \boldsymbol{C}_1(\boldsymbol{C}_1{}^H\boldsymbol{C}_1)^{-1}\boldsymbol{g}$
- $\rho^2 = t.(\|\boldsymbol{f}_c\|^2),\ \beta^2 = \rho^2 - \|\boldsymbol{f}_c\|^2, \boldsymbol{f}_a(0) = 0^T$
- $\lambda_0^1(n) = 0; \boldsymbol{f}_a(n) = \boldsymbol{0}_{N_a}; \overline{\boldsymbol{d}}(0) = \boldsymbol{B}^H(\zeta \boldsymbol{I}_{N_a})\boldsymbol{f}_c$

For $n = 1,2,\cdots,$ **do**

- $\boldsymbol{z}(n) = \boldsymbol{B}^H\boldsymbol{x}(n); N_aN_f$
- $y(n) = \boldsymbol{x}^H(n)\boldsymbol{f}_c + \boldsymbol{z}^H(n)\boldsymbol{f}_a(n-1); N_f + N_a$
- $\overline{\boldsymbol{z}}(n) = y(n)\boldsymbol{z}(n)$
- $e(n) = 1 - |y(n)|^2$
- $\overline{\boldsymbol{R}}_B(n) = \eta\overline{\boldsymbol{R}}_B(n-1) + \overline{\boldsymbol{z}}^H(n)\overline{\boldsymbol{z}}(n); N_a{}^2$
- $\boldsymbol{d}(n-1) = \overline{\boldsymbol{d}}(n-1) + \lambda_0^1(n-1)\boldsymbol{f}_a(n-1)$
- $\overline{\boldsymbol{d}}(n) = \eta\overline{\boldsymbol{d}}(n-1) - e(n)\overline{\boldsymbol{z}}(n) - \mu_{opt}(n-1)\overline{\boldsymbol{R}}_B(n)\boldsymbol{d}(n-1)$
- $\mu_{opt}(n) = \frac{\gamma \boldsymbol{d}^H(n-1)\boldsymbol{d}(n-1)}{\boldsymbol{d}^H(n-1)\overline{\boldsymbol{R}}_B(n)\boldsymbol{d}(n-1)+\upsilon}; N_a^2 + 2N_a$

  - *if* $\left(\mu_{opt}(n) > \frac{\beta\|\boldsymbol{f}_a(n-1)\|}{\sqrt{\|\boldsymbol{f}_a(n-1)\|^2\overline{\boldsymbol{d}}^H(n)\overline{\boldsymbol{d}}(n) - \overline{\boldsymbol{d}}^H(n)\boldsymbol{f}_a(n-1)\boldsymbol{f}_a^H(n-1)\overline{\boldsymbol{d}}(n)}}\right)$; (step-size upper bound)
  - $\mu_{opt}(n) = \frac{\beta\|\boldsymbol{f}_a(n-1)\|}{\sqrt{\|\boldsymbol{f}_a(n-1)\|^2\overline{\boldsymbol{d}}^H(n)\overline{\boldsymbol{d}}(n) - \overline{\boldsymbol{d}}^H(n)\boldsymbol{f}_a(n-1)\boldsymbol{f}_a^H(n-1)\overline{\boldsymbol{d}}(n)}}; N_a$
  - end

- $\overline{\boldsymbol{f}}_a(n) = \boldsymbol{f}_a(n-1) - \mu_{opt}(n)\overline{\boldsymbol{d}}(n)$

  - *if*$(\|\overline{\boldsymbol{f}}_a(n)\|^2 > \beta^2)$; (QI constraint not met)
    - $c = \overline{\boldsymbol{f}}_a^H(n)\overline{\boldsymbol{f}}_a(n) - \beta^2,\ b = -2\mu\overline{\boldsymbol{f}}_a^H(n)\boldsymbol{f}_a(n-1),$ $a = \mu(n)^2\|\boldsymbol{f}_a(n-1)\|; 2N_a$
    - $\lambda_0^1(n) = \frac{-b-(\sqrt{b^2-4ac})}{2a}$
    - $\boldsymbol{f}_a(n) = \overline{\boldsymbol{f}}_a(n) - \mu_{opt}(n)\lambda_0^1(n)\boldsymbol{f}_a(n-1)$
  - *else*; (QI constraint met)
    - $\boldsymbol{f}_a(n) = \overline{\boldsymbol{f}}_a(n)$
    - $\lambda_0^1(n) = 0$

End

6.4 BSCMA Algorithm

In order to establish the BSCMA, we must first extend the CM cost function (6.1) to admit block processing similar to [73] and [74]. If we take a block from the received vector signal with length $N_f \times M$, where M is the block length (that is, the number of data bits), then:

$$D_i = [\boldsymbol{x}((i-1)M) : \boldsymbol{x}((i-1)M+1) : \cdots : \boldsymbol{x}(iM-1)] \in R^{N_f \times M} \tag{6.26}$$

The block CMA objective function is defined as follows:

$$\Psi(f) = \frac{1}{4M}\sum_{l=0}^{N-1}\left[\left|\boldsymbol{w}^H \boldsymbol{x}((i-1)M+l)\right|^2 - 1\right]^2 \tag{6.27}$$

The length of $\boldsymbol{x}((i-1)M+l)$ is equal to the detector length N_f.

The objective function is amended to fulfill the real requirement of Shanno's algorithm as follows [51, 73, 74, 80]:

$$\Psi(\boldsymbol{f}) = \frac{1}{4M}\sum_{n=0}^{M-1}\left[\boldsymbol{f}^H \boldsymbol{X}(n)\boldsymbol{f} - 1\right]^2 \tag{6.28}$$

and

$$\nabla_f \Psi(\boldsymbol{f}) = \frac{1}{M}\sum_{n=0}^{M-1}\left[\boldsymbol{f}^H \boldsymbol{X}(n)\boldsymbol{f} - 1\right]\boldsymbol{X}(n)\boldsymbol{f} \tag{6.29}$$

where $\boldsymbol{X}(n) = \boldsymbol{x}^T(n)\boldsymbol{x}(n)$ for real data samples. If we adopt the approach proposed in [73] for complex data samples, then

$$\boldsymbol{f}(k) = \begin{bmatrix} \mathrm{Re}\{\boldsymbol{f}(k)\} \\ \mathrm{Im}\{\boldsymbol{f}(k)\} \end{bmatrix} \tag{6.30}$$

$$\boldsymbol{X}(n) = \boldsymbol{x}_f(n)\boldsymbol{x}_f^T(n) + \boldsymbol{x}_b(n)\boldsymbol{x}_b^T(n) \tag{6.31}$$

where

$$\boldsymbol{x}_f(n) = \begin{bmatrix} \mathrm{Re}\{\boldsymbol{x}(n)\} \\ -\mathrm{Im}\{\boldsymbol{x}(n)\} \end{bmatrix}, \quad \boldsymbol{x}_b(n) = \begin{bmatrix} \mathrm{Re}\{\boldsymbol{x}(n)\} \\ -\mathrm{Im}\{\boldsymbol{x}(n)\} \end{bmatrix} \tag{6.32}$$

The PLIC structure is invoked to identify the MAI interference by dividing the weight vector $\boldsymbol{f}(n)$ as follows:

$$\boldsymbol{f} = \boldsymbol{f}_{c(N_f \times 1)} - \boldsymbol{B}\boldsymbol{f}_{a(N_a \times 1, N_a = N_f - N_g)} \tag{6.33}$$

The Newton algorithm is used to updates the filter tap weights as follows [87, 88]:

$$\boldsymbol{f}_a(j) = \boldsymbol{f}_a(j-1) - \boldsymbol{H}_{newton}^{-1}(\boldsymbol{f}_a(j-1))\boldsymbol{g}(\boldsymbol{f}_a(j-1)) \quad (6.34)$$

where $\boldsymbol{H}_{newton}^{-1}(\boldsymbol{f}_a(j-1))$ is the Hessian of the objective function $\Psi(\boldsymbol{f})$ and $\boldsymbol{g}(\boldsymbol{f}_a(j-1))$ is the gradient of the cost function with respect to $\boldsymbol{f}_a$ evaluated at the iteration block index $j-1$. Shanno's approximation is used to approximate the inverse of the Hessian matrix with $O(M)$ complexity. The Shanno approximation matrix is implicitly determined by the gradients of the two most recent iterations, a search direction, and a step size based on the previous iteration. As a result, we can obtain $\boldsymbol{H}_{newton}^{-1}(\boldsymbol{f}_a(j-1))$ as follows:

$$\boldsymbol{H}_{newton}^{-1}(\boldsymbol{f}_a(j-1)) = \boldsymbol{I} - \frac{\boldsymbol{u}(j)\boldsymbol{d}^T(j-1) - \boldsymbol{d}^T(j-1)[c(j)\boldsymbol{d}^T(j-1) - \boldsymbol{u}^T(j)]}{\boldsymbol{d}^T(j-1)\boldsymbol{u}(j)} \quad (6.35)$$

where

$$c(j) = \delta(j-1) + \frac{|\boldsymbol{u}(j)|^2}{\boldsymbol{d}^T(j-1)\boldsymbol{u}(j)} \quad (6.36)$$

$$\boldsymbol{u}(j) = \boldsymbol{g}(\boldsymbol{f}_a(j)) - \boldsymbol{g}(\boldsymbol{f}_a(j-1)) \quad (6.37)$$

$$\boldsymbol{d}(j) = -\boldsymbol{H}^{-1}(\boldsymbol{f}_a(j-1))\boldsymbol{g}(\boldsymbol{f}_a(j-1)) \quad (6.38)$$

The step-size is $\delta(j-1)$, and hence the adaptive weight vector can be updated as follows [73, 74]:

$$\boldsymbol{f}_a(j) = \boldsymbol{f}_a(j-1) + \delta(j)\boldsymbol{d}(j) \quad (6.39)$$

The step-size should satisfy the following constraints to guarantee convergence:

$$\Psi(\boldsymbol{f}_a(j-1) + \delta(j)\boldsymbol{d}(j)) \le \Psi(\boldsymbol{f}_a(j-1)) + \alpha\delta(j)\boldsymbol{g}^T(\boldsymbol{f}_a(j-1))\boldsymbol{d}(j) \quad (6.40)$$

$$(\boldsymbol{g}(\boldsymbol{f}_a(j)))^T\boldsymbol{d}(j) \ge \sigma(\boldsymbol{g}(\boldsymbol{f}_a(j-1)))^T\boldsymbol{d}(j) \quad (6.41)$$

A procedure for selecting the step-size according to the above constraints is outlined by Wang and Dowling [73].

Substituting from (6.35)–(6.37) into (6.38) and after some manipulations, the following update equation for $\boldsymbol{d}(j)$ is obtained:

$$\boldsymbol{d}(j) = \boldsymbol{d}(j-1)e(j) + (a(j)-1)\boldsymbol{g}(\boldsymbol{f}_a(j-1)) \tag{6.42}$$

where

$$a(j) = \frac{\boldsymbol{u}(j)\boldsymbol{d}^T(j-1)}{\boldsymbol{d}^T(j-1)\boldsymbol{u}(j)} \tag{6.43}$$

$$e(j) = \frac{[\boldsymbol{u}^T(j) - c(j)\boldsymbol{d}^T(j-1)]\boldsymbol{g}(\boldsymbol{f}_a(j-1))}{\boldsymbol{d}^T(j-1)\boldsymbol{u}(j)} \tag{6.44}$$

If we set $a(j) = 0$, we get the block conjugate gradient CMA (referred to as BCGCMA), and if we set $a(j) = e(j) = 0$, we obtain the block gradient descent CMA (referred to as BGDCMA).

6.5 BSCMA with Quadratic Inequality Constraint

The QI constraint can be used on the block's adaptive weight portion $\boldsymbol{f}_a(j)$. Consequently, the robust weight vector can be obtained from the solution of the following constrained optimization problem [51]:

$$\overline{\Psi}(f) = \frac{1}{4M}\sum_{n=0}^{M-1}\left[\overline{\boldsymbol{f}}_a^T(j)\boldsymbol{Z}(n)\overline{\boldsymbol{f}}_a(j) - 1\right]^2 \; s.t. \; \overline{\boldsymbol{f}}_a^H(j)\overline{\boldsymbol{f}}_a(j) \leq \beta^2 \tag{6.45}$$

where the overbar indicates that a parameter is quadratically constrained.

Unfortunately, no closed-form solution can be obtained for the above optimization problem. Alternatively, the BSCMA can be used to update the weight vector and the Lagrange methodology to solve the QI constraint. The new cost function and gradient vector will be represented, respectively, as follows:

$$\Psi(\overline{\boldsymbol{f}}_a) = \frac{1}{4M}\sum_{n=0}^{M-1}\left[\overline{\boldsymbol{f}}_a^H(j)\,\boldsymbol{Z}(n)\overline{\boldsymbol{f}}_a(j) - 1\right]^2 + \frac{1}{2}\lambda.s(\overline{\boldsymbol{f}}_a^T(j)\overline{\boldsymbol{f}}_a(j) - \beta^2) \tag{6.46}$$

$$\overline{\boldsymbol{g}}(\overline{\boldsymbol{f}}_a(j-1)) = \frac{1}{M}\sum_{n=0}^{M-1}\left[\overline{\boldsymbol{f}}_a^T(j-1)\boldsymbol{Z}(n)\overline{\boldsymbol{f}}_a(j-1) - 1\right] \times \boldsymbol{Z}(n)\overline{\boldsymbol{f}}_a(j-1) + \lambda(j)\overline{\boldsymbol{f}}_a(j-1) \tag{6.47}$$

Therefore,

$$\bar{\boldsymbol{d}}(j) = -\boldsymbol{H}^{-1}(\bar{\boldsymbol{f}}_a(j-1))\left[\boldsymbol{g}(\bar{\boldsymbol{f}}_a(j-1)) + \lambda(j)\bar{\boldsymbol{f}}_a(j-1)\right] \quad (6.48)$$

In order to simplify the computation of the Lagrange multiplier $\lambda(j)$ and to avoid the computation of Hessian matrix, we have to update the QI term in (6.46) using the steepest-descent algorithm rather than the Newton algorithm, as follows:

$$\bar{\boldsymbol{d}}(j) = -\boldsymbol{H}^{-1}(\bar{\boldsymbol{f}}_a(j-1))\boldsymbol{g}(\bar{\boldsymbol{f}}_a(j-1)) + \lambda(j)\bar{\boldsymbol{f}}_a(j-1) \quad (6.49)$$

$$\bar{\boldsymbol{f}}_a(j) = \bar{\boldsymbol{f}}_a(j-1) - \delta(j)\boldsymbol{H}^{-1}(\bar{\boldsymbol{f}}_a(j-1))\boldsymbol{g}(\bar{\boldsymbol{f}}_a(j-1)) + \delta(j)\lambda(j)\bar{\boldsymbol{f}}_a(j-1) \quad (6.50)$$

$$\bar{\boldsymbol{f}}_a(j) = \boldsymbol{f}_a(j) + \delta(j)\lambda(j)\bar{\boldsymbol{f}}_a(j-1) \quad (6.51)$$

where

$$\boldsymbol{f}_a(j) = \bar{\boldsymbol{f}}_a(j-1) - \delta(j)\boldsymbol{H}^{-1}(\bar{\boldsymbol{f}}_a(j-1))\boldsymbol{g}(\bar{\boldsymbol{f}}_a(j-1)) \quad (6.52)$$

Substituting the update equation (6.51) of $\bar{\boldsymbol{f}}_a(j)$ into the constraint $\bar{\boldsymbol{f}}_a^H(j)\bar{\boldsymbol{f}}_a(j) \leq \beta^2$, we can solve for $\lambda(j)$ as follows [51]:

$$\lambda(j) = [b \pm \sqrt{b^2 - 4ac}]/2a \quad (6.53)$$

where

$$a = \delta^2(j)\|\bar{\boldsymbol{f}}_a(i, j-1)\|^2, N_a \quad (6.54)$$

$$b = 2\delta(j)\boldsymbol{f}_a^T(i,j)\bar{\boldsymbol{f}}_a(i, j-1), 2N_a \quad (6.55)$$

$$c = \|\boldsymbol{f}_a(i,j)\|^2 - \beta^2, N_a \quad (6.56)$$

By substituting from (6.54)–(6.56) and (6.52) into $b^2 - 4ac \geq 0$, an inequality similar to (5.27) is obtained:

$$\left[(\bar{\boldsymbol{f}}_a(j-1) + \delta(j)\boldsymbol{d}(j))^T\bar{\boldsymbol{f}}_a(j-1)\right]^2 \geq \|\bar{\boldsymbol{f}}_a(j-1)\|^2\left[(\bar{\boldsymbol{f}}_a(j-1) + \delta(j)\boldsymbol{d}(j))^T(\bar{\boldsymbol{f}}_a(j-1) + \delta(j)\boldsymbol{d}(j)) - \beta^2\right] \quad (6.57)$$

After some manipulations to (6.57), we get the following upper bound on the step-size.

$$\delta(j) \leq \frac{\beta\|\bar{\boldsymbol{f}}_a(j-1)\|}{\sqrt{\|\bar{\boldsymbol{f}}_a(j-1)\|^2\|\boldsymbol{d}(j)\|^2 - \boldsymbol{d}^T(j)\bar{\boldsymbol{f}}_a(j-1)\bar{\boldsymbol{f}}_a^T(j-1)\boldsymbol{d}(j)}} \quad (6.58)$$

Therefore, this upper bound on $\delta(j)$ guarantees real positive roots in (6.50) and consequently the optimal loading level can be obtained. Additionally, the two constraints in (6.40) and (6.41) must be considered to guarantee the convergence of the BSCMA algorithm.

6.6 Block Processing and Adaptive Implementation

The adaptive block algorithm based on the block Shanno algorithm and the simplified VL technique is summarized in Box 6.2. There are two main loops in the algorithm. The outer loop is for each block of data and the inner loop is repeated over the same block of data until a certain number of iterations has been completed or the norm of the gradient vector is sufficiently small. A block of data with length $N_f \times M$ is selected and the output of the lower branch from the PLIC structure $z(n)$ is commuted. The adaptive vector $\bar{\boldsymbol{f}}_a(1,1)$ is initialized with the required user delay, obtained from the corresponding column in the blocking matrix, or with its signature sequence. On the other hand, the final weight vector $\bar{\boldsymbol{f}}_a(i,j)$ of each block is used as an initial vector to the next block (that is, $\bar{\boldsymbol{f}}_a(i+1,1) = \bar{\boldsymbol{f}}_a(i,j)$). There is no need to put a constraint on the gradient vector while the quadratic constraint on $\boldsymbol{f}_a$ oversees the gradient vector norm increase. The gradient vector is estimated, taken into consideration the previous diagonal loading term $\lambda(j-1)\bar{\boldsymbol{f}}_a(i,j-1)$, to avoid computing it with $\boldsymbol{d}(j)$, which requires Hessian matrix computation, as shown in (6.48). The vector $\bar{\boldsymbol{d}}(j)$ is computed according to (6.42) and then the initial adaptive vector is updated using (6.39).

If the norm constraint on $\boldsymbol{f}_a(i,j)$ is not met, then the VL technique is used to fulfill the quadratic inequality constraint. Unfortunately, we will not be able to find an optimum step-size equation because the cost function (6.49) is not a quadratic equation. This means it has no global minimum with step size. Alternatively, the procedure provided by Wang and Dowling [73] can be used in addition to (6.58) to guarantee optimal diagonal loading, as shown in Box 6.2. After convergence of a block of data, the output of this block is calculated.

Box 6.2 Summary of the robust BSCMA receiver.

Initialization

- $\rho^2 = t.(\|\boldsymbol{f}_c\|^2)$, $\beta^2 = \rho^2 - \|\boldsymbol{f}_c\|^2$, $\varepsilon = 0.1$
- $\lambda(0) = 0$, $\boldsymbol{d}(0) = \boldsymbol{g}(f_a(0))$, $\alpha = 0.25$, $\sigma = 0.5$

For $i = 1,2,\cdots,\lceil N/M \rceil$ **outer loop on block basis**

- $D_i = [\boldsymbol{x}((i-1)M) \vdots \boldsymbol{x}((i-1)M+1) \vdots \cdots \vdots \boldsymbol{x}(iM-1)]$
- $\boldsymbol{Z}(i) = \boldsymbol{z}(n)\boldsymbol{z}^T(n)$, $\boldsymbol{z}(n) = \boldsymbol{B}^T\boldsymbol{x}(n)$
  - If $(i = 1)\overline{\boldsymbol{f}}_{\boldsymbol{a}}(1,1) = \boldsymbol{B}^T c_1$
  - Else $\overline{\boldsymbol{f}}_{\boldsymbol{a}}(i,1) = \overline{\boldsymbol{f}}_{\boldsymbol{a}}(i,j)$
- $j = 0$
- For $j = 1,2,\cdots$, until convergence or max number of iterations reached
  - $\overline{\boldsymbol{g}}(\overline{f}_a(i,j-1)) = \frac{1}{M}\sum_{n=0}^{M-1}[\overline{\boldsymbol{f}}_a^{\,T}(i,j-1)\boldsymbol{Z}(n)\overline{\boldsymbol{f}}_a(i,j-1) - 1]$ $\boldsymbol{Z}(n)\overline{\boldsymbol{f}}(i,j-1)_a - \lambda(j-1)\overline{\boldsymbol{f}}_a(i,j-1)$
  - If $j = 1$
    - o $\boldsymbol{d}(j) = \overline{\boldsymbol{g}}(\overline{\boldsymbol{f}}_a(i,j-1))$
  - Else
    - o $\boldsymbol{d}(j) = \boldsymbol{d}(j-1)e(j) + (a(j)-1)\overline{\boldsymbol{g}}(\overline{\boldsymbol{f}}_a(i,j-1))$
  - $\boldsymbol{f}_a(i,j) = \overline{\boldsymbol{f}}_a(i,j-1) + \delta(j)\boldsymbol{d}(j)$
  - $if(\|\boldsymbol{f}_a(i,j)\|^2 > \beta^2)$
    - o $c = \|\boldsymbol{f}_a(i,j)\|^2 - \beta^2, N_a$
    - o $b = 2\delta(j)\boldsymbol{f}_a^T(i,j)\overline{\boldsymbol{f}}_a(i,j-1), 2N_a$
    - o $a = \delta^2(j)\|\overline{\boldsymbol{f}}_a(i,j-1)\|^2, N_a$
    - o $\lambda(j) = -b - (\sqrt{b^2 - 4ac})/2a$
    - o $\overline{\boldsymbol{f}}_a(i,j) = \boldsymbol{f}_a(i,j) + \lambda(j)\delta(j)\overline{\boldsymbol{f}}_a(i,j-1)$
  - Else
    - o $\overline{\boldsymbol{f}}_{\boldsymbol{a}}(i,j) = \boldsymbol{f}_a(i,j)$; $\lambda(j) = 0$
  - End if

- If $\left[\Phi(\bar{f}_a(i,j)) \le \Phi(\bar{f}_a(i,j-1)) + \alpha\delta(j)(\bar{g}(\bar{f}_a(i,j-1)))^T\bar{d}(j)\right]$
  - o $\delta(j+1) = \delta(j) + \epsilon\delta(j)$
- Else If $\left[(\bar{g}(\bar{f}_a(j)))^T\bar{d}(j) \ge \sigma(\bar{g}(\bar{f}_a(j-1)))^T\bar{d}(j)\right]$
  - o $\delta(j+1) = \delta(j) - \epsilon\delta(j)$
- Else if
  - o $\delta(j) > \dfrac{\beta\| \bar{f}_a(i,j)\|}{\sqrt{\| \bar{f}_a(i,j)\|^2 \| d(j)\|^2 - d^T(j)\bar{f}_a(i,j)\bar{f}_a^T(i,j)\, d(j)}} = \delta\left(j+1\right)$
- Otherwise
  - o $\delta(j+1) = \delta(j)$
- End if

- End for j loop

$f(i,j) = f_c - B\bar{f}_a(i,j)$

$y(((i=1)M):(iM=1)) = x(((i=1)M):(iM=1))f(i,j)$

End for i loop

As shown in Box 6.2. the total amount of the required computation at each snapshot for the robust subroutine is about $O(4N_a)$. As the complexity of the BSCMA is about $O(28.6N_a + 11N_a/M)$ [74], the total complexity of the robust BSCMA is about $O(28.6N_a + 15N_a/M)$ per output point from every block of data with length M [51].

6.7 Simulation Results for Robust LCCMA

In this section, the performance of the robust detectors is investigated and compared with different CM-based detectors. Five asynchronous users in a multipath Rayleigh fading channel with five multipath components are simulated. Gold codes are used. Code length is 31 chips and detector length is assumed to be one bit length. The channel length (maximum delay spread) is assumed to be ten delayed components and multipath delays are randomly distributed. The detector is assumed to be synchronized to the required user. Users are assumed to have equal power, except for the required user, which is assumed to be 10 dB less than the other users to simulate the near–far effect. In addition, a signature mismatch of order 0.2 T_c is added to the channel vector to examine the detector's robustness against mismatch errors. Two performance measures are adopted here: the output SINR and BER versus bit iterations.

We will refer to the detectors, based on the cost functions (6.1) and (6.6), as LCCMA1 and LCCMA2, respectively. Three LCCMA1 based detectors are simulated. These detectors are:

- LCCMA1 without weighting (LCCMA1 w/t W)
- LCCMA1 with de-biased weighting (LCCMA1 w. W.) [104]
- LCCMA1 with weighting and QI constraint on the weight vector norm (LCCMA1 w. VL).

The LCCMA1-based detectors are adapted using the well-known stochastic gradient method. The simplified VL technique is used with LCCMA1 cost function to apply the QI constraint. Straightforward steps are required to compute the diagonal loading term and other update equations with the cost function given in (6.1).

The LCCMA2 cost function combined with the fast RSD algorithm is used to develop three detectors. The first detector without the QI constraint, the second detector using the SP technique to satisfy the QI constraint, and the third with the simplified VL technique. Additionally, the LCCMA2 function adapted with the RCG algorithm [103, 147] is simulated as well. The four detectors are referred to as:

- LCCMA2 w/t QI
- LCCMA2 w. SP
- LCCMA2 w. VL
- LCCMA2 w. CG

The initializations of the simulated detectors are summarized in Table 6.1. The selection of a constrained value is a compromise between robustness and performance optimality.

Figures 6.1 and 6.2 show the SINR and BER respectively for the seven detectors. We can readily see that the combination of a QI constraint with LCCMA2-based detectors substantially improves their performance and robustness.

Table 6.1 Initializations for simulated CM-based detectors.

Detector	Parameter					
	η	ζ	t	QI	γ	υ
LCCMA1 w/t W.	—	—	—	—	0.1	0.1
LCCMA1 w. W.	—	—	—	—	0.1	0.1
LCCMA1 w. VL	—	—	1.6	VL	0.1	0.1
LCCMA2 w/t QI	1	1	—	—	0.9	0.1
LCCMA2 w. SP	1	1	1.3	SP	0.001	0.1
LCCMA2 w. VL	1	1	1.3	VL	0.001	0.1
LCCMA2 w. CG	1	1	—	—	0.9	0.1

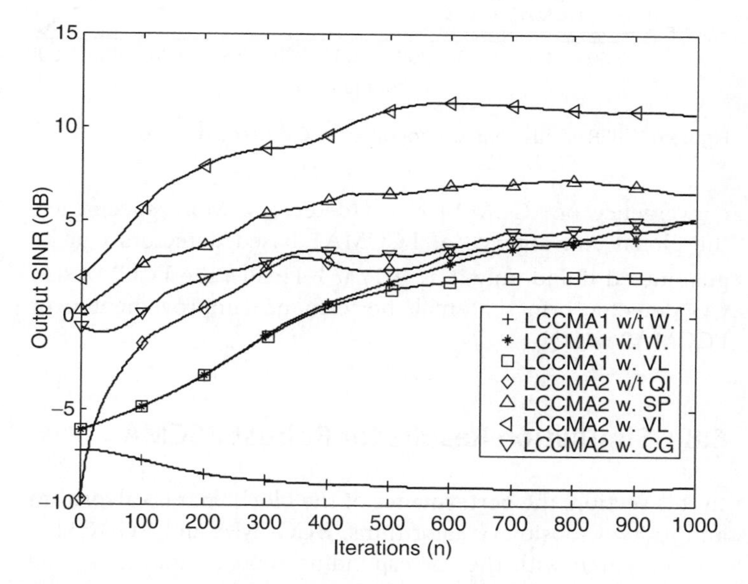

Figure 6.1 SINR versus iterations for different CM-based detectors.

More importantly, the simplified VL technique performs better than the SP technique. Furthermore, a quadratic constraint does not appear to offer any performance improvement for LCCMA1-based detectors. In addition to this, weighting is a mandatory initialization technique to guarantee global

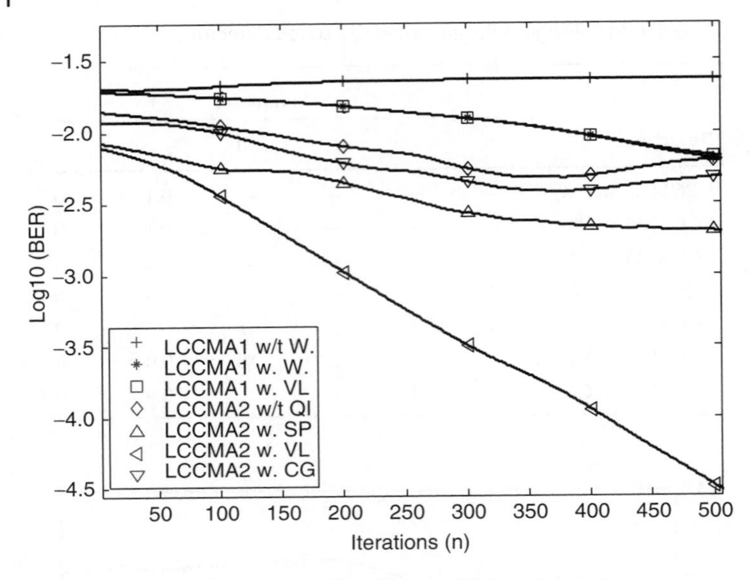

Figure 6.2 BER versus iterations for different CM-based detectors.

convergence of LCCMA1-based detectors. More specifically, the global convergence of LCCMA1 based detectors can be guaranteed if and only if $a_1 \geq 1/\sqrt{3}$. Finally, the LCCMA2 w. CG detector brings a small performance improvement over LCCMA2 w/t QI.

6.8 Simulation Results for Robust BSCMA

In this section, the performance of the block Shanno algorithm and its two subsidiary algorithms, BCGCMA and BGDCMA, are compared with the corresponding robust algorithms. The robust techniques will be referred respectively as:

- BSCMA w. VL
- BCGCMA w. VL
- BGDCMA w. VL.

Two scenarios are simulated: in the first scenario, the system deployed in Section 6.7 is used except that perfect power control (that is, equal user powers) is assumed. The second scenario is identical to the system that was utilized in Section 6.7 to model the near–far effect. The block length is 100 bits and the maximum number of iterations inside the block is 25. The performance of the six detectors is assessed in terms of output SINR and BER versus block iterations. Figures 6.3–6.6 show the SINR and BER for the two scenarios respectively. The figures show that the quadratic constraint brings a considerable improvement for the BSCMA algorithm and its variants, BCGCMA and BGDCMA. With perfect power control, the three robust algorithms perform almost equally well in terms of BER. The BGDCMA is, of course, the slowest converging algorithm. For both scenarios, the convergence of the robust algorithms is achieved after about 10 block iterations, which means four blocks of data are required for convergence. On the other hand, the non-robust algorithms require at least 150 block iterations to reach the same convergence point. Additionally, in the second scenario, the steady-state behavior of the presented robust algorithms is superior to the corresponding non-robust algorithms in terms of BER and SINR.

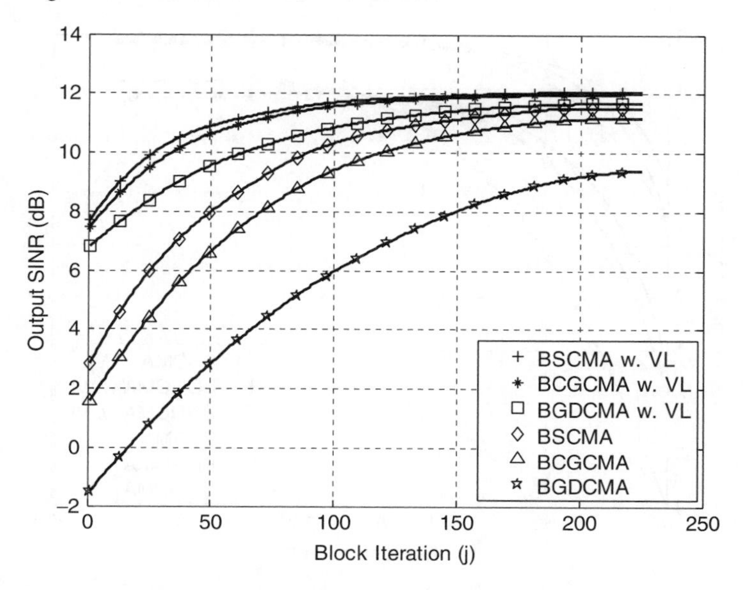

Figure 6.3 SINR versus block iterations for first scenario.

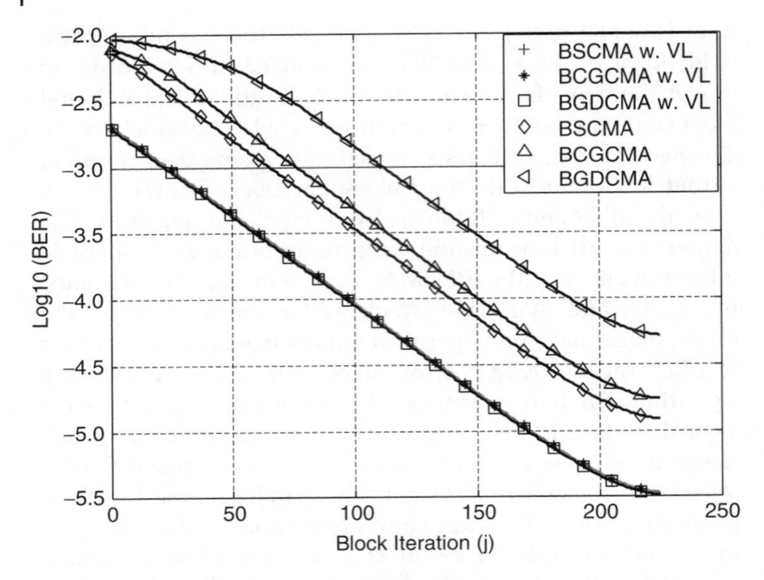

Figure 6.4 BER versus block iterations for first scenario.

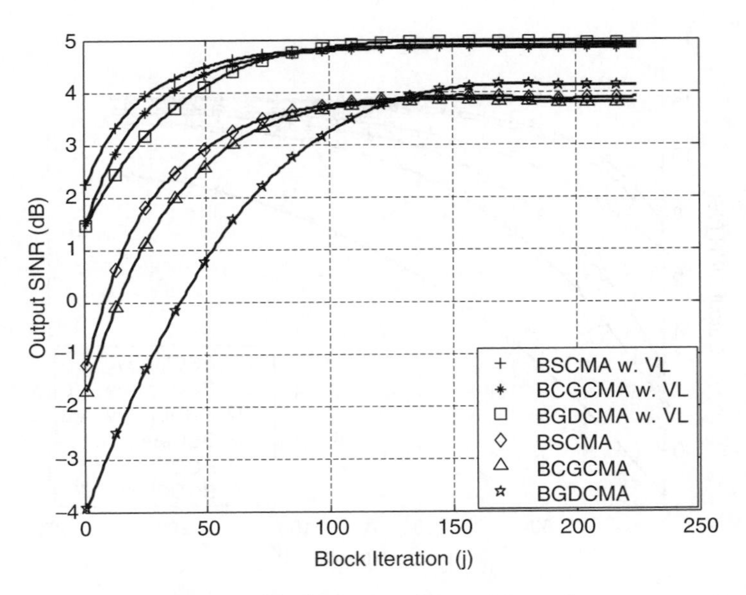

Figure 6.5 5 SINR versus block iterations for second scnario.

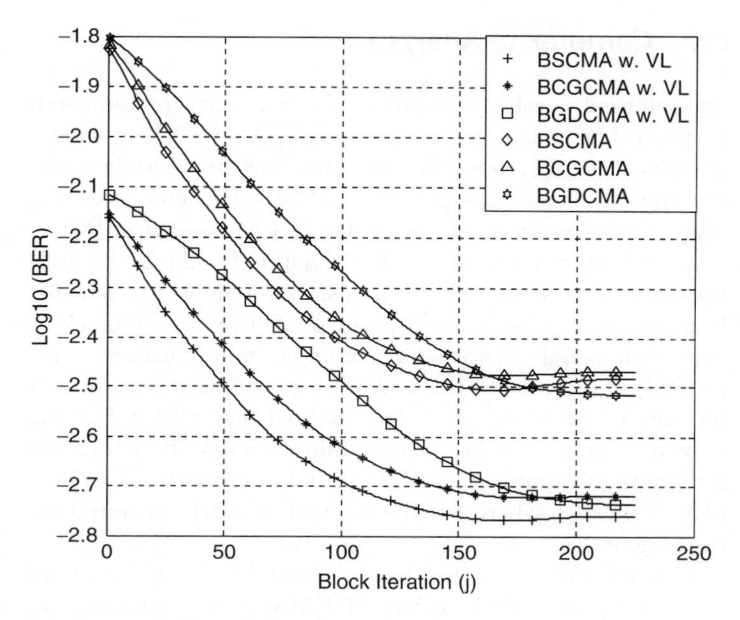

Figure 6.6 BER versus block iterations for second scenario.

Finally, it is interesting now to compare the performance of the LCCMA2 w. VL and BSCMA w. VL detectors. LCCMA2 w. VL performs better than BSCMA w. VL in terms of its steady state: there is an improvement of about 5 dB improvement in SINR. However, the BSCMA w. VL performance is optimal in terms of convergence speed, as the BSCMA w. VL detector requires almost 100 block-iterations to converge, whereas LCCMA2 w. VL requires at least 550 bit iterations to converge. In terms of complexity, we can generally say that LCCMA2 w. VL is of $O(N_a^2)$ complexity while BSCMA w. VL is of $O(N_a)$ complexity. A detailed complexity analysis is provided in Section 6.9. In addition, it is important to highlight that the LCCMA and BSCMA algorithms do not offer the peak SINR or the minimum BER offered by the MOE detectors in Chapters 4 and 5.

All Matlab scripts are provided with the book and the reader can fine tune and optimize these algorithms to achieve the same SINR and BER as the MOE algorithms.

6.9 Complexity Analysis

The required number of multiplications at each iteration step of the robust adaptive detector LCCMA2 w. VL is shown in Table 6.1. Therefore, the total number of multiplications required for each sample of the LCCMA2 w. VL detector during adaptive implementation is about $O(4L^2 + 4LN + 13L + 2N)$.

On the other hand, and as illustrated in Table 6.2, the multiplication complexity of the robust BSCMA w. VL detector is about $O(17L + 6LM)$ per output point from every block of data with length M. So the maximum multiplication complexity per block is $O((17L + 6LM)J)$, assuming the maximum number of iterations is always attained and where J is the maximum number of iterations within the block. Therefore, the total multiplication complexity up to convergence is $O((21L + 6LM + 4LN)JI)$, where I is the number of blocks required for convergence.

As demonstrated in Section 6.8, four blocks are required for convergence of the robust BSCMA w. VL detector and the number of iterations inside the block is 25. Therefore, the maximum multiplication complexities required for convergence of the BSCMA w. VL detector is about $((21L + 6LM + 4LN)100)$, while the LCCMA2 w. VL detector requires about $((4L^2 + 4LN + 13L + 2N)550)$ multiplications before convergence. By considering the simulation scenario in this chapter with $L = 21$, $N = 31$ and $M = 100$ bits, the robust BSCMA detector requires about 1,564,500 multiplications while the LCCMA2 w. VL detector requires about 2,586,650 multiplications; that is, $O(4L^2 + 4LN + 13L + 2N)|_{L=21,N=31}$. Therefore, the robust BSCMA w. VL detector offers a significant complexity reduction compared to the robust LCCMA2 w. VL.

The multiplication complexity versus the detector length (N) for the two detectors is demonstrated in Figure 6.7. It is concluded from the figure that the BSCAM w. VL detector has very low computational complexity compared to the LCCMA w. VL detector and it is evident that the complexity of the BSCMA w. VL detector increases almost linearly with the detector length while the complexity of LCCMA w. VL detector increases exponentially with the detector length.

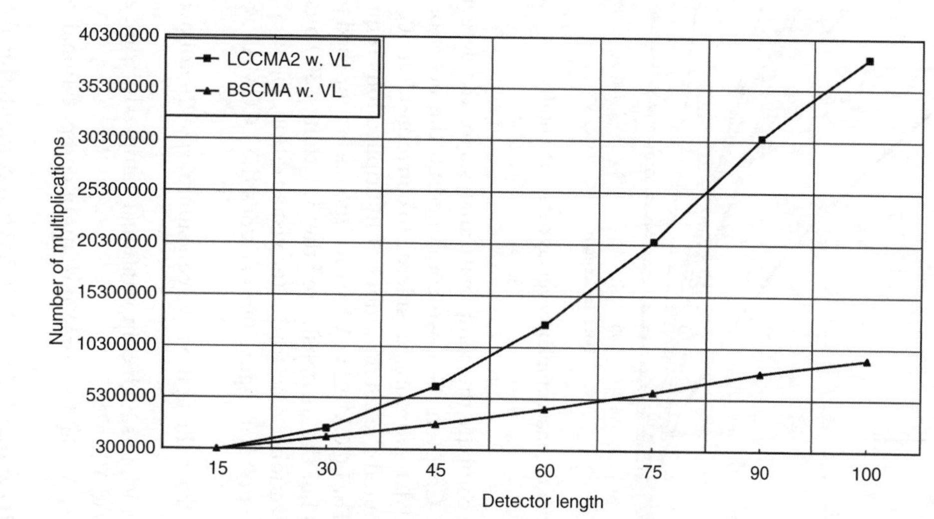

Figure 6.7 Computational complexity of the developed robust detectors versus detector length.

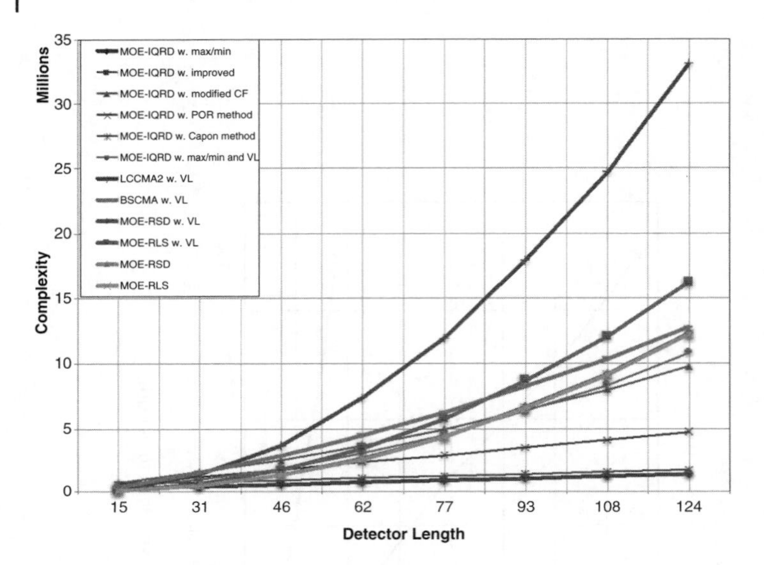

Figure 6.8 Computational complexity versus detector length.

In order to compare the complexity analysis for the detectors developed in Chapters 4–6, we combined all the detectors in Figure 6.8. All detectors are considered to converge after 300 bits and we computed the total number of multiplications required to converge. For LCCMA2 w. VL, convergence is supposed to be after 550 bits but the we considered the situation at 300 bits for better visualizations. Also for better visualization, Figure 6.9 is similar to the previous figure but excludes the LCCMA2 w. VL detector.

As evident from the figures, we can summarize the following:

- BSCMA w. VL has high computational complexity, similar to MOE-RLS w. VL.
- LCCMA2 w. VL is the worst in terms of computational complexity.
- MOE-IQRD w. max/min and VL has less computational complexity than MOE-RLS and MOE-RLS with VL.
- MOE-RSD and MOE-RSD w. VL have the same computational complexity and they offer very low computational complexity at small detector lengths.

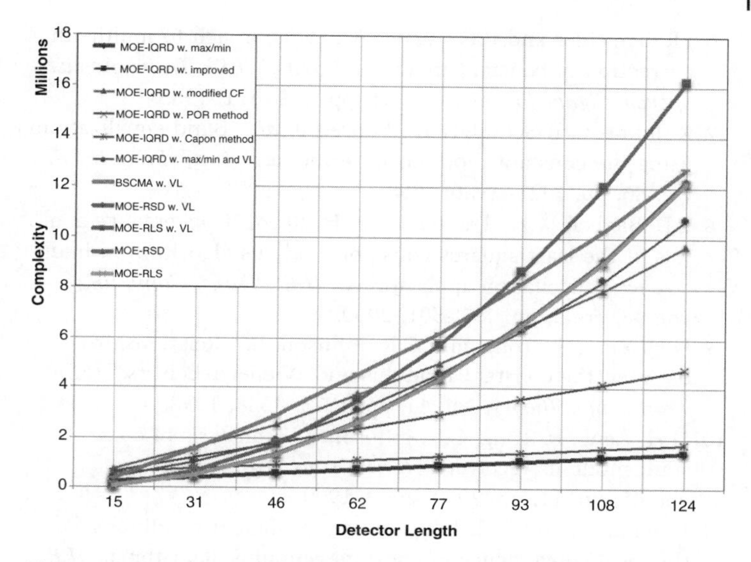

Figure 6.9 Computational complexity versus detector length.

References

1 J.K. Tugnait and T. Li, "Blind asynchronous multiuser CDMA receivers for ISI channels using code-aided CMA," *IEEE J. Sel. Areas Commun.*, vol. 19, no. 8, pp. 1520–1530, 2001.

2 S. Denno and T. Ohira, "Modified constant modulus algorithm for digital signal processing adaptive antenna with microwave analog beamforming," *IEEE Trans. Antennas Propag.*, vol. 50., no. 6, 2002.

3 P.K.P. Cheung and P.B. Rapajic, "CMA-based code acquisition scheme for DS-CDMA systems," *IEEE Trans. Commun.*, vol. 48, no. 5, pp. 852–862, 2000.

4 P. Schniter, "Minimum-entropy blind acquisition/equalization for uplink DS-CDMA." In: *Proceedings of the 36th Allerton Conference on Communications, Control, and Computing*, September 1998.

5 D.N. Godard, "Self-recovering equalization and carrier tracking in two-dimensional data communication systems," *IEEE Trans. Commun.*, vol. COMM-28, pp. 1867–1875, 1980.

6 J.R. Treichler and B.G. Agee, "A new approach to multipath correction of constant modulus signals," *IEEE Trans. Acoust., Speech, Signal Process.*, vol. 31, pp. 459–471, 1983.
7 R. Johnson, P. Schniter, T.J. Endres, *et al.*, "Blind equalization using the constant modulus criterion: a review," *Proc. IEEE*, vol. 86, pp. 1927–1950, 1998.
8 T.E. Biedka, W.H. Tranter, and J.H. Reed, "Convergence analysis of the least squares constant modulus algorithm in interference cancellation applications," *IEEE Trans. Commun.*, vol. 48, no. 3, pp. 491–501, 2000.
9 H.H. Zeng, L. Tong, and C.R. Johnson, Jr.," Relationships between the constant modulus and Wiener receivers," *IEEE Trans. Info. Theory*, vol. 44, pp. 1523–1538, 1998.
10 H.H. Zeng, L. Tong, C.R. Johnson, "An analysis of constant modulus receivers," *IEEE Trans. Signal Process.*, vol. 47, no. 11, pp. 2990–2999, 1999.
11 P. Schniter, and C.R. Johnson, Jr.,, "Sufficient conditions for the local convergence of constant modulus algorithms," *IEEE Trans. Signal Process.*, vol. 48, pp. 2785–2796, 2000.
12 Z. Xu and L. Ping, "Constrained CMA-based multiuser detection under unknown multipath." In: *12th IEEE International Symposium on Personal, Indoor and Mobile Radio Communications*, 30 Sept.-3 October 2001., vol. 1, pp. A-21–A-25.
13 D.J. Brooks, S. Lambotharan and J.A. Chambers, "Optimum delay and mean square error using a mixed crosscorrelation and constant modulus algorithm," *IEE Proc. Commun.*, vol. 147, no. 1, pp. 18–22, 2000.
14 L. Castedo, C.J. Escudero, and A. Dapena "A blind signal separation method for multiuser communications," *IEEE Trans. Signal Process.*, vol. 45, no. 5, pp. 1343–1348, 1997.
15 C.B. Papadias and A.J. Paularj, "A constant modulus algorithm for multiuser signal separation in presence of delay spread using antenna array," *IEEE Trans. Signal. Process. Lett.*, vol. 4, no. 6, pp. 178–181, 1997.
16 Y. Luo, J.A. Chambers and S. Lambotharan, "Global convergence and mixing parameter selection in the cross-correlation constant modulus algorithm for the multiuser environment," *IEE Proc.-Vis. Image Signal Process.*, vol. 148, no. 1, pp. 9–20, 2001.

17 Y. Luo, J.A. Chambers, "Steady-state mean-square error analysis of the cross-correlation and constant modulus algorithm in a MIMO conventional system," *IEE Proc.-Vis. Image Signal Process.*, 149, no. 4, 2002.
18 P. Schniter, R.A. Cassas, A. Touzni, and C.R. Johnson, Jr., "Performance analysis of Godard-based blind channel identification," *IEEE Trans. Signal. Process.*, vol. 49, no. 8, pp. 1757–1767, 2001.
19 R. Lopez-Valcarce and F. Perez-Gonzalez, "Efficient reinitialization of the prewhitened constant modulus algorithm," *IEEE Commun. Lett.*, vol. 5, no. 12, pp. 488–490, 2001.
20 S. Evans and L. Tong, "Online adaptive reinitialization of constant modulus algorithm," *IEEE Trans. Commun.*, vol. 48, pp. 537–539, 2000.
21 Z. Xu and L. Ping, "Blind multiuser detection by kurtosis maximization/minimization", *IEEE Signal Process. Lett.*, vol. 11, pp. 1–4, 2004.
22 O. Shalvi and E. Weinstein, "New criteria for blind deconvolution of nonminimum phase systems (channels)," *IEEE Info. Theory*, vol. IT-36, pp. 312–320, 1990.
23 J.K. Tugnait, "Identification and deconvolution of multichannel linear non-Gaussian process using higher order statistics and inverse filter criteria," *IEEE Trans. Signal. Process.*, vol., 45, no. 3, pp. 658–672, 1997.
24 C.B. Papadias, "Globally convergent blind source separation based on a multiuser kurtosis maximization criterion," *IEEE Trans. Signal. Process.*, vol. 48, no. 12, pp. 3508–3519, 2000.
25 Y. Li and Z. Ding, "Global convergence of fractionally spaced Godared (CMA) adaptive equalizers," *IEEE Trans. Signal Process.*, vol. 44, pp. 818–816, 1996.
26 M. Gu, L. Tong, "Domains of attraction of Shalvi-Weinstein receivers," *IEEE Trans. Signal. Process.*, vol. 49, no. 7, pp. 1397–1408, 2001.
27 P. Schniter, "Existence and performance of Shalvi-Weinstein estimators," *IEEE Trans. Signal. Process.*, vol. 49, no. 9, pp. 2031–2041, 2001.
28 J.-C. Lin, "Blind equalization technique based on an improved constant modulus adaptive algorithm," *IEE Proc. Commun.*, vol. 149, no. 1, pp. 45–50, 2002.

29 S. Abrar, A. Zerguine, M. Deriche, "Soft constraint satisfaction multimodulus blind equalization algorithms", *IEEE Signal Process. Lett.*, vol. 12, pp. 637–640, 2005.
30 S. Abrar, S.I. Shah, "New multimodulus blind equalization algorithm with relaxation", *IEEE Signal Process. Lett.*, vol. 13, pp. 425–428, 2006.
31 J. Migues and L. Castedo, "A linearly constrained constant modulus approach to blind adaptive multiuser interference suppression," *IEEE Commun. Lett.*, vol. 2, no. 8, pp. 217–219, 1998.
32 Z. Tang, Z. Yang, and Y. Yao, "Closed-form analysis of linearly constrained CMA-based blind multiuser detector", *IEEE Commun. Lett.*, vol. 4, no. 9, pp. 273–276, 2000.
33 C. Xu and G. Feng, "Non-canonically constrained CMA for blind multiuser detection," *Electronics Lett.*, vol. 36, no. 2, pp. 171–172, 2000.
34 C. Xu and G. Feng, and K.S. Kwak, "A modified constrained constant modulus approach to blind adaptive multiuser detection," *IEEE Trans. Commun.* vol. 49, no. 9, pp. 1642–1648, 2001.
35 C. Xu and G. Feng, Comments on "A linearly constrained constant modulus approach to blind adaptive multiuser interference suppression", *IEEE Commun. Lett.*, vol. 4, no. 9, pp. 2017–2019, 2000.
36 C. Xu and K.S. Kwak, Comments on "Closed-form analysis of linearly constrained CMA-based blind multiuser detector", *IEEE Commun. Lett.*, vol. 5, no. 7, pp. 290–291, 2001.
37 H. Jiang and K.S. Kwak, "A non-canonical linearly constrained constant modulus algorithm for a blind multiuser detector," *ETRI J.*, vol. 24, no. 3, pp. 239–246, 2002.
38 C. Xu, K. S. Kwak, "Comments on closed-form analysis of linearly constrained CMA-based blind multiuser detector", *IEEE Commun. Lett.*, vol. 5, pp. 290–291, 2001.
39 P. Liu and Z. Xu, "Blind MMSE-constrained multiuser detection", *IEEE Trans. Vehic. Techn.*, vol. 57, pp. 608–615, 2008.
40 L. Wang, R.C. de Lamare, "Adaptive constrained constant modulus algorithm based on auxiliary vector filtering for beamforming", *IEEE Trans. Signal Process.*, vol. 58, pp. 5408–5413, 2010.

41 J.K. Tugnait and T. Li, “Blind asynchronous multiuser CDMA receivers for ISI channels using code-aided CMA”, *IEEE J. Sel. Areas Commun.*, vol. 19, pp. 1520–1530, 2001.

42 Y. Cai and R.C. de Lamare, “Low-complexity variable step-size mechanism for code-constrained constant modulus stochastic gradient algorithms applied to CDMA interference suppression”, *IEEE Trans. Signal Process.*, vol. 57, pp. 313–323, 2009.

43 J.B. Whitehead and F. Takawira, “Performance analysis of the linearly constrained constant modulus algorithm-based multiuser detector”, *IEEE Trans. Signal Process.*, vol. 53, pp. 643–653, 2005.

44 S. Li and R.C. de Lamare, “Blind reduced-rank adaptive receivers for DS-UWB systems based on joint iterative optimization and the constrained constant modulus criterion”, *IEEE Trans. Vehic. Techn.*, vol. 60, pp. 2505–2518, 2011.

45 R.C. de Lamare and P.S.R. Diniz, “Blind adaptive interference suppression based on set-membership constrained constant-modulus algorithms with dynamic bounds”, *IEEE Trans. Signal Process.*, vol. 61, pp. 1288–1301, 2013.

46 J.P. de Villiers and L.P. Linde, “On the convexity of the LCCM cost function for DS-CDMA blind multiuser detection”, *IEEE Commun. Lett.*, vol. 8, pp. 351–353, 2004.

47 Y. Cai, R.C. de Lamare, M. Zhao, and J. Zhong, “Low-complexity variable forgetting factor mechanism for blind adaptive constrained constant modulus algorithms”, *IEEE Trans. Signal Process.*, vol. 60, pp. 3988–4002, 2012.

48 S. Zhang, J. Zhang, “New steady-state analysis results of variable step-size lms algorithm with different noise distributions,” *IEEE Signal Process. Lett.*, vol. 21, pp. 653–657, 2014.

49 F. Huang, J. Zhang, S. Zhang, “NLMS algorithm based on a variable parameter cost function robust against impulsive interferences,” *IEEE Trans. Circuits Syst II*, vol. 64, pp. 600–604, 2017.

50 R. C. de Lamare, R. Sampaio-Neto, and M. Haardt, “Blind adaptive constrained constant-modulus reduced-rank interference suppression algorithms based on interpolation and switched decimation,” *IEEE Trans. Signal Process.*, vol. 59, pp. 681–695, 2011.

51 A. Elnashar, S. Elnoubi, and H. Elmikati, "Sample-by-sample and block-adaptive robust constant modulus-based algorithms," *IET Signal Process.*, vol. 6, no. 8, pp. 805–813, 2012.

52 T.J. Endres, S.N. Hulyalkar, C.H. Strolle, and T.A. Schaffer, "Low-complexity and low-latency implementation of the Godared/CMA update," *IEEE Trans. Commun.*, vol. 49. no. 2, pp. 219–225, 2001.

53 S.R. Nelatury and S.S. Rao "Increasing the speed of convergence of the constant modulus algorithm for blind channel equalization" *IEEE Trans. Commun.*, vol. 50, no. 6, pp. 872–876, 2002.

54 A. Mathur, A.V. Keerthi, and J.J. Shynk, "A variable step-size CM array algorithm for fast fading channel" *IEEE Trans. Signal Process.*, vol. 45, no. 4, pp. 1083–1087, 1997.

55 L. He and M. Amin, "A dual-mode technique for improved blind equalization for QAM signals" *IEEE Signal Process. Lett.*, vol. 10, no. 2, pp. 29–31, 2003.

56 P. Arasaratnam, S. Zhu and A.G. Constantinides, "Fast convergent multiuser constant modulus algorithm for use in multiuser DS-CDMA environment." In: *IEEE International Conference on Acoustics, Speech, and Signal Processing*, (ICASSP '02), 13–17 May 2002, vol.3, pp. 2761–2764.

57 Z. Du, S. Zhou, P. Wan, and W. Wu, "Novel variable step size constant modulus algorithms for blind multiuser detection." In: *IEEE Vehicular Technology Conference* 2001, pp. 673–677.

58 P. Arasaratnam, S. Zhu and A.G. Constantinides, "Robust and fast convergent blind multi-user detectors for DS-CDMA systems in nonstationary multipath environments," *14th International Conference on Digital Signal Processing, DSP 2002*, 1–3 July 2002, vol.2, pp. 627–630.

59 S.-J. Chern and S.-M. Wang "Adaptive multiuser detector for DS-CDMA over multipath fading channel with linearly constrained constant modulus modified conjugate gradient algorithm." In: *ISPACS* 2003, December 2003, pp. A3–3.

60 P.S. Chang and A.N. Willson, Jr., "Analysis of conjugate gradient algorithms for adaptive filtering," *IEEE Trans. Signal Process.*, vol. 48, no. 2, pp. 409–418, 2000.

61 S. Haykin, *Adaptive Filter Theory*, 4th edn. Prentice-Hall, 2002

62 V. Krishnamurthy, "Averaged stochastic gradient algorithms for adaptive blind multiuser detection in DS/CDMA systems," *IEEE Trans. Commun.*, vol. 48, pp. 125–134, 2000.

63 D. Das and M.K. Varanasi, "Blind adaptive multiuser detection for cellular systems using stochastic approximation with averaging," *IEEE J. Sel. Areas Commun.*, vol. 20, no. 2, pp. 310–319, 2002.

64 B.T. Polyak and A.B. Juditsky, "Acceleration of stochastic approximation by averaging," *SIAM J. Contr. Optim.*, vol. 30, no. 4, pp. 838–855, 1992.

65 V.J. Mathews and Z. Xie, "A stochastic gradient adaptive filter with gradient adaptive step size," *IEEE Trans. Signal Process.*, vol. 41, no. 6, pp. 2075–2087, 1993.

66 H.J. Kushner and J. Yang, "Analysis of adaptive step-size SA algorithms for parameter tracking," *IEEE Trans. Autom. Control*, vol. 40, pp. 1403–1410, 1995.

67 P. Yuvapoositanon and J. Chambers, "Adaptive step-size constant modulus algorithm for DS-CDMA receivers in nonstationary environments," *Signal Process.*, vol. 82, pp. 311–315, 2002.

68 Z. Xu and P. Liu, "Code-constrained blind detection of CDMA signals in multipath channels," *IEEE Signal Process. Lett.*, vol. 9, no. 12, 2002.

69 R.C. de Lamare and R. Sampaio Neto, "Blind adaptive code-constrained constant modulus algorithms for CDMA interference suppression in multipath channels", *IEEE Commun. Lett.*, vol. 9. no. 4, pp. 334–336, 2005.

70 E. Dong, C. Zhu and L. Evans, "Blind multiuser detector based on reduced rank subspace and least square constant modulus algorithm." In: *8th International Conference on Signal Processing*, 2006.

71 R.C. de Lamare, M. Haardt and R. Sampaio-Neto, "Blind adaptive constrained reduced-rank parameter estimation based on constant modulus design for CDMA interference suppression," *IEEE Trans. Signal Process.*, vol. 56., no. 6, pp. 2470–2482, 2008.

72 R.C. de Lamare, R. Sampaio-Neto and M. Haardt, "Blind adaptive constrained constant-modulus reduced-rank interference suppression algorithms based on interpolation and switched decimation," *IEEE Trans. Signal Process.*, vol.59, no.2, pp. 681–695, 2011.
73 Z. Wang and E.M. Dowling, "Block Shanno constant modulus algorithm for wireless equalizations." In: *Proceedings of ICASSP,* Atlanta, GA, USA, *April 1996*, pp. 2678–2681.
74 U.G. Jani, E.M. Dowling, R.M. Golden, and Z. Wang, "Multiuser interference suppression using block Shanno constant modulus algorithm," *IEEE Trans. Signal. Process.*, vol., 48, no. 5, pp. 1503–1506, 2000.
75 P. Arasaratnam, S. Zhu and A.G. Constantinides, "Block-Shanno multi-user constant modulus algorithm for use in multi-user CDMA environment," *Electronics Lett.*,vol. 38, no. 9, pp. 423–425, 2002.
76 P. Arasaratnam, S. Zhu and A.G. Constantinides, "Fast convergent multiuser constant modulus algorithm for use in multiuser DS-CDMA environment." In: *IEEE International Conference on Acoustics, Speech, and Signal Processing*, (ICASSP02), 13–17 May 2002, vol.3, pp. 2761–2764.
77 P. Arasaratnam, S. Zhu and A.G. Constantinides, "Robust and fast convergent blind multi-user detectors for DS-CDMA systems in nonstationary multipath environments." In: *2002 14th International Conference on Digital Signal Process.*, DSP 2002, 1–3 July 2002, vol.2, pp. 627–630.
78 P. Daehyun, R.M. Golden, M. Torlak, and E.M. Dowling, "Blind adaptive CDMA processing for smart antennas using the block shanno constant modulus algorithm," *IEEE Trans. Signal Process.*, vol. 54, no. 5, pp. 1956–1959, 2006.
79 O. Diene, A. Bhaya, "Conjugate gradient and steepest descent constant modulus algorithms applied to a blind adaptive array", *Signal Process.*, vol. 90, pp. 2835, 2010.
80 T. Samir, S. Elnoubi, and A. Elnashar, "Block-Shanno minimum BER beamforming," *IEEE Trans. Vehic. Techn.*, vol. 57, no. 5, pp. 2981–2990, 2008.
81 C.-Y. Chi, C.-Y. Chen, C.-H. Chen, C.-C. Feng, "Batch processing algorithms for blind equalization using higher-order statistics", *IEEE Signal Process. Mag.*, vol. 20, pp. 25–49, 2003.

82 F.-B. Ueng, J.-D. Chen, S.-C. Tsai, P.-Y. Chen, "Blind adaptive DS/CDMA receivers for multipath and multiuser environments", *IEEE Trans. Circuits Syst I*, vol. 53, pp. 440–453, 2006.

83 A.-C. Chang, "Robust multiuser detection based on variable loading RLS technique", *Signal Process.*, vol. 90, pp. 579, 2010.

84 E.-L. Kuan and L. Hanzo, "Burst-by-burst adaptive multiuser detection CDMA: a framework for existing and future wireless standards", *Proc. IEEE*, vol. 91, pp. 278–302, 2003.

85 R. O'Donnell, "Prolog to: burst-by-burst adaptive multiuser detection CDMA: a framework for existing and future wireless standards", *Proc. IEEE*, vol. 91, pp. 275–277, 2003.

86 J.K. Tugnait, "Blind estimation and equalization of MIMO channels via multidelay whitening", *IEEE J. Sel. Areas Commun.*, vol. 19, pp. 1507–1519, 2001.

7

Robust Adaptive Beamforming

7.1 Introduction

The use of antenna arrays in a communication system can theoretically improve performance in terms of capacity and coverage [1–3]. In particular, a multi-element antenna receiver at the base station of a cellular communication system is able to compensate for signal degradations in the mobile-to-base link caused by co-channel interference. This is known to be the most important factor limiting the number of users that a system can handle [4]. Adaptive interference suppression techniques based on multi-user detection and antenna array beamforming have recently been considered as a powerful method for increasing the quality, capacity and coverage of cellular systems based on DS/CDMA or OFDM-MIMO [5–7]. They provide a superior, although computationally more expensive, alternative to conventional single-sensor/user (matched filter) detection, which is seriously limited by MAI, multipath propagation in frequency-selective channels, asynchronism, and accuracy, calibration and direction of arrival (DOA) errors. The use of antenna arrays offers the possibility to use the spatial characteristics of different user signals to augment the temporal discrimination provided by their signature sequences. This will lead to a spatial-temporal detector [8]. Combined multi-user beamforming algorithms for space-time processing can be found in the literature [7–15].

In Chapter 8, we will present a combined multi-user and beamforming processing approach for DS/CDMA and OFDM-MIMO systems in the context of minimum BER

Simplified Robust Adaptive Detection and Beamforming for Wireless Communications,
First Edition. Ayman Elnashar.

Companion website: www.wiley.com/go/elnashar49

detection. In this chapter, we will present robust adaptive beamforming techniques that are generally suitable for wireless communications and particularly for base stations in cellular systems. For simplicity, and without losing the generality, this chapter will consider simple beamforming formulations in order to develop robust beamforming techniques. The techniques in these chapters can be easily extended to OFDM-MIMO and other wireless systems, as well as other applications, such as sonar, radar, acoustic arrays, and medical imaging analysis.

Beamforming is a signal processing technique by means of an array of sensors that allows spatial filtering of signals [16, 17]. Adaptive beamforming has been implemented in wireless communications, radar, sonar, speech processing, and other areas. Recently, there has been a great effort to design adaptive beamforming techniques that are robust against mismatch and modeling errors, calibration errors, mutual coupling, and that enhance interference cancellation capability [18–26]. Adaptive beamforming is a versatile approach to detect and estimate the signal-of-interest (SOI) at the output of a sensor array. It has applications in wireless communications, radar, sonar, astronomy, seismology, medical imaging, and microphone array speech processing. Unfortunately, traditional adaptive array algorithms are known to be extremely sensitive to even slight mismatches between the presumed and the actual array responses to the desired signal direction [1]. Such mismatches can occur as a result of DOA errors, local scattering, near–far spatial signature mismatches, waveform distortion, source spreading, array calibration errors, distorted antenna shapes, and mutual coupling. Whenever a mismatch occurs, the adaptive beamformer tends to misconstrue the SOI components in the array observations as interference and hence suppresses them. The errors in array response to the SOI can be due to look-direction errors, uncertainty in array sensor positions, mutual coupling, imperfect array calibration, multipath propagation due to local and remote scattering, and limited sample support.

There are several existing approaches to robust adaptive beamforming. The so-called linearly constrained minimum variance (LCMV) beamformer, also known as Capon's method, has been a popular beamforming technique. In Capon's beamforming method, the weights are chosen to minimize the array output power subject to side constraint(s) in the desired look direction(s). The method assumes that the array manifold is accurately known, but unfortunately even a small discrepancy in the array manifold can substantially degrade its performance. This degradation in performance is particularly manifest at high signal-to-noise ratios (SNRs) [28]. Minimum variance beamforming, which uses a weight vector that maximizes the signal-to-interference-plus-noise ratio (SINR), is often sensitive to estimation error and uncertainty in the parameters, and also to steering-vector and covariance-matrix estimation errors. Robust beamforming attempts to systematically alleviate this sensitivity by explicitly incorporating a data uncertainty model in the optimization problem.

Diagonal loading is one of the most widely deployed robust techniques. It involves addition of a scaled identity matrix to the sample covariance matrix. Diagonal loading is a technique where the diagonal of the covariance matrix is augmented with a positive or negative constant prior to inversion [29]. Diagonal loading has been a widespread approach to improving robustness against mismatch errors and random perturbations [30, 31]. It has been suggested that diagonal loading desensitizes the system by compressing the noise eigenvalues of the covariance matrix so that the nulling capability against small interference sources is reduced [30]. Moreover, radar signal detectors, such as sample-matrix inversion (SMI), constant false alarm rate (CFAL), and generalized likelihood ratio test (GLRT), which utilize the inverse of the data covariance matrix, experience serious degradation when the sample support available for estimating the matrix is limited [32]. In this case, it is essential to improve the covariance matrix numerical conditioning. This problem can be tackled by diagonally loading the data covariance matrix prior to inversion. A detector combining the covariance matrix taper (CMT) approach and diagonal loaded SMI (SMI-DL), termed the SMI-CMT, preserves the minimal sample support properties and hence does not suffer from undernulled interference [33]. As described by Carlson [30], the diagonal loading controls the effective rank of colored noise by raising the white noise "floor". This corresponds to artificially varying the colored to white noise ratio (CWNR). Viewed in this way, the SMI-DL can be interpreted as an approximate principal component (PC) eigencanceler method to which a CMT can be applied. As a consequence, the performance of the SMI-CMT method can closely approximate the PC-CMT when the diagonal loading factor (DLF) is correctly chosen. Furthermore, it is well known that antenna sidelobes can be made small if the sample data correlation matrix is diagonally loaded before inversion is performed [34].

In digital communications, user information is coded into a sequence with the constant modulus (CM) in the baseband. Obviously, if the beamformer can cancel interference sufficiently well, the outputs of the beamformer should satisfy the constant modulus (CM) condition more precisely. Therefore, minimizing CM errors can act as a natural criterion for determining diagonal loading factor (DLF). Liu and Ding presented a cost function based on the CM criterion to determine the DLF [68]. They also proposed a direct search algorithm for DLF determination by minimizing the cost CM function. The CM property has also been used in the so-called blind beamformer [56, 69]. The blind beamformer does not require knowledge of the signal steering vector, so it does not rely on precise steering vector estimation. The non-blind beamformer is preferred, because the blind beamformer does not give a unique result in two cases [56]:

- If there is more than one signal source meeting the CM, one will get at least two beamformer weights that can satisfy the optimal problem.
- Solutions can be found only up to an arbitrary phase, since the CM property is phase-blind.

In addition, the cost function for the CM-blind beamformer is non-convex. There are local optimal points that make optimization difficult. What is different in Liu and Ding's approach [68] is that the proposed method for finding DLF is based on loading sample-matrix inversion (LSMI) beamforming with a method for determining the one-dimensional parameter, DLF, in terms of the CM condition. In this way, the solution space of the proposed method can be narrowed to a minimum, so that the non-uniqueness will no longer occur. Since the DLF is a scalar, its optimization is a one-dimensional problem: it can be solved easily, whether it is convex or not.

The main drawback of the conventional diagonal loading method is that the diagonal loading level is chosen in an ad-hoc way. Sometimes it is dependent on the noise level, and sometimes it is based on a norm constraint of the weight vector. The most important drawback of the diagonal loading technique is the difficulty involved in deriving a closed-form expression for the diagonal loading term, which relates the amount of diagonal loading with the upper bound of the mismatch constraint value or the required level of robustness. The asymptotic description given by Mestre and Lagunas [35] establishes the existence of an optimum loading factor, which can be estimated from the received data.

Cox *et al.* [19] suggested a new robust SMI beamformer which includes a quadratic inequality constraint on the array gain against uncorrelated noise (occasionally termed an QC-SMI beamformer). A simple scaled projection technique is proposed to satisfy the quadratic inequality constraint. The QC-SMI beamformer is similar to the minimum variance distortionless response (MVDR) beamformer, except that the covariance data matrix is diagonally loaded. The diagonal loading term is obtained by scaling the tentative projection weights to fulfill the quadratic inequality constraint. Unfortunately, the scaled projection technique does not yield the optimal diagonal loading [2, 31]. The performance of the QC-SMI beamformer has been further analyzed with a generalized side-loop canceller (GSC) formulation [36], and it has been demonstrated that the weight ratio can be related to the field scenario. The weight ratio is the ratio between the norm of the adaptive portion of the GSC beamformer and the norm of the quiescent vector. It is additionally recommended that setting the weight ratio to a value of about 0.5 engenders a robust QC-SMI beamformer.

A robust cyclic adaptive beamformer (R-CAB) was proposed by Wu and Wong [24]. This includes the diagonal loading form of the QC-SMI beamformer but the data covariance matrix and the steering vector are swapped, respectively, with the interference covariance matrix and cyclic adaptive beamformer weight vector. They proposed adjusting the diagonal loading term to the trace of the covariance matrix of the desired signal, which can be computed using the cyclostationarity of the signals (which is a practical value). However, the computation burden of R-CAB is $O(M^3)$ (where M is the number of sensors). Further discussion of the R-CAB beamformer, with emphasis on steering vector mismatches and appropriate loading strengths, is provided by He and Chen in [37]. Additionally, the R-CAB has been realized using an improved Hopfield neural network with a netlike parallel structure that can manage the data more rapidly [37].

A Monte Carlo simulation has been used to assess the CFAL detector with diagonal loading [32]. This technique, unfortunately, is scenario dependent. A modified GLRT detector with diagonal loading of the covariance matrix was suggested to improve the performance of the GLRT detector for cases when the interference and noise covariance matrix is poorly conditioned [32]. Unfortunately again, there is no closed form expression given for the amount of diagonal loading λ. Otherwise, it is suggested that the diagonal loading is bounded by $\sigma_v^2 < \lambda < \gamma_{\min}$, where $\gamma_{\min}$ is the minimum interference eigenvalue and σ_v^2 is the variance of the white noise.

Tian *et al.* [38–42] proposed a technique for implementing a quadratic inequality constraint on the beamformer weights to improve robustness against pointing errors and random perturbations in sensor parameters. They derived a closed-form solution for the amount of loading level. This approach provides a robust LCMV beamformer based on the recursive least squares (RLS) algorithm and a variable loading (VL) technique. This VL technique is based on a Taylor series expansion approximation, but unfortunately it is applicable only for small loading levels. It fails with large loading levels due to the breach of the Taylor series approximation. More specifically, the solution may lead to complex roots and hence the optimal loading level cannot be computed. This drawback is covered in the literature [2, 31, 43–46]. These papers provide a new VL technique capable of precisely computing the required amount of diagonal loading level without approximation and with low computational complexity. In these approaches, a quadratic inequality constraint is applied on the weight vector in preference to the steering vector. Therefore, these approaches can be interpreted as general robust techniques; the constrained value is not directly connected to uncertainties in the steering vector.

Recently, robust beamforming approaches derived from standard LCMV beamforming with a spherical or ellipsoidal uncertainty constraint have been developed [48–52]. The uncertainty constraint can be applied on the weight vector or the steering vector as well. Some approaches [48, 49] reformulate robust adaptive beamforming as a convex second-order cone programming (SOCP) problem. The SOCP approach can be interpreted as a diagonal loading technique in which the optimal value of diagonal loading is computed based on the known upper bound on the norm of the signal steering vector mismatch [49]. The SeDuMi optimization Matlab toolbox [53] has been used to compute the weight vector in the SOCP approach. Unfortunately, the computational burden of this software seems to be cumbersome, which limits the practical implementation of the technique. Although several efficient convex optimization software tools are currently accessible, the SOCP-based method does not provide any closed-form solution, and hence does not have simple on-line adaptive implementations [54]. In addition, the technique can be regarded as a batch algorithm rather than an adaptive scheme; the weight vector of the beamformer is individually computed in each step and cannot be recursively updated.

The robust Capon beamformer proposed in [51, 52] computes precisely the diagonal loading level based on the ellipsoidal uncertainty set of the array steering vector. The essence behind this technique is the estimation of the signal of interest power (SOIP) rather than the signal itself. An eigendecomposition batch algorithm is proposed to compute the diagonal loading level, which would also hit the wall of the computational complexity burden. Alternatively, this approach can be recursively implemented using subspace-tracking algorithms via tracking signal and noise subspaces. However, it is difficult to apply this approach to wireless communications, where the dimensions of signal and noise may be ambiguous or not exactly known. Besides, the eigendecomposition approach is based on continuous diagonal loading even without mismatches. This means that the data autocorrelation matrix will be always diagonally loaded even without DOA mismatch errors, such as in a moving source scenario. More specifically, this approach does not check for ellipsoidal/uncertainty constraints at each adaptive step and accordingly applies diagonal loading only if it is needed (that is, the ellipsoidal or uncertainty constraint is not met).

Many approaches have been proposed to improve the robustness of traditional beamforming methods. A survey of these approaches can be found in the literature [55, 56]. Among them, the worst-case (WC) performance optimization is a powerful technique that yields a beamformer with robustness against an arbitrary signal steering vector mismatch, data non-stationarity problems, and small sample support [47, 49, 56–59]. The WC approach explicitly models an arbitrary (but bounded in norm) mismatch in the desired signal array response and uses WC performance optimization to improve the robustness of the minimum variance distortionless response (MVDR) beamformer [47]. A theoretical analysis of this class of robust beamformers, in terms of SINR in the presence of random steering vector errors, is presented in the literature [60, 61]. These papers derive closed-form expressions for the SINR.

Unfortunately, the natural formulation of WC performance optimization involves minimization of a quadratic function subject to infinity nonconvex quadratic constraints [47]. Some approaches [47, 57] reformulate WC optimization as SOCP, and solve it efficiently via the well-established interior point method [53]. As highlighted earlier, the SOCP method does not provide a closed-form solution for the beamformer weights and it cannot be implemented online because the weight vector needs to be recomputed completely with the arrival of a new array observation. To alleviate the computational cost of conventional SOCP techniques, an iterative reweighted least-squares procedure, which also converges to the support vector machine SVM solution, has been proposed [62]. Robust beamforming via worst-case SINR maximization – that is, the problem of finding a weight vector that maximizes the worst-case SINR over the uncertainty model – was proposed by Kim *et al.* [64]. With a general convex uncertainty model, the WC SINR maximization problem can be solved using convex optimization. In particular, when the uncertainty model can be represented by linear matrix inequalities, the WC SINR maximization problem can be solved via semidefinite programming approaches, such as SeDuMi or CXA. The convex formulation result allows more general uncertainty models to be handled, rather than using a special form of uncertainty model.

The problem of robust beamforming for antenna arrays with arbitrary geometry and magnitude response constraints is one of considerable importance. Due to the presence of the non-convex magnitude response constraints, conventional convex optimization techniques cannot be applied directly. A new approach based on iteratively linearizing the non-convex constraints [85] reformulates the non-convex problem as a series of convex subproblems, each of which can be optimally solved using SOCP. The basic idea of the proposed approach is to linearize the non-convex magnitude squared response constraints in a neighborhood of the complex array weights in each iteration. For this linearization, it is shown that the problem of finding the optimal updates around the previous iterates is a convex SOCP problem that can be efficiently solved. Moreover, in order to obtain a beamformer that is more robust against array imperfections, the proposed method is further extended by optimizing its WC performance, again using SOCP. Different from some conventional methods, which are restricted to linear arrays, this method is applicable to arbitrary array geometries since the weight vector, rather than its autocorrelation sequence, is used as the variable. This suggests a general framework for the design of beamformers of arbitrary array geometries that can satisfy different common sets of robustness requirements.

Recently, another approach to the design of robust adaptive beamforming, based on probability-constrained optimization, has been introduced [71]. Unlike the WC performance optimization approach, which is very conservative, the probability-constrained-based approach aims at providing robustness against steering vector errors with a certain preselected probability. The approach introduced [71] specifies the parameters of the uncertainty region in terms of the beamformer outage probability. It therefore leads to a better robust beamformer design by providing robustness against steering vector errors that are more likely to occur while discarding errors that occur with low probability. In this context, a new approach to the design of robust adaptive beamformers estimates the difference between the actual and presumed steering vectors in the sense that the output SINR is maximized [72]. It uses this difference to correct the erroneous presumed steering vector. The estimation process is performed iteratively, with a quadratic convex optimization problem solved at each iteration. In contrast to the WC performance-based [47, 57] and the probability-constrained-based approaches [71], this approach does not make any assumptions on either the norm of the mismatch vector or its probability distribution. Hence, it avoids the need for estimating their values.

Attractive approaches based on eigendecomposition of the sample covariance matrix have been introduced in several papers [49, 54, 58]. These approaches developed a closed-form solution for a WC robust detector using the Lagrange method, which incorporates the estimation of the norm of the weight vector and/or the Lagrange multiplier. A binary search algorithm followed by a Newton-like algorithm is proposed to estimate the norm of the weight vector after dropping the Lagrange multiplier. Although these approaches have provided closed-form solutions for the WC beamformer, they unfortunately have several difficulties. First, eigendecomposition for the sample covariance matrix is required with the arrival of a new array observation. Second, the inverse of the diagonally loaded sample covariance matrix is required to estimate the weight vector. Third, some difficulties are encountered during algorithm initialization and a stopping criterion is necessary to prevent negative solutions of the Newton-like algorithm.

The robust Capon beamformer (RCB) suffers from mutual coupling between the array sensors, which is difficult to neglect in practice. An autocalibration algorithm has been proposed to improve the RCB [63], by recursively calibrating the uncertainty due to mutual coupling. The proposed method is better able to estimate the angle-of-arrival and power of the signal of interest than the RCB. Conventional beamforming methods typically aim at maximizing the SINR [64]. However, this does not guarantee a small mean-squared error (MSE), so that on average the resulting signal estimate can be far from the true signal. Elder *et al.* consider strategies that attempt to minimize the MSE between the estimated and unknown signal waveforms [65]. The suggested methods maximize the SINR but at the same time are designed to have good MSE performance. Since the MSE depends on the signal power, which is unknown, we develop competitive beamforming approaches that minimize a robust MSE measure.

In the presence of large steering vector mismatches, the uncertainty set has to expand to accommodate the increased error. This degrades the output SINRs of these beamformers since their interference-plus-noise suppression abilities are weakened. In [124], an iterative robust minimum variance beamformer (IRMVB) is proposed. It uses a small uncertainty sphere (and a small flat ellipsoid) to search for the desired array steering vector iteratively. This preserves the interference-plus-noise suppression ability of the proposed beamformer by preserving its degrees of freedom and by using the corrected desired array steering vector. As a result, the IRMVB beamformer achieves higher output SINR than other beamformers [47–54]. The IRMVB applies two stopping criteria. Due to one of them, the steering vector calculated by the IRMVB is not allowed to converge to the steering vectors of interferences.

Fully parameter-free robust adaptive beamformers are scarce. One example is the HKB (after its developers, Hoerl, Kennard, and Baldwin [72]), which uses ridge regression based on the GSC parameterization of the standard Capon beamformer (SCB). This approach can be extended to other well-investigated methods in the literature, such as principal component regression (PCR) [73] and partial least squares (PLS) [74]. Another example uses general linear combination (GLC) shrinkage-based covariance matrix estimation [75], which has been demonstrated to be useful in the case of small sample sizes. Note that both HKB and GLC can be seen as diagonal loading algorithms. Yang *et al.* reformulated a general form of the loading matrix with automatic parameter determination [76]. This can be seen as a generalized loading approach in which a Hermitian matrix is loaded on a sample covariance matrix instead of a diagonal matrix. All the existing techniques –diagonal loading methods, PCR, and PLS and so on – can be seen as special cases of the general form. Based on this formulation, we also propose two special generalized loading algorithms. Simulation results show that the proposed methods are more robust to errors in the steering vector and sample covariance matrix than other tested parameter-free methods. The approach of Yang *et al.* [76] is extended to a fully automatic robust linear receiver technique for joint space-time decoding and interference rejection in multi-access MIMO systems. This uses orthogonal space-time block codes and erroneous channel state information (CSI) [77]. The proposed receiver does not need any *a-priori* knowledge of channel estimation errors and has a simple closed form.

The HKB algorithm may have an inherent problem in choosing an appropriate DL level, which makes its usefulness somewhat limited [82]. An alternative and simple approach to the fully automatic computation of the DL level is proposed by Ledoit and Wolf [82]. The conventional sample covariance matrix used in SCB is replaced by an enhanced estimate obtained via a shrinkage method (see [83] and the references therein). The enhanced estimate is obtained in turn by a GLC of the sample covariance matrix and the identity matrix in an optimal MSE sense. The idea of shrinkage-based covariance matrix estimation has also been applied in knowledge-aided signal processing, where prior knowledge is utilized to construct the enhanced covariance matrix estimate [84, 85]. The issue of DL and the performance comparisons of the GLC-based user parameter-free DL with other DL approaches, including HKB and the conventional DL methods are analyzed by Ledoit and Wolf [82].

This problem of robust beamforming for general-rank signal models with norm bounded uncertainties in the desired and received signal covariance matrices as well as positive semidefinite constraints on the covariance matrices is addressed by Xing *et al.* [78]. Two novel minimum variance robust beamformers are derived in closed-form. The first is a closed-form solution to the iterative beamforming algorithm of Shahbazpanahi *et al.* [54], thus offering the same performance as the iterative algorithm but with a much lower complexity. The second one provides even better performance than the first one since fewer approximations are made in the derivation. Note that both of the proposed beamformers have the advantage of low complexity due to the simple closed-form solutions and thus are attractive for use in practical systems.

Two new approaches to adaptive beamforming in sparse subarray-based partly calibrated sensor arrays are developed by Lei *et al.* [79]. Each subarray is assumed to be well calibrated, so that the steering vectors of all subarrays are exactly known. However, the intersubarray gain and/or phase mismatches are known imperfectly or remain completely unknown. The first approach is based on a WC beamformer design which, in contrast to the existing WC designs [47, 48, 58, 59], exploits a specific structured ellipsoidal uncertainty model for the signal steering vector rather than the commonly used unstructured uncertainty models. The second approach is based on estimating the unknown intersubarray parameters by maximizing the output power of the minimum variance beamformer subject to a suitable constraint that helps to avoid trivial solutions of the resulting optimization problem.

The Bayesian approach is a robust beamforming method [27, 80]. The desired DOA is assumed to be a discrete random variable of which there is *a-priori* knowledge [27]. However, when the Bayesian model does not hold, which means that the true DOA is deterministic and is not included in the prior, the Bayesian optimality is lost [80]. A Bayesian beamforming approach to deal with situations where the Bayesian model does not hold has been developed [81]. A support vector machine (SVM), which was first used to solve the sidelobe control problem [125], is used to obtain the optimal weights. The Bayesian beamforming problem has also been reformulated with an additional inequality constraint [81].

Traditional robust wideband beamforming is mainly achieved by extending narrowband robust beamforming methods to the wideband case through frequency sampling, with robust constraints imposed on different frequency points separately. The resultant beamformer may have different responses at different frequencies, causing intolerable distortion of the desired signal. To improve the frequency-response consistency of the beamformer against arbitrary array manifold errors, a novel robust beamformer is proposed with a closed-form solution. To extend the robustness to the general case with arbitrary array manifold errors, a robust beamformer with robust frequency-invariance constraints and WC performance optimization has been proposed [85]. In this approach, the frequency invariance property of the (mismatched) desired signal in the presence of arbitrary manifold errors is preserved. In addition, for the RB-WC, constraints with a fixed maximum norm value for the error vector are imposed on different sampled frequency points separately. However, the actual norm bound with respect to different frequencies could vary.

As a result, the beamformer will be over constrained for some frequencies. For the RB-RFI-WC, worst-case performance optimization is realized by bounding an error matrix including all sampled frequency-angle points. A higher optimum output SINR can be achieved due to increased number of degrees of freedom for interference suppression.

Khabbazibasmenj *et al.* have rethought the notion of robustness and present a unified approach to MVDR robust adaptive beamforming (RAB) design; that is, to use a standard MVDR beamformer in tandem with steering vector estimation based on prior information and data covariance matrix estimation [87]. Motivated by this unified approach, a new technique is developed. This uses as little as possible of the imprecise, and easy-to-obtain prior information about the desired signal/source, the antenna array, and the propagation media. The new RAB technique involves estimation of the steering vector through beamformer output power maximization under the requirement that the estimate does not converge to any of the interference steering vectors and their linear combinations. The only prior information used is the imprecise knowledge of the angular sector of the desired signal and the antenna array geometry, while knowledge of the presumed steering vector is not needed. Such MVDR RAB techniques can be mathematically formulated as a non-convex (due to an additional steering vector normalization condition) quadratically constrained quadratic programming (QCQP) problem. This problem has been considered in the literature for decades and can be solved using the semi-definite programming relaxation technique. Moreover, this specific optimization problem allows for an exact solution using, for example, the duality theory [88] or the iterative rank reduction technique [89]. In the optimization context, several new results are developed when answering the questions of:

- how to obtain a rank-one solution from a general-rank solution of the relaxed problem algebraically
- when it is guaranteed that the solution of the relaxed problem is rank-one.

The latter question, for example, is important because it has been observed that the probability of obtaining a rank-one solution for the class of problems similar to the one considered by Khabbazibasmenj *et al.* [87] is close to 1, while the theoretical upper bound suggests a significantly smaller probability [90]. The result of Khabbazibasmenj *et al.* proves the correctness of the experimental observations about the high probability of a rank-one solution for the relaxed problem.

In a recent study [91], a new beamformer algorithm that is robust against DOA mismatches and gives excellent performance, even with small sample sizes, is proposed. In this method, the desired degree of freedom and the range of the angular location of the desired signal (DS) are first defined. Based on the desired degree of freedom, the array is divided into subarrays. Subsequently, the subarray data are combined with a weighting vector, which is orthogonal to the subspace of possible steering vectors of the DS. As a result, the new data are free of the DS. It is shown that this DS blocking process is equivalent to a simple linear transformation. The subarray covariance matrix is estimated using this new data and then the subarray beamforming vector is computed. Ultimately, the final beamforming vector is computed from the subarray beamforming vector. The proposed method decreases the degree of freedom and gives a high convergence rate. Based on the proposed subarray technique, this algorithm is applicable to arrays where the structure can be synthesized by scrolling the subarrays. The main drawback is the degree of freedom selection, which can significantly degrade the performance if it not properly selected.

In this chapter, we will present four approaches for robust adaptive beamforming. We first develop an improved recursive realization for robust LCMV beamforming. This includes an uncertainty ellipsoidal constraint on the steering vector. The robust recursive implementation presented here is based on a combination of the ellipsoidal constraint formulation and the variable diagonal loading technique outlined in Chapter 5. As a consequence, an accurate technique for computing the diagonal loading level without eigendecomposition or SOCP is developed. The geometrical interpretation of the diagonal loading technique will be demonstrated and compared with the eigendecomposition approach. Note that this approach adopts a spherical constraint on the steering vector to optimize the beamformer output power [2]. Unfortunately, the adaptive beamformer developed here is apt to noise enhancement at low SNR and an additional constraint is required to bear the ellipsoidal constraint [2].

The second approach is the development of a joint constraint approach for a joint robustness beamformer. A joint constraint approach has previously been presented for joint robustness against steering vector mismatches and unstationarity of interferers [66, 67]. An alternative approach involves imposing an ellipsoidal uncertainty constraint and a quadratic constraint on the steering vector and the beamformer weights, respectively. We introduce a new, simple approach to get the corresponding diagonal loading value. The quadratic constraint is invoked as a cooperative constraint to overcome noise enhancement at low SNR. The performance of the presented robust adaptive schemes as well as other robust approaches are demonstrated in scenarios involving steering vector mismatch and moving jammers.

In the third approach, the robust MVDR beamformer with a single WC constraint is implemented using an iterative gradient minimization algorithm. It uses a simple technique to estimate the Lagrange multiplier instead of the Newton-like algorithm. As well as its simplicity, and low computational load, this simple algorithm does not need either sample-matrix inversion or eigendecomposition. A geometric interpretation of the robust MVDR beamformer is demonstrated to supplement the theoretical analysis.

In the last approach, a robust linearly constrained minimum variance (LCMV) beamformer with multiple-beam WC (MBWC) constraints is developed using a novel multiple WC constraint formulation. The Lagrange method is exploited to solve this optimization problem. The solution of the robust LCMV beamformer with MBWC constraints entails solving a set of nonlinear equations. As a consequence, a Newton-like method is mandatory to solve the system of nonlinear equations, which yields a vector of Lagrange multipliers.

The rest of this chapter is organized as follows. In Section 7.2, the standard MVDR and LCMV beamformers with single and multiple constraints are summarized in the context of a single point source and a source with multipath rays, respectively. In

Section 7.3, the WC optimization formulation is introduced by summarizing general and special formulations for the steering vector uncertainty set. Efficient implementations of single- and multiple-WC formulations are derived and analyzed in Section 7.4. A geometric illustration for the single-WC implementation is presented. Simulations and a performance analysis are provided in Section 7.5. Conclusions and points for future work are encapsulated in Section 7.6.

7.2 Beamforming Formulation

7.2.1 Capon Beamforming

In this section, the conventional Capon or MVDR beamformer will be presented. Let us consider a linear array with M sensors. The array received signal vector is an $M \times 1$ vector and is given by:

$$\mathbf{x}(k) = \sum_{i=0}^{L-1} \mathbf{a}_i \mathbf{s}_i(k) + \mathbf{v}(k) \tag{7.1}$$

where $\mathbf{s}_i(k)$ is the kth sample transmitted by ith user, L is the number of sources (users), $\boldsymbol{a}_i$ is the $M \times 1$ complex array steering vector of the ith user, and $\boldsymbol{n}(k)$ is the complex vector of ambient channel noise samples, assumed to be independent and identically distributed (iid) random variables with a Gaussian distribution. That is:

$$\boldsymbol{v}(k) = N(0, \sqrt{\sigma_v^2/2}) + iN(0, \sqrt{\sigma_v^2/2}) \tag{7.2}$$

where σ_v^2 is the noise power, and i and N stand for the complex term and the Gaussian random generator, respectively. The antenna array response vector $\boldsymbol{a}_i(k)$ can be estimated as follows [2]:

$$\boldsymbol{a}_i(\theta) = \left[1 \;\; e^{-j2\pi \sin\theta(k)\frac{d}{\lambda}} \;\; \cdots \;\; e^{-j(N-1)2\pi \sin\theta(k)\frac{d}{\lambda}}\right]^T \tag{7.3}$$

where $\theta(k)$ is the incident angle (the DOA of the user of interest), d is the array element spacing, $\lambda = c/f_c$ is the wavelength for a given carrier frequency f_c and c is the speed of light. $[.]^T$ stands for the transpose. The beamformer output is a linear combination of the array sampled received signals at each sensor. This leads to:

$$y(k) = \boldsymbol{w}^H(k)\boldsymbol{x}(k) \tag{7.4}$$

where the complex vector $\boldsymbol{w}^H(n)$ is the $M \times 1$ beamformer weights and $(\cdot)^H$ stands for the Hermitian transpose. The beamformer output energy is given by

$$E\{|y(k)|^2\} = E\{|\boldsymbol{w}^H\boldsymbol{x}(k)|^2\} = \boldsymbol{w}^H\boldsymbol{R}_{xx}(k)\boldsymbol{w} \tag{7.5}$$

where $\boldsymbol{R}_{xx}(k) = E\{\boldsymbol{x}(k)\boldsymbol{x}^H(k)\}$ is the $M \times M$ autocorrelation matrix of the array data signals $\boldsymbol{x}(k)$. In practical applications, $\boldsymbol{R}_{xx}$ can be obtained using one of the following methods:

Sample covariance method

$$\boldsymbol{R}_{xx}(k) = \frac{1}{N}\sum_{n=1}^{N}\boldsymbol{x}(k)\boldsymbol{x}^H(k) \tag{7.6}$$

where N is the number of snapshots.

Exponentially decaying data window

$$\boldsymbol{R}_{xx}(k) = \eta\boldsymbol{R}(k-1) + \boldsymbol{x}(k)\boldsymbol{x}^H(k) \tag{7.7}$$

and η is the usual forgetting factor with $0 \ll \eta \leq 1$. Refer to Chapter 2 for more details.

The conventional LCMV beamformer can be obtained by minimizing the output energy of the beamformer subject to a certain number of constraints. To preclude the cancellation of a desired signal source during the minimization of the beamformer output energy, a linear constraint of the form $\boldsymbol{w}^H a_0(\theta_0) = 1$ is imposed throughout the optimization, where a_0 and θ_0 are the array steering vector and the DOA of the desired source, respectively. This is expressed mathematically as follows:

$$\min_{w} \boldsymbol{w}^H\boldsymbol{R}_{xx}\boldsymbol{w} \quad \text{subject to} \quad \boldsymbol{w}^H\boldsymbol{a}_0(\theta_0) = 1 \tag{7.8}$$

The well-known Capon beamformer is the solution to (7.8) and can be obtained using Lagrange method as follows [2]:

$$\boldsymbol{w}_0 = \frac{\boldsymbol{R}_{xx}^{-1}\boldsymbol{a}_0(\theta_0)}{\boldsymbol{a}_0^H(\theta_0)\boldsymbol{R}_{xx}^{-1}\boldsymbol{a}_0(\theta_0)} \tag{7.9}$$

A major drawback of conventional beamforming is its sensitivity to mismatches between presumed and genuine steering vectors. This mismatch can have one of the following causes:

- DOA mismatch
- array calibration errors
- the near–far effect
- other mismatch and modeling errors.

In such scenarios, a robust adaptive beamforming technique is required to improve the robustness of conventional beamforming against mismatch errors. In the following section, a modified constraint formulation will be imposed to improve the robustness of the conventional beamformer.

7.2.2 LCMV Beamforming

Consider an array comprising M uniformly spaced sensors, which receives a narrowband signal $\boldsymbol{s}_d(k)$. Initially, we assume that the desired signal is a point source with a time-invariant wavefront. The $M \times 1$ vector of array observations can be modeled as [55, 56, 97]:

$$\boldsymbol{x}(k) = \boldsymbol{a}_d(\varphi)s_d(k) + \boldsymbol{i}(k) + \boldsymbol{n}(k) \tag{7.10}$$

where k is the time-index, $s_d(k)$ is the complex signal waveform of the desired signal, and $\boldsymbol{a}_d(\varphi)$ is its $M \times 1$ steering vector, where φ is the angle-of-incidence (AOI) and $\boldsymbol{i}(k)$ and $\boldsymbol{n}(k)$ are the statically independent components of the interference and the noise, respectively.

A generalized model with multipath propagation can be expressed as follows:

$$\boldsymbol{x}(k) = s_d(k)\sum_{n=1}^{L} \gamma_n \boldsymbol{a}_d(\varphi + \phi_n) + \boldsymbol{i}(k) + \boldsymbol{n}(k) \tag{7.11}$$

where L is the number of multipaths, and each path has a random complex gain γ_n and an angular deviation ϕ_n from the nominal AOI φ. The scattered signals associated with the multipath propagation from a single source arrive at the base station (BS) from several directions, within an angular region called the angular spread. The angular spread arises due to the multipath propagation, both from local scatterers near to the source and near to the

BS, and from remote scatters as well. It varies according to the cell morphological type (dense urban, urban or rural), cell radius, BS location, and antenna height. It can vary from few degrees in rural road cells to 360° in microcellular and indoor environments due to the reflecting surfaces surround the BS antenna.

We assume that the time delays of the different multipath components are small compared to the inverse of the signal bandwidth (that is, a narrowband channel model) and therefore the delay can be modeled as a phase shift in the complex gain γ_n [98]. The angular spread is used here to describe the angular region associated with entire multipaths. However, each of the rays itself may be composed of a large number of "mini-rays", with roughly equal angles and delays but with arbitrary phases due to scattering close to the source [99]. Here the model is simplified by using the nominal AOI of each ray group and multipath delays are modeled as small angles in the complex gain.

The beamformer output signal can be written as

$$y(k) = \boldsymbol{w}^H(k)\boldsymbol{x}(k) \tag{7.12}$$

where $\boldsymbol{x}(k) = [x_1(k), \ldots, x_M(k)]^T$ is a $M \times 1$ complex vector of the array observations, $\boldsymbol{w}(k) = [w_1(k), \ldots, w_M(k)]^T$ is a $M \times 1$ complex vector of the beamformer weights, and $(\cdot)^T$ and $(\cdot)^H$ stand for the transpose and Hermitian transpose, respectively.

Let us consider the simplified model in (7.11) with the point source. The optimal weight vector seeks maximization of the output SINR [49, 55, 56, 60]:

$$SINR = \frac{\sigma_d^2|\boldsymbol{w}^H\boldsymbol{a}_d|^2}{\boldsymbol{w}^H\boldsymbol{R}_{i+n}\boldsymbol{w}} \tag{7.13}$$

where $\boldsymbol{R}_{i+n} \triangleq E\{(\boldsymbol{i}(k) + \boldsymbol{n}(k))(\boldsymbol{i}(k) + \boldsymbol{n}(k))^H\}$ is the interference-plus-noise covariance matrix and σ_d^2 is the desired source power. The optimal solution of $\boldsymbol{w}$ that maximizes the output SINR in (7.13) can be obtained by maintaining a distortionless response to the desired source while minimizing the output interference-plus-noise power $(\boldsymbol{w}^H\boldsymbol{R}_{i+n}\boldsymbol{w})$. In practical applications, the interference-plus-noise covariance matrix can be replaced by the sample covariance matrix [2, 3, 56], which can be estimated using the first-order recursion:

$$\widehat{\boldsymbol{R}}(n) = \sum_{i=1}^{n} \eta^{n-i}\boldsymbol{x}(i)\boldsymbol{x}^H(i) = \eta\widehat{\boldsymbol{R}}(n-1) + \boldsymbol{x}(n)\boldsymbol{x}^H(n) \tag{7.14}$$

where η is a forgetting factor which satisfies $0 \ll \eta \leq 1$. The minimum variance beamformer with single-beam constraint (SBC) can be formulated as follows:

$$\min_{\boldsymbol{w}} \boldsymbol{w}^H \widehat{\boldsymbol{R}} \boldsymbol{w} \text{ subject to } \boldsymbol{w}^H \boldsymbol{a}_d = 1 \tag{7.15}$$

The solution of (7.15) gives the standard MVDR beamformer with SBC and can be easily derived as:

$$\boldsymbol{w}_{SBC} = \frac{\widehat{\boldsymbol{R}}^{-1} \boldsymbol{a}_d}{\boldsymbol{a}_d^H \widehat{\boldsymbol{R}}^{-1} \boldsymbol{a}_d} \tag{7.16}$$

By considering the generalized received signal model in (7.11), the optimum MVDR beamformer can be obtained using multiple constraints to provide a multiple-beam constraint (MBC) beamformer:

$$\min_{\boldsymbol{w}} \boldsymbol{w}^H \widehat{\boldsymbol{R}} \boldsymbol{w} \text{ subject to } \boldsymbol{w}^H \Lambda(\theta_0) = \boldsymbol{v} \tag{7.17}$$

where $\theta_0 = [\theta_1 \ \cdots \ \theta_L]$, $\Lambda(\theta_0) = [\boldsymbol{a}(\theta_1) \ \cdots \ \boldsymbol{a}(\theta_L)]$ is the $M \times L$ spatial constraint matrix consisting of the steering vectors corresponding to the AOIs of the multipath components associated with the desired source, and $\boldsymbol{v}$ is a vector of constrained values (the gain vector). The latter can be set to all one vector for equal gain combining or alternatively it can be optimized using the maximal ratio combining (MRC) technique. Accordingly, the optimal weight vector of (7.18), termed the linearly constrained minimum variance (LCMV) beamformer, is given by [49]:

$$\boldsymbol{w}_{MBC} = \widehat{\boldsymbol{R}}^{-1} \Lambda (\Lambda^H \widehat{\boldsymbol{R}}^{-1} \Lambda)^{-1} \Lambda \tag{7.18}$$

7.3 Robust Beamforming Design

A generalization of (7.8) is based on the minimization of the weighted power output of the array in the presence of uncertainties in the steering vector as follows [48]:

$$\min_{\boldsymbol{w}} \boldsymbol{w}^H \boldsymbol{R}_{xx} \boldsymbol{w} \text{ subject to } |\boldsymbol{w}^H \boldsymbol{c}| \geq 1 \quad \forall \boldsymbol{c} \in \boldsymbol{A}(\varepsilon) \tag{7.19}$$

The genuine steering vector $\boldsymbol{a}_0(\theta_0)$ belongs to $A(\varepsilon)$

$$\boldsymbol{A}(\varepsilon) = \{\boldsymbol{c} | \boldsymbol{c} = \overline{\boldsymbol{a}}_0 + e, \|e\| \leq \varepsilon\} \tag{7.20}$$

where $\overline{\boldsymbol{a}}$ is the presumed steering vector and ε is an ellipsoid that covers the possible range of values of $\boldsymbol{a}_0(\theta)$ due to imprecise knowledge of the array steering vector (that is, uncertainty in the DOA) and other mismatch errors. Unfortunately, according to the constraint $|\boldsymbol{w}^H\boldsymbol{c}| \geq 1$, the solution of (7.19) represents a nonlinear and nonconvex optimization problem since there are an infinite number of vectors $\boldsymbol{c}$ in $A(\varepsilon)$ (that is, there is an infinite number of constraints that satisfy $|\boldsymbol{w}^H\boldsymbol{c}| \geq 1$). Vorobyov *et al.* [48] reformulate (7.19) as a convex SOCP problem as follows:

$$\min_{\boldsymbol{w}} \boldsymbol{w}^H\boldsymbol{R}_{xx}\boldsymbol{w} \text{ subject to } |\boldsymbol{w}^H\overline{\boldsymbol{a}}_0| \geq \varepsilon\|\boldsymbol{w}\| + 1$$
$$\text{Im}\{\boldsymbol{w}^H\overline{\boldsymbol{a}}_0\} = 0 \tag{7.21}$$

Other well-known MVDR RAB techniques are summarized in Table 7.1 [87]. The notions of robustness and prior information used by each techniques are listed. It can be seen from Table 7.1 that the main problem for any MVDR RAB technique is to estimate the steering vector while avoiding its convergence to any of the interferences and their linear combinations. This is achieved in different techniques by exploiting different prior information and solving different optimization problems. The complexity of these steering vector estimation problems can vary from the complexity of eigenvalue decomposition to the complexity of solving QCQP programming problems. All known MVDR RAB techniques require knowledge of the presumed steering vector, which implies that the antenna array geometry, propagation media, and desired source characteristics, such as the presumed angle of arrival, are known.

Although, the problem in (7.21) is much simpler and is converted to a positive semi-definite convex problem, it requires $O(M^6)$ flops using SeDuMi software [53]. This prohibitive complexity restricts the practical implementation of this technique. Additionally, the Matlab script provided by the authors in [143] requires a very long time to produce the robust detector. Moreover, the weight vector of this beamformer cannot be easily updated and has to be recomputed in each step. Therefore, this technique can be interrupted as a batch algorithm rather than as an adaptive scheme.

Table 7.1 MVDR RAB techniques, their notions of robustness and prior information used.

RAB	Robustness concept	Prior information	Limitation
Eigenspace-based beamformer	The projection of the presumed steering vector to signal-plus-interference subspace	The presumed steering vector and the number of interfering signals	High probability of subspace swap and incorrect estimation of the signal-plus-interference subspace dimension at low SNR
Worst-case and doubly constrained robust beamformer	The presumed steering vector and its uncertainty region	The presumed steering vector and the uncertainty bounding value	Difficulty to obtain the uncertainty bound value in practice
Probabilistically constrained robust beamformer	The non-outage probability; the steering vector is modeled as presumed vector plus a random mismatch vector with known or worst-case distribution	The presumed steering vector, the preselected non-outage probability value, and possibly distribution of the random mismatch vector	The value of probability constrained value and mismatch distribution may not be known.
Eigenvalue robust beamformer	General and the steering vector lies in a known signal subspace and the rank of the signal correlation matrix is known	The linear subspace in which the desired signal is located and the rank of the desired signal covariance matrix	Very specific modeling of the covariance matrix. The signal subspace has to be known.

Li *et al.* [51, 52] have proposed a different approach that gives robust Capon beamforming with $O(M^3)$ flops. This robust Capon beamforming can be obtained by maximizing the beamformer output power after the interference has been rejected. Consequently, the following max/min formulation for the robust Capon beamformer is obtained (see also Tsatsanis and Xu [100] for similar approach with multiuser detection):

$$\max_{\hat{\boldsymbol{a}}_0} \min_{\boldsymbol{w}} \boldsymbol{w}^H \boldsymbol{R}_{xx} \boldsymbol{w} \text{ subject to } \boldsymbol{w}^H \hat{\boldsymbol{a}}_0(\theta_0) = 1;$$
$$(\hat{\boldsymbol{a}}_0(k) - \overline{\boldsymbol{a}}_0)^H \boldsymbol{C}^{-1} (\hat{\boldsymbol{a}}_0(k) - \overline{\boldsymbol{a}}_0) \leq 1 \tag{7.22}$$

This optimization problem can be simplified without loss of generality by setting $\boldsymbol{C}^{-1} = \frac{1}{\varepsilon}\boldsymbol{I}$ and solving the inner optimization problem via injection of the standard robust Capon beamformer (7.9) into (7.22), to get

$$\max_{\hat{\boldsymbol{a}}_0} \frac{1}{\hat{\boldsymbol{a}}_0^H(k) \boldsymbol{R}_{xx}^{-1}(k) \hat{\boldsymbol{a}}_0(k)} \text{ subject to } (\hat{\boldsymbol{a}}_0(k) - \overline{\boldsymbol{a}}_0)^H$$
$$(\hat{\boldsymbol{a}}_0(k) - \overline{\boldsymbol{a}}_0) \leq \varepsilon \tag{7.23}$$

where $\overline{\boldsymbol{a}}_0$ and ε are given and $\hat{\boldsymbol{a}}_0(k)$ is the estimated (genuine) array response vector with mismatch and other errors encountered. The optimization problem in (7.23) can be rewritten as follows:

$$\min_{\hat{a}} \hat{\boldsymbol{a}}_0^H(k) \boldsymbol{R}_{xx}^{-1}(k) \hat{\boldsymbol{a}}_0(k) \text{ subject to } \|\hat{\boldsymbol{a}}_0(k) - \overline{\boldsymbol{a}}_0\|^2 \leq \varepsilon \tag{7.24}$$

The ellipsoidal uncertainty set is reduced to a spherical constraint. This problem can be solved using the Lagrange method by forming the following cost function

$$\Psi_{\hat{\boldsymbol{a}}}(k) = \hat{\boldsymbol{a}}_0^H(k) \boldsymbol{R}_{xx}^{-1}(k) \hat{\boldsymbol{a}}_0(k) + \lambda t(\|\hat{\boldsymbol{a}}_0(k) - \overline{\boldsymbol{a}}_0\|^2 - \varepsilon) \tag{7.25}$$

where $t()$ is the step function and Lagrange multiplier λ is a real scalar determined from the spherical constrained value ε. The problem is converted from a constrained minimization to an unconstrained minimization problem and the solution is given by [51, 52]:

$$\hat{\boldsymbol{a}}_0 = \left(\frac{\boldsymbol{R}_{xx}^{-1}(k)}{\lambda} + \boldsymbol{I} \right)^{-1} \overline{\boldsymbol{a}}_0 \tag{7.26}$$

It should be noted that the step function is introduced to guarantee that the term $(\|\widehat{\boldsymbol{a}}_0(k) - \overline{\boldsymbol{a}}_0\|^2 - \varepsilon)$ is positive. Therefore, during optimization the step function will not be differentiated. Additionally, the diagonal loading term λ should be positive to absolutely guarantee the positive definiteness of matrix $\left(\frac{\boldsymbol{R}_{xx}^{-1}(k)}{\lambda} + \boldsymbol{I}\right)$.

The robust Capon beamformer[1] $\widehat{\boldsymbol{w}}_0$ can be obtained by injecting (7.26) into (7.9); that is:

$$\widehat{\boldsymbol{w}}_0 = \frac{\boldsymbol{R}_{xx}^{-1}\left(\frac{\boldsymbol{R}_{xx}^{-1}(k)}{\lambda} + \boldsymbol{I}\right)^{-1}\overline{\boldsymbol{a}}_0}{\left(\frac{\boldsymbol{R}_{xx}^{-1}(k)}{\lambda} + \boldsymbol{I}\right)^{-1}\overline{\boldsymbol{a}}_0\boldsymbol{R}_{xx}^{-1}\left(\frac{\boldsymbol{R}_{xx}^{-1}(k)}{\lambda} + \boldsymbol{I}\right)^{-1}\overline{\boldsymbol{a}}_0} \tag{7.27}$$

Regrettably, the value of the diagonal loading term λ cannot be easily estimated where there is no closed-form expression for the optimal loading level. Li *et al.* have described a batch algorithm for solving this problem based on eigendecomposition to $\boldsymbol{R}_{xx}(k)$ [51]. They apply the matrix inversion lemma to (7.26). This yields:

$$\widehat{\boldsymbol{a}}_0 = \overline{\boldsymbol{a}}_0 - (\boldsymbol{I} + \lambda\boldsymbol{R}_{xx})^{-1}\overline{\boldsymbol{a}}_0 \tag{7.28}$$

Therefore, based on the spherical constraint in (7.24), the Lagrange multiplier λ can be obtained as the solution to the following constraint equation:

$$g(\lambda) \triangleq \|(I + \lambda\boldsymbol{R}_{xx})^{-1}\overline{\boldsymbol{a}}\|^2 = \varepsilon \tag{7.29}$$

Applying eigendecomposition to $\boldsymbol{R}_{xx} = \boldsymbol{U}\boldsymbol{\Gamma}\boldsymbol{U}^H$, where $\boldsymbol{U}$ contains the eigenvectors and the diagonal of Γ contains the eigenvalues denoted by $\gamma_1 \geq \gamma_2 \geq \cdots \geq \gamma_m$, and letting $\boldsymbol{z} = \boldsymbol{U}^H\overline{\boldsymbol{a}}_0$ where z_m denotes the mth element of $\boldsymbol{z}$, allows (7.29) to be reformulated as:

$$g(\lambda) \triangleq \sum_{j=1}^{M} \frac{|z_m|^2}{(1 + \lambda\gamma_m)^2} = \varepsilon \tag{7.30}$$

The diagonal loading term can be computed by solving (7.30) using a Newton-like algorithm. To prevent negative, zero,

1 This robust beamformer can be regarded as a Capon beamformer because it includes steering vector optimization.

or complex solutions of (7.30), the following assumption is mandatory:

$$\|\bar{\boldsymbol{a}}_0\|^2 > \varepsilon \tag{7.31}$$

This technique has the following major limitations. First, eigendecomposition has a high computational burden, of order $O(M^3)$. Additionally, the recursive implementation updates both the covariance matrix and its inverse to compute the diagonal loading value and the robust detector, respectively. Second, the technique is based on a batch algorithm and there is no clear vision for its recursive implementation, even with subspace tracking techniques. This is because subspace tracking algorithms adaptively update the eigenvectors and then, at a certain snapshot, the current estimate is not the optimal eigenvectors and therefore the diagonal loading term will not be precisely calculated. This may lead to error accumulation. Furthermore, the rank of signal and noise may be uncertain or not exactly known and need to be estimated in advance [101]. As a consequence, this technique is expected to fail in a dynamic interference scenario. Third, the technique reduces the inequality in (7.24) to an equality, which is not always acceptable. More specifically, the covariance matrix will be diagonally loaded even without mismatch. However, the assumption given in (7.31) is generally acceptable for most practical cases, but in the general ellipsoidal constraint ($\boldsymbol{C}^{-1} = \boldsymbol{D}^H\boldsymbol{D}/\varepsilon$) it cannot be guaranteed. Therefore, dropping this assumption is technically interesting.

7.3.1 Adaptive Implementation

The first robust approach in this chapter is to reformulate the diagonal loading approach presented in Chapter 5 [2, 31, 43–46] to manage the ellipsoidal uncertainty constraint on the steering vector and hence to obtain the genuine steering vector from the presumed steering vector. The optimal diagonal loading term λ will be precisely computed with low computational complexity using a variable loading approach instead of the eigendecomposition or SOCP batch approaches. The methods of steepest descent (SD) or conjugate gradient (CG) [102, 103] are used to adaptively update the genuine steering vector that minimizes the Lagrangian functional (7.25); that is, respectively:

$$\widehat{\boldsymbol{a}}_0(k) = \widehat{\boldsymbol{a}}_0(k-1) - \mu_{SD}(n)\boldsymbol{g}(k) \tag{7.32}$$

$$\widehat{\boldsymbol{a}}_0(k) = \widehat{\boldsymbol{a}}_0(k-1) - \mu_{CG}(n)\boldsymbol{p}(n) \tag{7.33}$$

where $\mu_{sd}(n), \mu_{cg}(n)$ are the step-sizes of the SD and CG algorithm respectively and $\boldsymbol{g}(k)$ is the conjugate derivative of $\boldsymbol{\Psi}_{\hat{a}}(k)$ with respect to the real and imaginary parts of $\widehat{\boldsymbol{a}}_0^H(n)$ and is given by:

$$\boldsymbol{g}(k) = \boldsymbol{R}_{xx}^{-1}(k)\widehat{\boldsymbol{a}}_0(k-1) + \lambda(\widehat{\boldsymbol{a}}_0(k-1) - \overline{\boldsymbol{a}}) \tag{7.34}$$

The direction vector $\boldsymbol{p}(n)$ is given by [102]

$$\boldsymbol{p}(k+1) = \boldsymbol{g}(k) + \omega(k)\boldsymbol{p}(k) \tag{7.35}$$

where

$$\omega(k) = \frac{(\boldsymbol{g}(k) - \boldsymbol{g}(k-1))^H \boldsymbol{g}(k)}{\boldsymbol{g}(k-1)^H \boldsymbol{g}(k-1)} \tag{7.36}$$

The optimal step sizes of the SD and CG algorithms can be obtained, respectively, by substituting from (7.32) or (7.33) into (7.25). The constructed cost function is a quadratic function in the step size and hence it has a global minimum. As a consequence, the step-sizes of the SD and CG algorithms are respectively given by [2, 102]:

$$\mu_{SD}(k) = \frac{\alpha \boldsymbol{g}^H(k)\boldsymbol{g}(k)}{\boldsymbol{g}^H(k)\boldsymbol{R}_{xx}^{-1}(k)\boldsymbol{g}(k) + \sigma} \tag{7.27}$$

$$\mu_{CG}(n) = \frac{\alpha \boldsymbol{p}^H(k)\boldsymbol{g}(k-1)}{\boldsymbol{p}^H(k)\boldsymbol{R}_{xx}^{-1}(k)\boldsymbol{p}(k) + \sigma} \tag{7.28}$$

where α, σ are two positive constants added to improve the numerical stability of the algorithm and to avoid dividing by zero, respectively. The constant α acts as dominant controller of the step-size range and it should be suitably selected. The constant σ can be set to a very small value (i.e. 10^{-6}).

The adaptive implementation of $\widehat{\boldsymbol{a}}_0(k)$ using th SD method can be obtained by substituting from (7.34) into (7.32) as follows:

$$\widehat{\boldsymbol{a}}_0(k) = \widehat{\boldsymbol{a}}_0(k-1) - \mu_{SD}(k)(\boldsymbol{R}_{xx}^{-1}(k)\widehat{\boldsymbol{a}}_0(k-1) + \lambda(\widehat{\boldsymbol{a}}_0(k-1) - \overline{\boldsymbol{a}}_0)) \tag{7.39}$$

Correspondingly, for the CG method, the adaptive implementation of $\widehat{\boldsymbol{a}}_0(k)$ is obtained by updating $\boldsymbol{p}(n+1)$ using (7.35) and hence updating $\widehat{\boldsymbol{a}}_0(k+1)$ in the next iteration.

The spherical constraint in (7.25) should be satisfied at each iteration step: $\|\widehat{\boldsymbol{a}}_0(k) - \overline{\boldsymbol{a}}_0\|^2 \leq \varepsilon$. Assuming the constraint was satisfied on the previous snapshot and using (7.39) we get:

$$((\tilde{\boldsymbol{a}}_0(k) - \mu_{SD}(k)\lambda(\widehat{\boldsymbol{a}}_0(k-1) - \overline{\boldsymbol{a}}_0)) - \overline{\boldsymbol{a}}_0)^H \\ ((\tilde{\boldsymbol{a}}_0(k) - \mu_{SD}(k)\lambda(\widehat{\boldsymbol{a}}_0(k-1) - \overline{\boldsymbol{a}}_0)) - \overline{\boldsymbol{a}}_0) \leq \varepsilon \quad (7.40)$$

where

$$\tilde{\boldsymbol{a}}_0(k) = \widehat{\boldsymbol{a}}_0(k-1) - \mu_{SD}\boldsymbol{R}_{xx}^{-1}(k)\widehat{\boldsymbol{a}}_0(k-1) \quad (7.41)$$

or using CG method:

$$\tilde{\boldsymbol{a}}_0(k) = \widehat{\boldsymbol{a}}_0(k-1) - \mu_{CG}\boldsymbol{p}(k) \quad (7.42)$$

Let us define the following two new vectors:

$$\overline{\boldsymbol{d}}_0(k) = \tilde{\boldsymbol{a}}_0(k) - \overline{\boldsymbol{a}} \quad (7.43)$$

$$\boldsymbol{d}_0(k) = \widehat{\boldsymbol{a}}_0(k) - \overline{\boldsymbol{a}} \quad (7.44)$$

The two vectors $\overline{\boldsymbol{d}}_0(k), \boldsymbol{d}_0(k)$ represent the unconstrained and the constrained difference vectors, respectively. Subsequently, the constrained difference vector satisfies $\|\boldsymbol{d}_0(k)\|^2 \leq \varepsilon$. Equation (7.30) is then reduced to the following simple form:

$$(\overline{\boldsymbol{d}}_0(k) - \mu_{SD|CG}\lambda\boldsymbol{d}_0(k-1))^H(\overline{\boldsymbol{d}}_0(k) - \mu_{SD|CG}\lambda\boldsymbol{d}_0(k-1)) \leq \varepsilon \quad (7.45)$$

Therefore, it is easily solved [2]. Subsequently, the value of λ that satisfies the ellipsoidal constraint is:

$$\lambda = \frac{b \pm \sqrt{b^2 - ac}}{a} \quad (7.46)$$

where

$$a = \mu^2\|\boldsymbol{d}_0(k-1)\|^2 > 0 \quad (7.47a)$$

$$b = \mu_{sd}(n)\mathrm{Re}\{\overline{\boldsymbol{d}}_0^H(k)\boldsymbol{d}_0(k-1)\} \quad (7.47b)$$

$$c = \|\overline{\boldsymbol{d}}_0(k)\|^2 - \varepsilon > 0 \quad (7.47c)$$

Similar to the approach adopted in Chapter 5 [2], to avoid a complex-roots solution to (7.45), we impose the following inequality constraint on the step size:

$$[\mathrm{Re}\{(\boldsymbol{d}_0(k-1)-\mu_{SD}\overline{\boldsymbol{g}}(k))^H(\boldsymbol{d}_0(k-1))\}]^2 \geq \|\boldsymbol{d}_0(k-1)\|^2 \\ [(\boldsymbol{d}_0(k-1)-\mu_{SD}\overline{\boldsymbol{g}}(k))^H(\boldsymbol{d}_0(k-1)-\mu_{SD}\overline{\boldsymbol{g}}(k))-\varepsilon] \tag{7.48}$$

where

$$\overline{\boldsymbol{g}}(k)=\boldsymbol{R}_{xx}^{-1}(k)\widehat{\boldsymbol{a}}_0(k-1) \tag{7.49}$$

and

$$\boldsymbol{g}(k)=\overline{\boldsymbol{g}}(k)+\lambda(\widehat{\boldsymbol{a}}_0(k-1)-\overline{\boldsymbol{a}}) \tag{7.50}$$

In order to simplify (7.48), which contains complex values, we have to convert it to a real-valued inequality by setting:

$$\breve{\boldsymbol{d}}_0(k-1)\triangleq[\mathrm{Re}\{\boldsymbol{d}_0(k-1)\}^T,\mathrm{Im}\{\boldsymbol{d}_0(k-1)\}^T] \tag{7.51}$$

$$\breve{\boldsymbol{g}}(k)=[\mathrm{Re}\{\overline{\boldsymbol{g}}(k)\}^T,\mathrm{Im}\{\overline{\boldsymbol{g}}(k)\}^T] \tag{7.52}$$

Therefore, substituting (7.51) and (7.52) into (7.48) and converting it to real-valued inequality, we get:

$$[(\breve{\boldsymbol{d}}_0(k-1)-\mu_{SD}\breve{\boldsymbol{g}}(k))^T(\breve{\boldsymbol{d}}_0(k-1))]^2 \geq \|\breve{\boldsymbol{d}}_0(k-1)\|^2 \\ [(\breve{\boldsymbol{d}}_0(k-1)-\mu_{SD}\breve{\boldsymbol{g}}(k))^T(\breve{\boldsymbol{d}}_0(k-1)-\mu_{SD}\breve{\boldsymbol{g}}(k))-\varepsilon] \tag{7.53}$$

After some manipulations to (7.53), the step-size upper bound inequality can be obtained as in Chapter 5:

$$\mu_{SD} \leq \sqrt{\frac{\varepsilon\|\breve{\boldsymbol{d}}_0(k-1)\|^2}{\|\breve{\boldsymbol{d}}_0(k-1)\|^2\|\breve{\boldsymbol{g}}(k)\|^2-\breve{\boldsymbol{g}}^T(k)\breve{\boldsymbol{d}}_0(k-1)\breve{\boldsymbol{d}}_0(k-1)^T\breve{\boldsymbol{g}}(k)}} \tag{7.54}$$

and similarly, for CG algorithm:

$$\mu_{CG} \leq \sqrt{\frac{\varepsilon\|\breve{\boldsymbol{d}}_0(k-1)\|^2}{\|\breve{\boldsymbol{d}}_0(k-1)\|^2\|\breve{\boldsymbol{p}}(k)\|^2-\breve{\boldsymbol{p}}^H(k)\breve{\boldsymbol{d}}_0(k-1)\breve{\boldsymbol{d}}_0(k-1)^H\breve{\boldsymbol{p}}(k)}} \tag{7.55}$$

where $\breve{\boldsymbol{p}}(k)=[\mathrm{Re}\{\boldsymbol{p}(k)\}^T,\mathrm{Im}\{\boldsymbol{p}(k)\}^T]$. Therefore, this upper bound on $\mu_{SD|CG}$ guarantees real positive roots for (7.55) and consequently the optimal diagonal loading level can be obtained. Moreover, based on the well-known Cauchy-Schwarz inequality [104], it is easily verified that:

$$\|\boldsymbol{d}_0(k-1)\|^2\|\overline{\boldsymbol{g}}(k)\|^2 \geq \overline{\boldsymbol{g}}^H(k)\boldsymbol{d}_0(k-1)(\overline{\boldsymbol{g}}^H(k)\boldsymbol{d}_0(k-1))^H \tag{7.56}$$

The equality in (7.56) represents the two positive equal real roots solution to (7.45); that is, $b^2 - 4ac = 0$. In this case, the optimal loading level is given by:

$$\lambda = \frac{\mathrm{Re}\{\overline{\boldsymbol{d}}_0^H(k)\boldsymbol{d}_0(k-1)\}}{\mu_{SD|CG}\|\boldsymbol{d}_0(k-1)\|^2} \tag{7.57}$$

In addition to (7.54) and (7.55), another condition on the step size is required to guarantee positive diagonal loading (that is, $b \geq 0$). That condition is:

$$\mu_{SD|CG}\mathrm{Re}\{\overline{\boldsymbol{d}}_0^H(k)\boldsymbol{d}_0(k-1)\} \geq 0 \tag{7.58}$$

Using (7.47) and (7.49) into (7.58), we get:

$$\mu_{SD|CG} \leq \mathrm{Re}\left[\frac{\|\boldsymbol{d}_0(k-1)\|^2}{\overline{\boldsymbol{g}}^H(k)\boldsymbol{d}_0(k-1)}\right] \tag{7.59}$$

It is easily verified that, the right-hand side of (7.59) is positive and real valued. Therefore, the selection of the step size according to the inequality (7.59) guarantees positive diagonal loading.

More interestingly, the presented implementation and the simplified VL technique can be extended to include the generalized flat ellipsoidal constraint [52]. Even the optimization problem in (7.19) can be solved using the Lagrange method [47], so it can be implemented on-line using this approach.

7.4 Cooperative Joint Constraint Robust Beamforming

The max/min approach outlined in Section 7.3 is notorious for exhibiting performance loss at low SNR [105–110]. This can be understood from (7.22); that is, "the maximization of the power after the interference has been mitigated":[2]

$$\max_{\hat{a}_0} \hat{w}_0^H R_{xx} \hat{w}_0 \tag{7.60}$$

where $\hat{w}_0$ is the optimal robust Capon beamformer with an ellipsoidal constraint on the steering vector. The array correlation matrix R_{xx} can be expressed as:

$$R_{xx} = \left(\sum_{i=1}^{L} \rho_i s_i s_i^H + R_n \right) \tag{7.61}$$

where s_i and ρ_i signify the steering vector and the power of source i, respectively, and $R_n = \sigma^2 I$ denotes the component of the array correlation matrix due to random noise. Substituting (7.61) into (7.60), we get:

$$\max_{\hat{a}_0} \left(\hat{w}_0^H \sum_{i=1}^{L} \rho_i s_i s_i^H \hat{w}_0 + \sigma^2 \hat{w}_0^H \hat{w}_0 \right) \tag{7.62}$$

Therefore, the optimization process includes maximization for the noise constituent $\sigma^2 \hat{w}_0^H \hat{w}_0$. In order to mitigate the effect of noise enhancement, we can impose a quadratic constraint on the weight vector norm. In addition to this, the presented robust beamformer may gain other useful features of the quadratic constraint, such as robustness to pointing errors and random perturbations in detector parameters, and small training-sample sizes [38]. To start the formulation of this constraint, we directly impose the quadratic constraint on the robust Capon beamformer as follows:

$$\min_{w} w^H R_{xx} w \text{ subject to } w^H \hat{a}_0(\theta_0) = 1 \ \varsigma\ \hat{w}_0^H \hat{w}_0 \le \tau \tag{7.63}$$

where $\hat{a}_0(\theta_0)$ is the actual estimated steering vector and can be obtained using (7.32) or (7.33). The solution to this optimization problem is given by [2, 38]:

$$\tilde{w}_0 = \frac{(R_{xx} + \upsilon I)^{-1} \hat{a}_0(k)}{\hat{a}_0^H(k)(R_{xx} + \upsilon I)^{-1} \hat{a}_0(k)} \tag{7.64}$$

2 For simplicity of notation, the ellipsoidal side constraint is omitted.

Two variable loading procedures are presented in Chapter 5 [2, 31, 38]. These give the optimum diagonal loading term. Regrettably, these approaches are mainly developed using the GSC architecture, which is not appropriate here due to the first optimization problem (7.22). More specifically, this optimization problem includes genuine steering vector optimization, which means the upper quiescent vector of the GSC structure needs to be optimized concurrently with the lower section. This dual optimization problem seems to be problematical with the GSC structure. Therefore, we have to adopt a direct-form realization here. Unfortunately, a direct solution of (7.64) seems also to be difficult to obtain, due to the existence of the diagonal loading term in both numerator and denominator. To overcome this difficulty, we can remove the diagonal loading term from the denominator of (7.64). This suggestion can be understood because the denominator of (7.64) is already optimized via (7.22). As a result, the optimum solution can be expressed as:

$$\breve{\boldsymbol{w}}_0 = \frac{(\boldsymbol{R}_{xx} + \upsilon \boldsymbol{I})^{-1}\hat{\boldsymbol{a}}_0(k)}{\hat{\boldsymbol{a}}_0^H(k)\boldsymbol{R}_{xx}^{-1}\hat{\boldsymbol{a}}_0(k)} \tag{7.65}$$

Therefore, a closed-form solution to the robust joint constraint beamformer can be attained by substituting from (7.16) into (7.65). This yields:

$$\breve{\boldsymbol{w}}_0 = \frac{(\boldsymbol{R}_{xx} + \upsilon \boldsymbol{I})^{-1}\boldsymbol{R}_{xx}(\boldsymbol{R}_{xx} + \boldsymbol{I}/\lambda)^{-1}\overline{\boldsymbol{a}}_0}{\overline{\boldsymbol{a}}_0^H(k)(\boldsymbol{R}_{xx} + \boldsymbol{I}/\lambda)^{-1}\boldsymbol{R}_{xx}(\boldsymbol{R}_{xx} + \boldsymbol{I}/\lambda)^{-1}\overline{\boldsymbol{a}}_0} \tag{7.66}$$

The closed-form solution contains two positive diagonal loading terms. The first diagonal loading term λ can be obtained using the VL technique proposed in Section 7.3.1. The Taylor series approximation used previously [18, 38] can be invoked to obtain the second diagonal loading term υ as follows:

$$\breve{\boldsymbol{w}}_0 = \frac{(\boldsymbol{I} + \upsilon \boldsymbol{R}_{xx}^{-1})^{-1}\boldsymbol{R}_{xx}^{-1}\hat{\boldsymbol{a}}_0(k)}{\hat{\boldsymbol{a}}_0^H(k)\boldsymbol{R}_{xx}^{-1}\hat{\boldsymbol{a}}_0(k)} \tag{7.67}$$

$$\breve{\boldsymbol{w}}_0 \approx \frac{(\boldsymbol{I} - \upsilon \boldsymbol{R}_{xx}^{-1})\boldsymbol{R}_{xx}^{-1}\hat{\boldsymbol{a}}_0(k)}{\hat{\boldsymbol{a}}_0^H(k)\boldsymbol{R}_{xx}^{-1}\hat{\boldsymbol{a}}_0(k)} \tag{7.68}$$

$$\breve{\boldsymbol{w}}_0 \approx \hat{\boldsymbol{w}}_0 - \upsilon \boldsymbol{\mathscr{w}}_0 \tag{7.69}$$

where

$$\boldsymbol{w}_0 = \boldsymbol{R}_{xx}^{-1}\widehat{\boldsymbol{w}}_0$$

Therefore, utilizing the quadratic constraint in (7.63) with the approximate update equation (7.69), we get the diagonal loading term υ using a simple quadratic equation analogous to the approach of Tian *et al.* [38]. It is interesting to highlight the drawback of the Taylor series approximation (7.68), namely that the diagonal loading term υ will not precisely calculated due to this approximation and even real diagonal loading cannot be guaranteed as it is in the first approach. Nevertheless, consolidating the two approaches generates an extremely robust beamformer, especially at low SNR, as we will see in the simulation section.

7.4.1 Adaptive Implementation

The matrix $\boldsymbol{R}_{xx}^{-1}(k)$ is updated using the well-known RLS algorithm:

$$\boldsymbol{q}(k) = \frac{\eta^{-1}\boldsymbol{R}_{xx}^{-1}(k-1)\boldsymbol{x}(k)}{1+\eta^{-1}\boldsymbol{x}^H(k)\boldsymbol{R}_{xx}^{-1}(k-1)\boldsymbol{x}(k)} \tag{7.70}$$

$$\boldsymbol{R}_{xx}^{-1}(k) = \eta^{-1}\boldsymbol{R}_{xx}^{-1}(k-1) - \eta^{-1}\boldsymbol{q}(k)\boldsymbol{x}^H(k)\boldsymbol{R}_{xx}^{-1}(k-1) \tag{7.71}$$

The unconstrained steering vector $\tilde{\boldsymbol{a}}_0(k)$ is updated using (7.32) or (7.33). If $\tilde{\boldsymbol{a}}_0(k)$ does not fulfill the ellipsoidal constraint, then the optimal loading level is calculated using (7.46) and the constrained vector $\widehat{\boldsymbol{a}}_0(k)$ is calculated according to:

$$\widehat{\boldsymbol{a}}_0(k) = \tilde{\boldsymbol{a}}_0(k) - \mu_{SD|CG}(k)\lambda(\widehat{\boldsymbol{a}}_0(k-1) - \overline{\boldsymbol{a}}) \tag{7.72}$$

This VL technique depends only on the difference between the previous updated and the presumed steering vector. Consequently, the total amount of computation required to calculate the optimal diagonal loading term in (7.65) is $O(M)$. Compared with previously published diagonal loading techniques, the VL technique presented here has considerably lower complexity; calculation of the diagonal loading terms in other approaches involves, for example, $O(M^6)$ [47] or $O(M^3)$ [53] flops.

$$\boldsymbol{R}_{xx}^{-1}(0) = \boldsymbol{I},\ \widehat{\boldsymbol{a}}_0(0) = \overline{\boldsymbol{a}},\ \eta = 1,$$

$$\lambda(0) = 0,\ \boldsymbol{g}(0) = \boldsymbol{p}(0) = \overline{\boldsymbol{a}},$$

$$\alpha = 0.01$$

Substituting into (7.27) we get the robust beamformer $\widehat{\boldsymbol{w}}_0$ with a single ellipsoidal constraint, which is referred to as the "first proposed". Alternatively, this robust beamformer can be obtained using the update equation of the RLS algorithm in conjunction with the look direction constraint [111].

The joint constraint beamformer $\breve{\boldsymbol{w}}_0$ can be obtained by solving a similar quadratic equation and for simplicity it is only summarized in Box 7.1, alongside the single constraint beamformer. As shown in Box 7.1, the total number of the multiplications required at each snapshot for the single and joint constraint beamformers is about $O(3M^2 + 10M)$ and $O(4M^2 + 13M)$ respectively.

The convergence of the presented algorithm depends on the eigenvalue spread of the diagonally loaded autocorrelation matrix. If the eigenvalues of $\left(\frac{\boldsymbol{R}_{xx}^{-1}(k)}{\lambda} + \boldsymbol{I}\right)$ are denoted by ρ_i, then the updated steering vector $\widehat{\boldsymbol{a}}_0(k)$ converges to its optimal value $\left(\frac{\boldsymbol{R}_{xx}^{-1}(k)}{\lambda} + \boldsymbol{I}\right)^{-1}\overline{\boldsymbol{a}}$ if [2, 31, 112]:

$$0 < \mu_{SD|CG}(k) < \frac{1}{\rho_{\max}} \tag{7.73}$$

7.5 Robust Adaptive MVDR Beamformer with Single WC Constraint

The beamforming formulations in (7.15) and (7.17) assume that the array response to the desired source (that is, the steering vector $\boldsymbol{a}_d$ of the point source or the spatial matrix Λ of a source with multipath rays) is precisely known. However, in practice, knowledge of the desired source steering vector or spatial matrix may be imprecise. In this section, the recently formulated rigorous approach to robust MVDR beamforming based on WC performance optimization is considered [3, 47, 48, 58, 59, 64, 70, 71].

We first consider the formulation of the standard MVDR beamformer in (7.15) with SBC and, following the approach in [3], to add robustness to the standard MVDR beamformer in (7.16), we minimize the WC weighted power output of the array in the presence of uncertainties in the steering vector:

$$\min_{\boldsymbol{w}} \boldsymbol{w}^H \widehat{\boldsymbol{R}} \boldsymbol{w} \;\text{ subject to } |\boldsymbol{w}^H \boldsymbol{z}| \geq 1 \quad \forall \boldsymbol{z} \in \varepsilon \tag{7.74}$$

Box 7.1 Flowchart of the robust Capon beamformers with Spherical and QI constraint

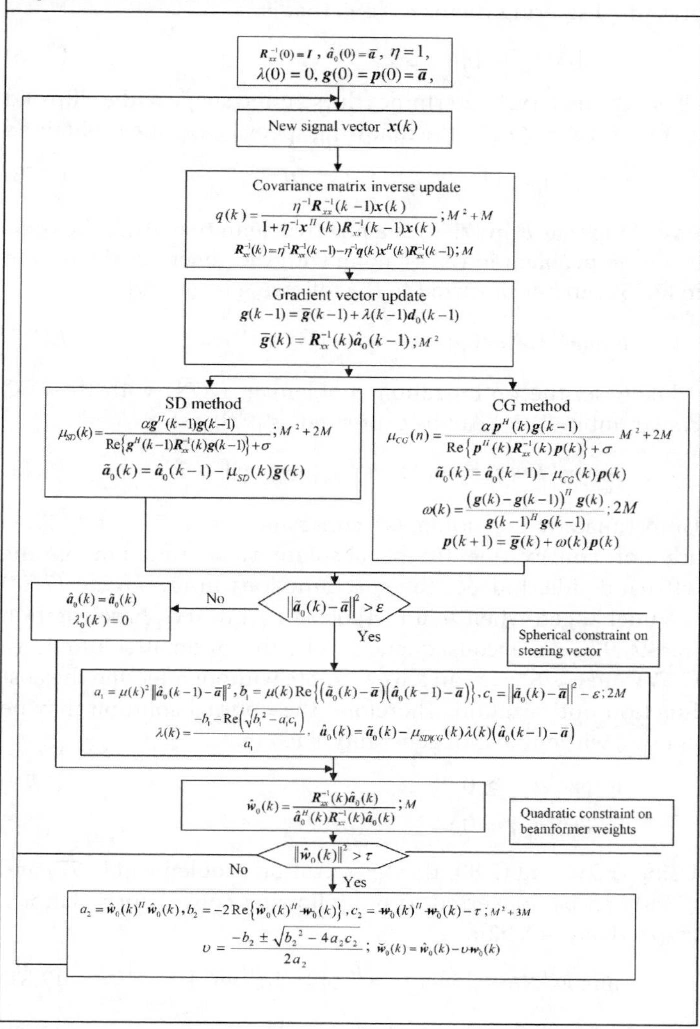

where ε is an ellipsoid that covers the possible range of the imprecise steering vector $\boldsymbol{z}$. We assume that ε is centered at the presumed steering vector $\boldsymbol{a}_d$ [49], that is:

$$\varepsilon = \{\boldsymbol{A}\boldsymbol{u} + \boldsymbol{a}_d \,|\, \|\boldsymbol{u}\| \leq 1\} \tag{7.75}$$

where the matrix $\boldsymbol{A}$ determines the size and shape of the ellipsoid ε. If we set $\boldsymbol{A} = \zeta \boldsymbol{I}$ [3], the following special case of ε is obtained

$$\varepsilon = \{\boldsymbol{e} + \boldsymbol{a}_d \,|\, \|\boldsymbol{e}\| \leq \zeta\}, \quad \boldsymbol{e} = \zeta \boldsymbol{u} \tag{7.76}$$

Assuming that $\widehat{\boldsymbol{R}}$ in (7.74) is a positive definite matrix, the optimization problem in (7.74) along with the generalized ellipsoid in (7.75) can be converted to the following form [49]

$$\min_{\boldsymbol{w}} \boldsymbol{w}^H \widehat{\boldsymbol{R}} \boldsymbol{w} \text{ subject to } |\boldsymbol{w}^H \boldsymbol{a}_d| \geq \|\boldsymbol{A}^H \boldsymbol{w}\| + 1 \tag{7.77}$$

Likewise, the optimization problem in (7.74) with the WC constraint in (7.76) can be expressed as [97]:

$$\min_{\boldsymbol{w}} \boldsymbol{w}^H \widehat{\boldsymbol{R}} \boldsymbol{w} \text{ subject to } |\boldsymbol{w}^H \boldsymbol{a}_d| \geq \zeta \|\boldsymbol{w}\| + 1 \tag{7.78}$$

Unfortunately, the nonlinear constraints in (7.77) and (7.78) are non-convex due to the absolute value function on the left-hand side. Indeed, the cost functions in (7.77) and (7.78) are unchanged when $\boldsymbol{w}$ undergoes an arbitrary phase rotation [3, 49, 97]. As a consequence, with the optimal solutions of (7.77) and (7.78), we can always rotate without affecting the cost function optimization. Therefore, the optimal solution may be chosen, without loss of generality, such that:

$$\text{Re}\{\boldsymbol{w}^H \boldsymbol{a}_d\} \geq 0 \tag{7.79}$$

$$\text{Im}\{\boldsymbol{w}^H \boldsymbol{a}_d\} = 0 \tag{7.80}$$

Using (7.79) and (7.80) the optimization problems in (7.77) and (7.78) can be converted to the following convex formulations, respectively [49, 97]:

$$\min_{\boldsymbol{w}} \boldsymbol{w}^H \widehat{\boldsymbol{R}} \boldsymbol{w} \text{ subject to } \boldsymbol{w}^H \boldsymbol{a}_d \geq \|\boldsymbol{A}^H \boldsymbol{w}\| + 1 \tag{7.81}$$

$$\min_{\boldsymbol{w}} \boldsymbol{w}^H \widehat{\boldsymbol{R}} \boldsymbol{w} \text{ subject to } \boldsymbol{w}^H \boldsymbol{a}_d \geq \zeta \|\boldsymbol{w}\| + 1 \tag{7.82}$$

The constraints in (7.81) and (7.82) are called second-order cone constraints. Two SOCP approaches are proposed in [57, 97] for real and complex formulations respectively.

7.5.1 Lagrange Approach

We start by forming the following Lagrange function:

$$J(\boldsymbol{w}, \lambda) = \boldsymbol{w}^H \widehat{\boldsymbol{R}} \boldsymbol{w} - \lambda t(\boldsymbol{w}^H \boldsymbol{a}_d - \zeta \|\boldsymbol{w}\| - 1) \tag{7.83}$$

where $t(.)$ is a step function which guarantees that $\boldsymbol{w}^H \boldsymbol{a}_d \geq \zeta \|\boldsymbol{w}\| + 1$ and λ is the Lagrange multiplier. The inequality constraint in (7.82) is satisfied by the equality if the cost function in (7.83) is minimized. This fact can be proved by contradiction [56, 59] and hence the step-function in (7.83) is dispensable. By differentiating (7.83) and equating the result to zero, we get [56]:

$$\widehat{\boldsymbol{R}} \boldsymbol{w} + \lambda \zeta \frac{\boldsymbol{w}}{\|\boldsymbol{w}\|} = \lambda \boldsymbol{a}_d \tag{7.84}$$

By solving for $\boldsymbol{w}$, the following closed-form solution is obtained:

$$\boldsymbol{w}_{WC} = \lambda \left(\widehat{\boldsymbol{R}} + \frac{\lambda \zeta}{\|\boldsymbol{w}_{WC}\|} \boldsymbol{I} \right)^{-1} \boldsymbol{a}_d \tag{7.85}$$

The WC robust MVDR beamformer in (7.85) has three problems: the estimation of the weight vector norm, the estimation of the Lagrange multiplier which achieves (7.85), and the computational load of computing the inverse of the diagonally loaded sample covariance matrix. In the following two sections, we summarize two techniques for computing $\boldsymbol{w}_{WC}$.

7.5.2 Eigendecomposition Method

Several eigendecomposition approaches have been developed to solve the WC performance optimization problem. The optimization problems in (7.81) and (7.82) have been solved, respectively, in [49] and [56, 59] using eigendecomposition methods. For sake of comparison, let us briefly review the approach in the latter pair of papers. Using the fact that multiplying $\boldsymbol{w}_{WC}$ in (7.84) by any arbitrary constant does not affect BER performance of the beamformer [56, 59], a scaled version of the WC beamformer can be obtained as follows [56]:

$$\tilde{\boldsymbol{w}}_{WC} = \left(\widehat{\boldsymbol{R}} + \frac{\zeta}{\|\tilde{\boldsymbol{w}}_{WC}\|} \boldsymbol{I} \right)^{-1} \boldsymbol{a}_d \tag{7.86}$$

A binary search algorithm followed by Newton-Raphson iterations has been proposed to compute the norm of the WC beamformer $\|\tilde{\boldsymbol{w}}_{WC}\|$ [59]. The eigendecomposition approach accurately estimates the norm of the robust detector $\|\tilde{\boldsymbol{w}}_{WC}\|$ and hence we can obtain the optimal weight vector using the closed-form in (7.86).

7.5.3 Taylor Series Approximation Method

By applying Taylor series expansion to (7.86), analogous to the approach of Tian *et al.* [38], we obtain the following:

$$\begin{aligned}\tilde{\boldsymbol{w}}_{WC} &= \left(\boldsymbol{I} + \frac{\zeta}{\|\tilde{\boldsymbol{w}}_{WC}\|}\widehat{\boldsymbol{R}}^{-1}\right)^{-1}\widehat{\boldsymbol{R}}^{-1}\boldsymbol{a}_d \\ &\approx \overline{\boldsymbol{w}}_{SBC} - \frac{\zeta}{\|\tilde{\boldsymbol{w}}_{WC}\|}\widehat{\boldsymbol{R}}^{-1}\overline{\boldsymbol{w}}_{SBC},\end{aligned} \tag{7.87}$$

where $\overline{\boldsymbol{w}}_{SBC} = \widehat{\boldsymbol{R}}^{-1}\boldsymbol{a}_d$ is a biased version of the standard MVDR beamformer in (7.16). By introducing a new vector $\tilde{\boldsymbol{w}}_{SBC} = \widehat{\boldsymbol{R}}^{-1}\overline{\boldsymbol{w}}_{SBC}$ and substituting into (7.87), this yields:

$$\tilde{\boldsymbol{w}}_{WC} \approx \overline{\boldsymbol{w}}_{SBC} - \kappa\tilde{\boldsymbol{w}}_{SBC}, \tag{7.88}$$

where $\kappa = \zeta/\|\tilde{\boldsymbol{w}}_{WC}\|$ is a parameter related to the weight vector norm of the WC robust beamformer and can be estimated by substituting (7.88) into the WC constraint in (7.82). This approach is almost similar to the eigendecomposition approach, where low complexity is introduced at the expense of the weight vector norm estimation accuracy as a result of the Taylor series approximation.

7.5.4 Adaptive MVDR Beamformer with Single WC Constraint

In this section, efficient adaptive implementations of robust adaptive beamformers based on WC performance optimization are developed. We efficiently implement WC performance optimization MVDR beamforming using iterative gradient minimization algorithms with an ad-hoc technique to satisfy the WC constraint.

The adaptive beamformer can be found by searching for a weight vector $\boldsymbol{w}$ that minimizes the cost function (7.83). In order to find the target beamformer in an iterative manner, the weight vector can be updated as follows:

$$\boldsymbol{w}(k+1) = \boldsymbol{w}(k) - \mu(k)\nabla(k) \tag{7.89}$$

where k is the snapshot index, $\nabla(k)$ is the gradient vector of the Lagrange function $J(\boldsymbol{w}, \lambda)$ in (7.83) with respect to $\boldsymbol{w}$ determined at snapshot k, and $\mu(k)$ is an adaptive step size, which determines the convergence speed of the algorithm.

The gradient vector of the cost function in (7.83) is given by:

$$\nabla = \partial J(\boldsymbol{w}, \lambda)/\partial \boldsymbol{w} = \widehat{\boldsymbol{R}}\boldsymbol{w} - \lambda(\boldsymbol{a}_d - \zeta \boldsymbol{w}/\|\boldsymbol{w}\|) \tag{7.90}$$

The step function is dropped due to the ad-hoc adaptive implementation. Therefore, the adaptive weight vector can be obtained by substituting (7.90) into (7.89), which yields:

$$\begin{aligned}\boldsymbol{w}(k+1) = \boldsymbol{w}(k) &- \mu(k)\widehat{\boldsymbol{R}}(k)\boldsymbol{w}(k) \\ &+ \mu(k)\lambda(\boldsymbol{a}_d - \zeta \boldsymbol{w}(k)/\|\boldsymbol{w}(k)\|)\end{aligned} \tag{7.91}$$

For simplicity, we introduce two new vectors: $\widehat{\boldsymbol{w}}(k+1) = \boldsymbol{w}(k) - \mu(k)\widehat{\boldsymbol{R}}(k)\boldsymbol{w}(k)$ (referred to as the unconstrained MV weight vector) and $\boldsymbol{\pi}(k) = \boldsymbol{a}_d - \zeta \boldsymbol{w}(k)/\|\boldsymbol{w}(k)\|$. Therefore, the weight vector of the robust WC adaptive beamformer can be updated as follows:

$$\boldsymbol{w}(k+1) = \widehat{\boldsymbol{w}}(k+1) + \mu(k)\lambda\pi(k) \tag{7.92}$$

7.5.4.1 Lagrange Multiplier Estimation

Assuming that the weight vector $\boldsymbol{w}(k)$ satisfies the WC constraint in (7.82), then, $\boldsymbol{w}(k+1)$ should also satisfy the WC constraint. The weight vector $\widehat{\boldsymbol{w}}(k+1)$ represents the minimization of the unconstrained MV cost function ($\boldsymbol{w}^H\widehat{\boldsymbol{R}}\boldsymbol{w}$), which leads to a trivial zero solution if the additional WC constraint is not imposed. In order to fully satisfy the inequality constraint in (7.82), we first compute the weight vector $\widehat{\boldsymbol{w}}(k+1)$ and then verify if $\widehat{\boldsymbol{w}}(k+1)$ achieves the WC constraint in (7.82). Consequently, if $\widehat{\boldsymbol{w}}(k+1)$ satisfies the WC constraint, the weight vector is accepted and the algorithm continues with a new array observation. Otherwise, we substitute (7.92) into the inequality constraint in (7.82) to estimate the Lagrange multiplier as follows:

$$\begin{aligned}&\mathrm{Re}\{(\widehat{\boldsymbol{w}}(k+1)+\mu(k)\lambda\pi(k))^{H}\boldsymbol{a}_{d}\}\\&\quad\geq\zeta\|(\widehat{\boldsymbol{w}}(k+1)+\mu(k)\lambda\pi(k))\|+1,\end{aligned}\tag{7.93}$$

where Re{•} is inserted to make sure that (7.79) and (7.80) are always met during adaptive implementation. After arranging and boosting both sides of (7.93) by a power of two, we get

$$\begin{aligned}&(\mathrm{Re}\{(\widehat{\boldsymbol{w}}(k+1)+\mu(k)\lambda\boldsymbol{\pi}(k))^{H}\boldsymbol{a}_{d}\}-1)^{2}\\&\quad\geq\zeta^{2}\|(\widehat{\boldsymbol{w}}(k+1)+\mu(k)\lambda\boldsymbol{\pi}(k))\|^{2}\end{aligned}\tag{7.94}$$

By rearranging (7.94), we have:

$$\begin{aligned}&((\mathrm{Re}\{\widehat{\boldsymbol{w}}(k+1)^{H}\boldsymbol{a}_{d}\}-1)+\mu(k)\lambda\mathrm{Re}\{\boldsymbol{\pi}(k)^{H}\boldsymbol{a}_{d}\})^{2}\\&\quad\geq\zeta^{2}(\widehat{\boldsymbol{w}}(k+1)+\mu(k)\lambda\boldsymbol{\pi}(k))^{H}(\widehat{\boldsymbol{w}}(k+1)+\mu(k)\lambda\boldsymbol{\pi}(k))\end{aligned}\tag{7.95}$$

We need now to estimate the Lagrange multiplier λ that achieves the WC constraint in (7.82). During ad hoc implementation, we will solve (7.95) only if the WC constraint is not met. If this is the case, the inequality in (7.95) is replaced by an equality and after some manipulations to (7.95), we have:

$$\begin{aligned}&\chi^{2}+2\mu(k)\lambda\chi\boldsymbol{\pi}(k)^{H}\boldsymbol{a}_{d}+\mu(k)^{2}\lambda^{2}(\boldsymbol{\pi}(k)^{H}\boldsymbol{a}_{d})^{2}\\&\quad=\zeta^{2}\|\widehat{\boldsymbol{w}}(k+1)\|^{2}+2\mu(k)\zeta^{2}\lambda\mathrm{Re}\{\boldsymbol{\pi}(k)^{H}\widehat{\boldsymbol{w}}(k+1)\}\\&\quad\quad+\mu(k)^{2}\zeta^{2}\lambda^{2}\|\boldsymbol{\pi}(k)\|^{2}\end{aligned}\tag{7.96}$$

where $\chi=\mathrm{Re}\{\widehat{\boldsymbol{w}}(k+1)^{H}\boldsymbol{a}_{d}\}-1$. Therefore, the Lagrange multiplier λ can be computed as the solution to the following quadratic equation:

$$\begin{aligned}&\mu(k)^{2}\lambda^{2}((\mathrm{Re}\{\boldsymbol{\pi}(k)^{H}\boldsymbol{a}_{d}\})^{2}-\zeta^{2}\|\boldsymbol{\pi}(k)\|^{2})\\&\quad+2\mu(k)\lambda(\chi\mathrm{Re}\{\boldsymbol{\pi}(k)^{H}\boldsymbol{a}_{d}\}-\zeta^{2}\mathrm{Re}\{\boldsymbol{\pi}(k)^{H}\widehat{\boldsymbol{w}}(k+1)\})\\&\quad+\chi^{2}-\zeta^{2}\|\widehat{\boldsymbol{w}}(k+1)\|^{2}=0\end{aligned}\tag{7.97}$$

Therefore, the value of λ that achieves the WC constraint in (7.82) has the following form:

$$\lambda=-B\pm\sqrt{B^{2}-AC}/A\tag{7.98}$$

where

$$\begin{aligned} A &= \mu(k)^2\left(\left(\text{Re}\left\{\boldsymbol{\pi}(k)^H\boldsymbol{a}_d\right\}\right)^2 - \zeta^2\|\boldsymbol{\pi}(k)\|^2\right) \\ B &= \mu(k)(\chi\text{Re}\{\boldsymbol{\pi}(k)^H\boldsymbol{a}_d\} - \zeta^2\text{Re}\{\boldsymbol{\pi}(k)^H\widehat{\boldsymbol{w}}(k+1)\}) \\ C &= \chi^2 - \zeta^2\|\widehat{\boldsymbol{w}}(k+1)\|^2 \end{aligned} \tag{7.99}$$

The Lagrange multiplier estimation is only executed when the WC constraint is not achieved: $\chi < \zeta\|\widehat{\boldsymbol{w}}(k+1)\|$. Therefore, $C < 0$ and the roots of (7.97) fall under one of the following categories:

$A > 0$ and $B > 0$: Equation 7.97 has two real roots, one positive and one negative, resulting from the pluse and minus signs in (7.98) respectively. The positive root is selected to make sure that $(\widehat{\boldsymbol{R}} + \lambda\zeta/\|\boldsymbol{w}_{RMV}\|\boldsymbol{I})^{-1}$ is a positive definite matrix.

$A < 0$ and $B > 0$: Equation 7.97 has only one real positive root resulting from the plus sign in (7.98) if $B^2 > AC$.

$A > 0$ and $B < 0$: Equation 7.97 has two positive real solutions. In this case, the smaller root is selected to guarantee algorithm stability.

$A < 0$ and $B < 0$: Equation (7.97) is guaranteed to have one real positive solution if $B^2 > AC$.

7.5.4.2 Recursive Implementation

The optimum step-size of minimizing $\boldsymbol{w}^H\widehat{\boldsymbol{R}}\boldsymbol{w}$ is the best estimate of the optimum step size that minimizes (7.83). As a consequence, the optimum step size can be obtained by substituting (7.89) into (7.83) and differentiating with respect to the adaptive step size. Equating the result to zero, the following optimum step-size is obtained:

$$\mu_{opt}(k) = \left.\frac{\alpha\widehat{\nabla}^H(k)\widehat{\nabla}(k)}{\widehat{\nabla}^H(k)\widehat{\boldsymbol{R}}(k)\widehat{\nabla}(k)}\right|_{\nabla=\widehat{\boldsymbol{R}}(k)\boldsymbol{w}(k)} \tag{7.100}$$

The parameter α is added to improve the numerical stability of the algorithm. For a practical system, it should be adjusted during initial tuning of the system and it should satisfy $0 < \alpha < 1$ [113, 114].

The WC robust adaptive beamformer algorithm is summarized in Box 7.2.

Box 7.2 Robust WC adaptive beamformer.

Step 0) Initialize: $\widehat{\boldsymbol{R}}(0) = \boldsymbol{I}$, $\boldsymbol{w}(0) = \boldsymbol{a}_d$, $\alpha = 0.1$, $\eta = 0.97$

Step 1) Pick a new sample from array observations and compute the sample covariance matrix: $\widehat{\boldsymbol{R}}(k) = \eta\widehat{\boldsymbol{R}}(k-1) + \boldsymbol{x}(k)\boldsymbol{x}^H(k)$; M^2.

Step 2) Compute the optimum step-size using (7.100); $M^2 + 2M$.

Step3) Update the unconstrained MV weight vector: $\widehat{\boldsymbol{w}}(k+1) = \boldsymbol{w}(k) - \mu(k)\widehat{\boldsymbol{R}}(k)\boldsymbol{w}(k)$; the matrix vector multiplication $\widehat{\boldsymbol{R}}(k)\boldsymbol{w}(k)$ is computed in Step 2.

Step 4) ***If*** $\chi < \zeta\|\widehat{\boldsymbol{w}}(k+1)\|$, compute λ using (7.98); $5M$.
Else $\lambda = 0$ and $\boldsymbol{w}(k+1) = \widehat{\boldsymbol{w}}(k+1) \rightarrow$ go to Step 1.

Step 5) Update the WC weight vector as: $\boldsymbol{w}(k+1) = \widehat{\boldsymbol{w}}(k+1) + \mu(k)\lambda\boldsymbol{\pi}(k) \rightarrow$ go to Step 1.

As shown in the above implementation, the total multiplication complexity of the presented algorithm is about $O(2M^2 + 7M)$. More interestingly, the WC optimization step with ad-hoc implementation requires $O(M)$ complexity, while it requires $O(M^3)$ complexity using SOCP. In the eigendecomposition method the estimation of the norm vector of the WC weight vector alone requires $O(M^3)$ complexity [3].

7.6 Robust LCMV Beamforming with MBWC Constraints

The majority of the robust techniques in the beamforming literature are based on a single constraint in the desired look direction [2, 3]. Therefore, if the desired source experiences multipath propagation and impinges on the antenna array from different angles associated with the dominant multipath rays, the robust technique with a single constraint is not capable of gathering all the multipath components, especially if there is a large angular spread. Alternatively, a robust technique may concentrate only on the nominal AOI and neglect other components scattered in different multipaths, which is not optimal in terms of maximizing the output SINR. As a consequence, it is worthwhile to generalize the WC robust technique to include multiple constraints. As a result, we can formulate a robust LCMV beamformer with MBWC constraints, analogous to the standard LCMV beamformer with MBC in (7.18). A generalization for (7.82) with MBWC constraints can be expressed as:

$$\min_{\boldsymbol{w}} \breve{\boldsymbol{w}}^H \widehat{\boldsymbol{R}} \breve{\boldsymbol{w}} \text{ subject to } \breve{\boldsymbol{w}}^H \boldsymbol{\Lambda} \geq \boldsymbol{\upsilon} \|\breve{\boldsymbol{w}}\| + \boldsymbol{i} \quad (7.101)$$

where $\boldsymbol{\Lambda}$ is a $M \times N$ spatial matrix of the desired source, $\boldsymbol{\upsilon}$ is a $1 \times N$ vector of the WC constrained values, and $\boldsymbol{i}$ is an $1 \times N$ all-ones vector, where N is the number of WC constraints (that is, the dominant multipath components, $N \leq L$). A generalized cost function corresponding to (7.83) can then be expressed as:

$$\Theta(\boldsymbol{w}, \boldsymbol{\tau}) = \breve{\boldsymbol{w}}^H \widehat{\boldsymbol{R}} \breve{\boldsymbol{w}} - (\breve{\boldsymbol{w}}^H \boldsymbol{\Lambda} - \boldsymbol{\upsilon} \|\breve{\boldsymbol{w}}\| - \boldsymbol{i})\boldsymbol{\tau} \quad (7.102)$$

where $\boldsymbol{\tau}$ is a $N \times 1$ vector of Lagrange multipliers. The step-function is dropped due to the ad-hoc implementation. The following equations correspond to (7.90) and (7.92), respectively:

$$\nabla = \partial\Theta(\breve{\boldsymbol{w}}, \boldsymbol{\tau})/\partial\breve{\boldsymbol{w}} = \widehat{\boldsymbol{R}}\breve{\boldsymbol{w}} - (\boldsymbol{\Lambda} - \boldsymbol{\upsilon}(\breve{\boldsymbol{w}}/\|\breve{\boldsymbol{w}}\|))\boldsymbol{\tau} \quad (7.103)$$

$$\breve{\boldsymbol{w}}(k+1) = \tilde{\boldsymbol{w}}(k+1) + \mu(k)\overline{\boldsymbol{\Lambda}}\boldsymbol{\tau}(k) \quad (7.104)$$

where $\overline{\boldsymbol{\Lambda}} = \boldsymbol{\Lambda} - \boldsymbol{\upsilon}(\breve{\boldsymbol{w}}(k)/\|\breve{\boldsymbol{w}}(k)\|)$ is a $M \times N$ matrix and $\tilde{\boldsymbol{w}}(k+1) = \breve{\boldsymbol{w}}(k) - \mu(k)\widehat{\boldsymbol{R}}(k)\breve{\boldsymbol{w}}(k)$ is similar to $\widehat{\boldsymbol{w}}(k+1)$. The vector of Lagrange multipliers is obtained by substituting (7.104) into the set of WC constraints in (7.101), giving the following set of nonlinear equations:

$$\begin{aligned}&(\tilde{\boldsymbol{w}}(k+1) + \mu(k)\overline{\boldsymbol{\Lambda}}\boldsymbol{\tau}(k)^H)\boldsymbol{\Lambda}\\ &\quad - \boldsymbol{\upsilon}\|\tilde{\boldsymbol{w}}(k+1) + \mu(k)\overline{\boldsymbol{\Lambda}}\boldsymbol{\tau}(k)\| - \boldsymbol{i} = \boldsymbol{0}\end{aligned} \quad (7.105)$$

where $\boldsymbol{0}$ is an all-zero vector. Unfortunately, a closed-form solution similar to (7.85) cannot be obtained because (7.105) is a system of nonlinear equations. In this case, a Newton-like method is obligatory to find the optimum vector of Lagrange multipliers $\boldsymbol{\tau}(k)$ that satisfies the set of WC constraints in (7.101).

The trust region method [115, 116] is adopted to solve the system of nonlinear equations in (7.105). A minor drawback of this technique is that we have to solve for all WC constraints via reducing them to equality if any of the WC constraints is not achieved.

The algorithm of the robust LCMV beamforming with MBWC consists is summarized in Box 7.3

Box 7.3 Robust LCMV beamforming with MBWC.

Step 0) Initialize $\widehat{\mathbf{R}}(0) = \mathbf{I}$, $\mathbf{w}(0) = \mathbf{a}_d$, $\alpha = 0.5$, $\eta = 0.97$

Step 1) Pick a new sample from array observations and compute the sample covariance matrix as $\widehat{\mathbf{R}}(k) = \eta\widehat{\mathbf{R}}(k-1) + \mathbf{x}(k)\mathbf{x}^H(k); M^2$.

Step 2) Compute the optimum step size using (7.100), where $\widehat{\nabla} \Leftarrow \widehat{\mathbf{R}}(k)\breve{\mathbf{w}}(k); M^2 + 2M$.

Step 3) Update the unconstrained MV weight vector as $\tilde{\mathbf{w}}(k+1) = \breve{\mathbf{w}}(k) - \mu(k)\widehat{\mathbf{R}}(k)\breve{\mathbf{w}}(k)$; the matrix vector multiplication $\widehat{\mathbf{R}}(k)\mathbf{w}(k)$ is computed in Step 2.

Step 4) If *any*($\tilde{\mathbf{w}}(k+1)^H\boldsymbol{\Lambda} \geq \upsilon\|\tilde{\mathbf{w}}(k+1)\| + \mathbf{i}$); $(MN+M)$
Compute $\boldsymbol{\tau}(k)$ by solving (7.105); $(2MN+M)(M+1)R$, where R is the required number of iterations for the trust region method convergence.
Else, $\boldsymbol{\tau}(k) = \mathbf{0}$ and $\breve{\mathbf{w}}(k+1) = \tilde{\mathbf{w}}(k+1) \rightarrow$ go to Step 1.

Step 5) Update the weight vector of robust LCMV beamformer using (7.104) $\rightarrow$ Go to Step 1.

As demonstrated in the above implementation, the robust LCMV beamformer with MBWC constraints requires $O(M^2(2NR+R+2)+M(2NR+N+R+3))$ complexity. Indeed, it requires a higher computational load, but it cannot be compared with single WC beamformers. Several simulation scenarios demonstrate that the trust region algorithm requires 4–12 iterations for convergence.

7.7 Geometric Interpretation

7.7.1 Ellipsoidal Constraint Beamforming

To elucidate the presented robust adaptive Capon beamformer algorithm with spherical constraint, let us consider a simple

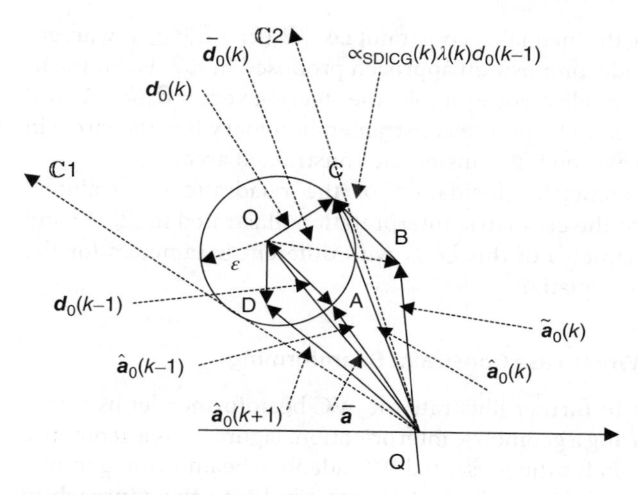

Figure 7.1 Geometric representation for robust Capon beamforming with ellipsoidal constraint.

2-D case, as shown in Figure 7.1. The center point Q is the array reference point. The steering vector $\bar{\boldsymbol{a}}$ $(\overrightarrow{QO})$ represents the presumed steering vector and the uncertainty constraint $\|\hat{\boldsymbol{a}}_0(k) - \bar{\boldsymbol{a}}\|^2 \leq \varepsilon$ is bounded by the circle centered at the origin O. The area bounded by the circle and the two lines ℂ1, ℂ2 give the possible steering vector area within the constrained value ε[3]. Assuming the previous steering vector update $\hat{\boldsymbol{a}}_0(k-1)$ $(\overrightarrow{QA})$ is sited inside the constrained vicinity. After next iteration, the steering vector $\tilde{\boldsymbol{a}}_0(k)$ $(\overrightarrow{QB})$ is obtained which may be outside the constrained region. Then, the variable loading technique is evoked and the diagonal loading term is computed using (7.36) and the amendment term $\mu_{SD|CG}(k)\lambda(k)\boldsymbol{d}_0(k-1)(\overrightarrow{BC})$ is appended to the steering vector to obtain the new constrained steering vector $\hat{\boldsymbol{a}}_0(k)(\overrightarrow{QC})$ fulfilling the spherical constraint (i.e., $\|\hat{\boldsymbol{a}}_0(k) - \bar{\boldsymbol{a}}\|^2 = \varepsilon$).

There are two solutions for the diagonal loading term $\lambda(k)$; the smaller solution is preferred to assure more algorithm stability. If the recent steering vector update $\hat{\boldsymbol{a}}_0(k+1)$ $(\overrightarrow{QD})$ is located inside the constrained area, the algorithm will continue without carrying out the VL algorithm. Therefore, this technique is fully

3 In a 3-D case this area is bounded by a sphere centered at origin O and a cone with nape at Q

attaining the inequality constraint i.e., $\|\hat{\boldsymbol{a}}_0(k) - \overline{\boldsymbol{a}}\|^2 \leq \varepsilon$, whereas the eigendecomposition approach proposed in [52] is diminishing it to equality. For example, the steering vector $\hat{\boldsymbol{a}}_0(k+1)$ will be dragged to be on the constrained boundary (i.e. the circle in the figure) even if it is inside the constrained area.

The geometric elucidation of the quadratic constraint is similar to the geometric interpretation illustrated in [2, 38] and as in Chapter 5 of this book with different parameters for the quadratic equation.

7.7.2 Worst-case Constraint Beamforming

In order to further illustrate the WC beamformer, let us exemplify it using a geometric interpretation. Figure 7.2 is a geometric illustration for the presented WC adaptive beamforming implementation using a simple 2-D case similar to the approach in [117]. The vector $\overrightarrow{OA}$ represents the presumed steering vector $\boldsymbol{a}_d$. The vector $\overrightarrow{OB}$ represents the WC robust beamformer at snapshot k. The concentric ellipses represent the unconstrained MV cost function ($\boldsymbol{w}^H\hat{\boldsymbol{R}}\boldsymbol{w}$) and the center of these ellipses is the minimal point (the trivial zero solution) that minimizes this cost function. We assume the WC weight vector $\boldsymbol{w}(k)$ satisfies the WC constraint; that is, $\overrightarrow{OB}^H\overrightarrow{OA} \geq \zeta\|\overrightarrow{OB}\| + 1$. The forthcoming update of the unconstrained MV weight vector is computed as $\hat{\boldsymbol{w}}(k+1) = \boldsymbol{w}(k) - \mu(k)\hat{\boldsymbol{R}}(k)\boldsymbol{w}(k)$; that is, $\overrightarrow{OC} = \overrightarrow{OB} + \overrightarrow{BC}$. As depicted in Figure 7.2, the vector $\overrightarrow{BC}$ represents the gradient of the MV cost function $-u(k)\hat{\boldsymbol{R}}(k)\boldsymbol{w}(k)$, which is perpendicularly inward inside the contours and towards the center of the ellipse. When the subsequent vector $\overrightarrow{OC}$ does not satisfy the WC constraint (that is, $\overrightarrow{OC}^H\overrightarrow{OA} < \zeta\|\overrightarrow{OC}\| + 1$), the condition in Step 4 in the algorithm is met and subsequently the vector $\overrightarrow{AE} = -\zeta\boldsymbol{w}(k)/\|\boldsymbol{w}(k)\|$, which is parallel to the vector $\overrightarrow{BO}$, is added to $\overrightarrow{OA}$ to estimate $\boldsymbol{\pi}(k)$; that is, $\overrightarrow{OE} = \overrightarrow{OA} + \overrightarrow{AE}$. Then the WC weight vector $\overrightarrow{OD} = \boldsymbol{w}(k+1)$ is generated by adding the vector $\overrightarrow{CD} = \mu(k)\lambda\boldsymbol{\pi}(k)$, which is parallel to the vector $\overrightarrow{OE}$, to the vector $\overrightarrow{OC}$; that is, $\overrightarrow{OD} = \overrightarrow{OC} + \overrightarrow{CD}$. Consequently, the ensuing weight vector $\overrightarrow{OD}$ satisfies the WC constraint. In a nutshell, the

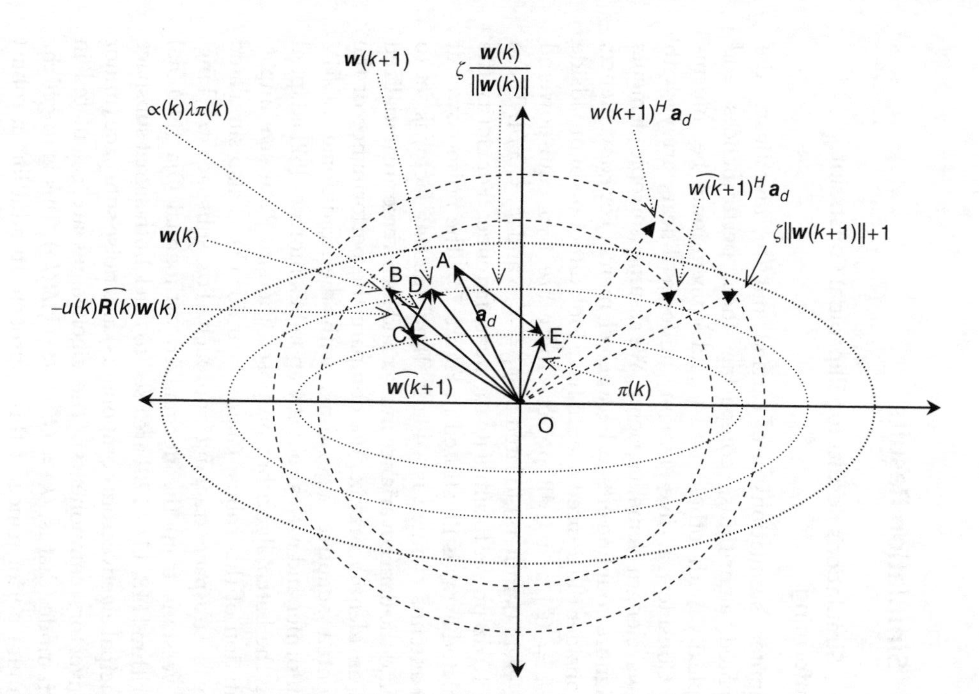

Figure 7.2 Geometric interpretation of the robust WC beamformer.

WC constraint prevents the weight vector from reaching the trivial zero solution by maintaining a distortionless response to a set of possible steering vectors. This is controlled by the WC constraint.

7.8 Simulation Results

7.8.1 Simulations Results for Ellipsoidal Constraint Beamforming

Computer simulations are carried out to investigate the presented algorithms, comparing their performances and complexity with other robust and conventional beamformers. Consider a five-element uniform linear array with half-wavelength element spacing. We assume isotropic elements and ignore coupling effects between elements. For convenience, we consider the desired user to have look direction of broadside (DOA = 0°) and 0 dB power. There are two equi-powered interferers (jammers) located at $\phi_1 = \pi/6$ and $\phi_2 = \pi/4$. These have 10 dB power to simulate the near–far effect. For clarity, the simulated array is illustrated in Figure 7.3. The noise-power at each antenna element is −40 dB. The forgetting factor is set to one and the data covariance matrix and its inverse are initialized with an identity matrix. We compare the performance of the presented adaptive algorithms with the traditional LCMV beamformer and with a robust Capon beamformer [52] adapted using the batch algorithm proposed therein. Recursive implementation of this robust Capon beamformer is also simulated using a subspace tracking approach. For subspace tracking, two versions from the normalized orthogonal Oja (NOOja) algorithm [113, 118–120] are used to track both signal subspace (principal eigenvectors components) and noise subspace (minor eigenvectors components). The eigenvalues are calculated in batch mode (that is, $(k) = \boldsymbol{U}^H(k)\boldsymbol{R}_{xx}(k)\boldsymbol{U}(k)$). This is to get the optimum performance of this approach. In addition, a robust beamformer based on SOCP [47] is also incorporated in our simulation.

The SeDuMi optimization Matlab toolbox [53] has been used to solve the SOCP problem. SeDuMi is a good piece of software

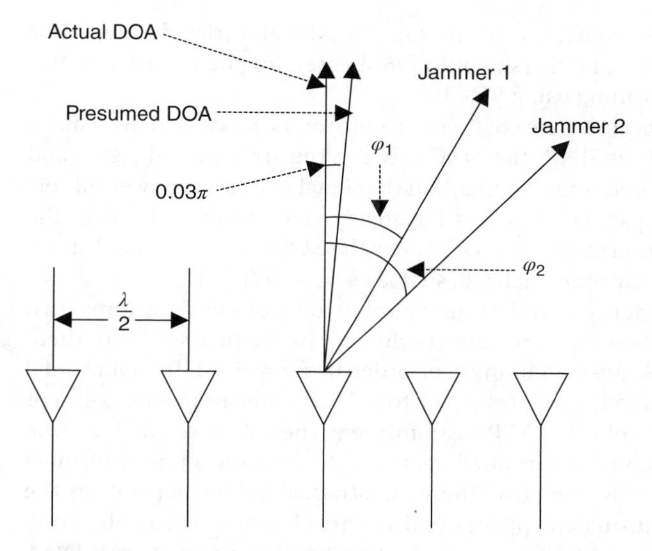

Figure 7.3 Five-element uniform linear array with one source and two jammers.

for optimization over symmetric cones. It was developed by Jos F. Sturm, who passed away in 2003. Currently SeDuMi is hosted and maintained by the CORAL Lab at the Department of Industrial and Systems Engineering at Lehigh University, but it may still give errors with recent releases of Matlab. For this reason, the simulations in this book use the CVX software package [95]. CVX is a modeling system for constructing and solving disciplined convex programs (DCPs). CVX supports a number of standard problem types, including linear and quadratic programs (LPs/QPs), second-order cone programs (SOCPs), and semidefinite programs (SDPs). CVX can also solve much more complex convex optimization problems, including many involving non-differentiable functions, such as L1 norms. CVX is convenient for formulating and solving constrained norm minimization, entropy maximization, determinant maximization, and many other convex programs. CVX supports two free SQLP solvers, SeDuMi [53] and SDPT3 [93, 94]. These solvers are included with the CVX distribution.

The SOCP beamformer is obtained using SeDuMi algorithm [53]. The SeDuMi algorithm in the CVX package is used to solve the SOCP problem [95]. The Matlab script developed by Vorobyov and Gershmanin [96] is used for the robust adaptive beamforming using SOCP.

The performance of these beamformers is assessed in terms of the output SINR, the MSE between the array output signal and the desired source signal, and the signal of interest power, all versus snapshots as well as the steady-state beam patterns of the antenna array against DOA. For the SOCP, the presumed steering vector is normalized, such as $\overline{\boldsymbol{a}} = M(\overline{\boldsymbol{a}}/\|\overline{\boldsymbol{a}}\|)$.

In order to corroborate the robustness of the algorithms, two simulation scenarios are conducted. In the first scenario, there is DOA mismatch angle of order $0.03\pi = 5.4°$. The ellipsoidal constrained value (ε) is set to 0.5 for all beamformers, except for the robust SOCP beamformer, where it is set to 1.5.[4] The selection of constrained values is the best for all beamformers within this scenario. These constrained values depend on the DOA mismatch upper bound and are obtained empirically using numerous simulation runs. In the second scenario, there is DOA mismatch angle of order $0.06\pi = 10.8°$. The constrained values are set to 1 for all beamformers and to 2 for SOCP. The quadratic inequality constraint value is set to $\tau = 0.2$ in the two scenarios. In the second scenario, the noise power at each antenna element is set to -20 dB. The first scenario is adapted using the SD method, while the CG method is used in the second scenario. All figures are obtained by averaging 50 independent runs.

Figures 7.4–7.6 show the output SINR, MSE, and SOIP for the first scenario. The robust beamformer with the ellipsoidal constraint only is referred to as the first algorithm [2]. The second algorithm is the robust beamformer with the joint constraints [2]. Two samples from steady-state beampatterns are shown in Figures 7.7 and 7.8. For clarity, the joint constraint beamformer is excluded from the beampattern figures because it is similar to the single constraint beamformer in steady-state performance, as shown in Figure 7.4. It is clear from this figure that the presented algorithms give an improvement of about 2 dB in the steady-state SINR over the SOCP approach. The recursive implementation using subspace tracking has considerable performance degradation compared to the batch algorithm

4 This is due to the normalization step required by SOCP algorithm [120].

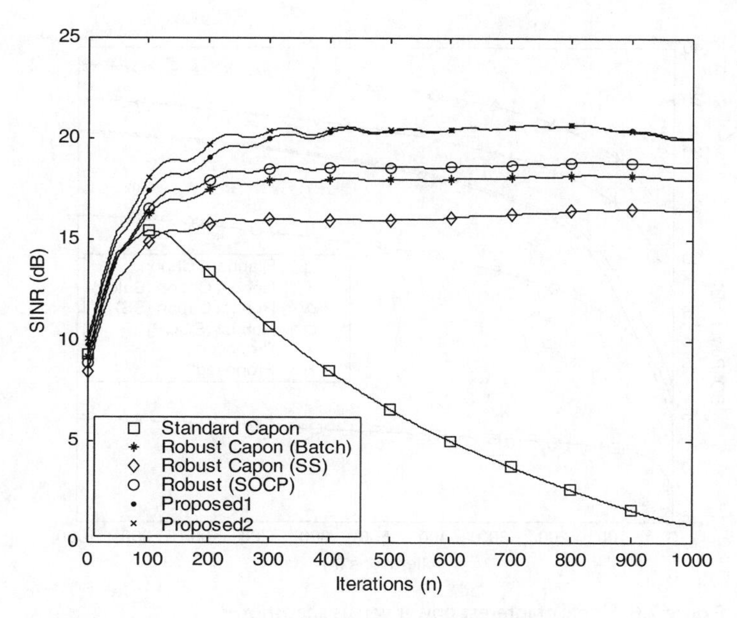

Figure 7.4 Output SINR versus snapshot for first scenario.

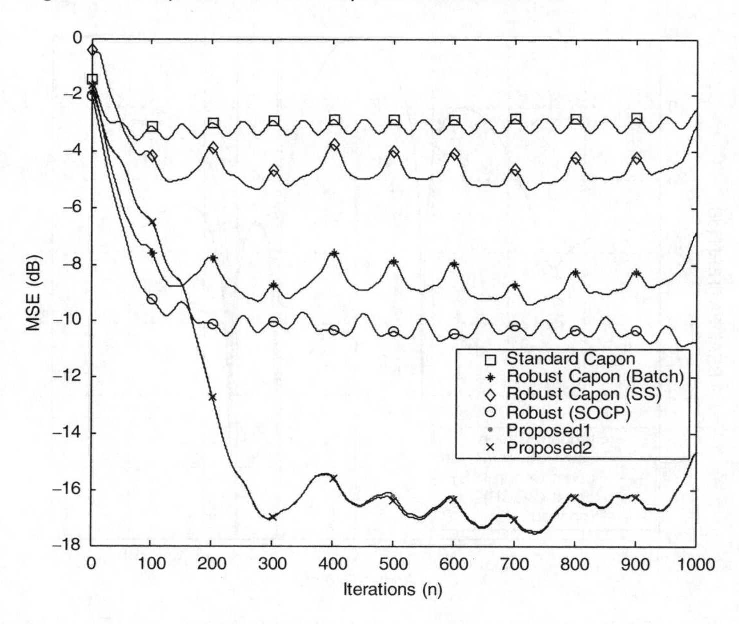

Figure 7.5 Mean squared error versus snapshots.

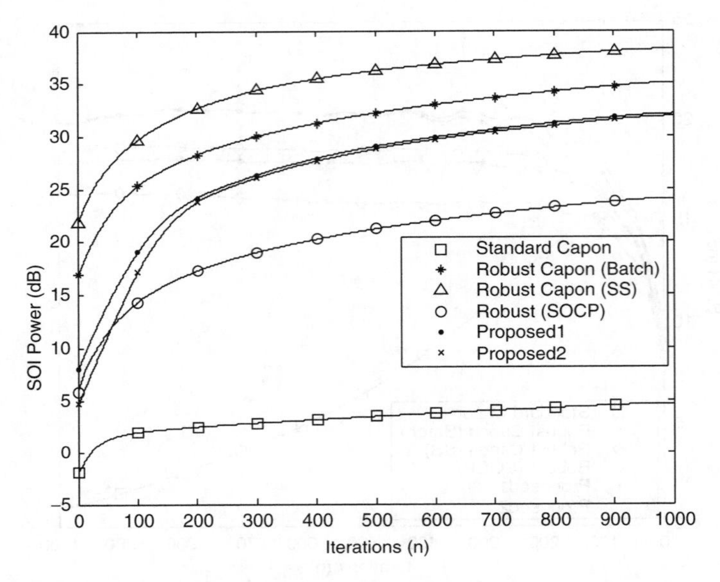

Figure 7.6 Signal of interest power versus snapshot.

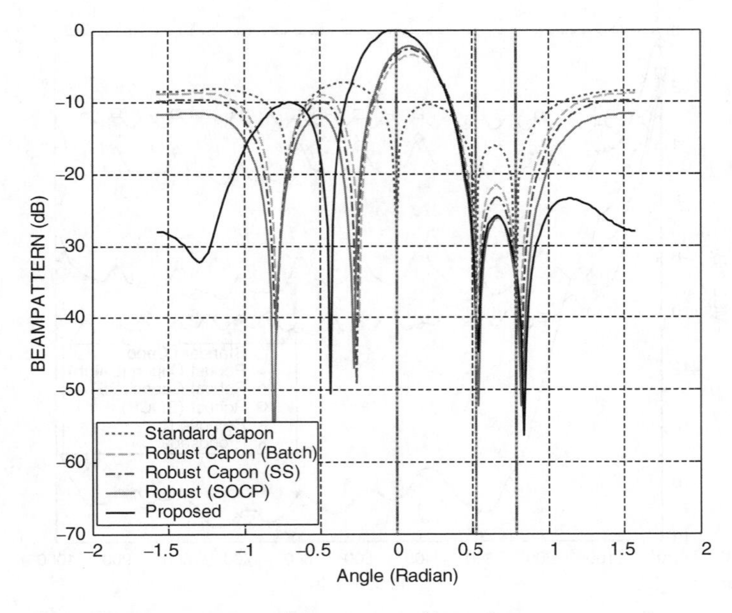

Figure 7.7 Beampatterns of the presented beamformers.

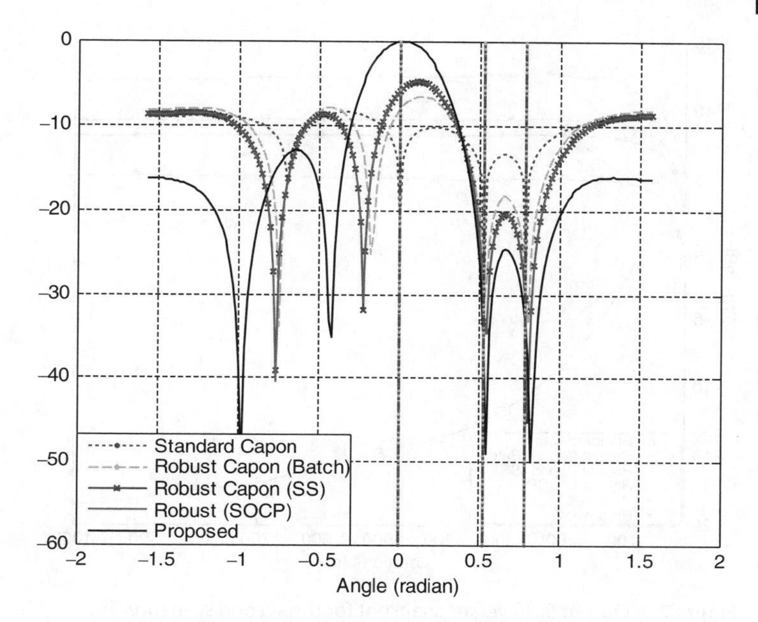

Figure 7.8 Beampatterns of the presented beamformers.

in spite of the deployment of a fast subspace tracking algorithm. The eigendecomposition approach (batch and subspace tracking) offer the maximum SOIP, as shown in Figure 7.6. This is, not surprisingly, due to the always diagonal loading technique utilized by these approaches. However, the SOIP is not the accurate performance measure. Figure 7.5 shows that the presented algorithms exhibit the minimum MSE performance and hence they have the best signal-tracking capability. It is clear from Figures 7.7 and 7.8 that the beampattern of the presented approach generates its maximum gain along the direction of target user, while the gain towards each interference is relatively much lower. Moreover, the presented approach exhibits the best sidelobe suppression. The performance of the standard Capon beamformer is substantially degraded because of the DOA mismatch. As a final point, the single and joint constraints approaches perform almost as well in the first scenario with high SNR.

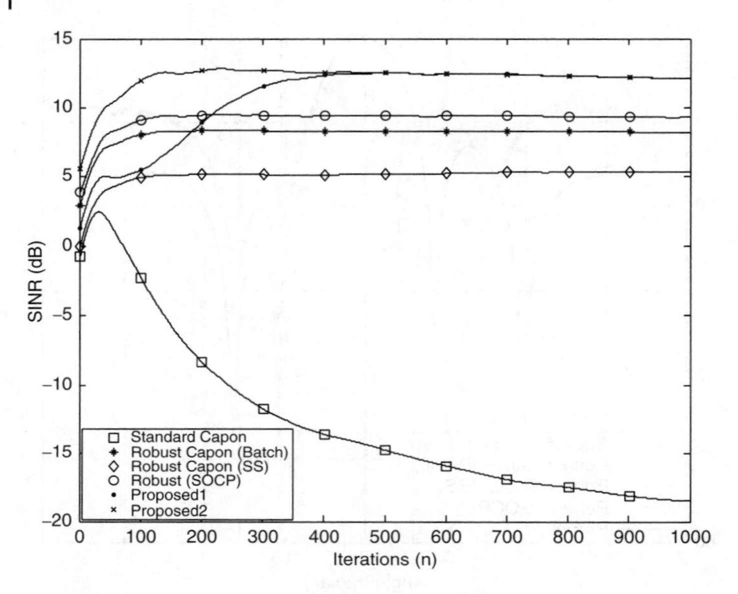

Figure 7.9 Output SINR versus snapshot for the second scenario.

Figure 7.9 illustrates the SINR for the second scenario.[5] The general view of Figure 7.9 is similar to Figure 7.4 except that the single constraint beamformer has an initial performance degradation in the form of a slow convergence rate due to low SNR (that is, antenna element noise power). The joint constraint approach overcomes this shortcoming and appears to be the superior approach. It is interesting now to distinguish between the quadratic inequality constraint solution [38–41] and the solution proposed in Section 7.3 [2]; it was seen from several simulation scenarios that this technique [2] provides the required robustness against noise enhancement over a wide range of the constraint value τ. On the other hand, the selection of the constraint value [13] depends on the particular environment and must be properly selected. This can be verified from the assumption made in (7.52) which yields a cooperative robust approach. More specifically, the quadratic constraint will have graceful effect at low SNR and in the initial phase of

5 We only illustrate the SINR in the second scenario as the other figures are comparable to the first scenario.

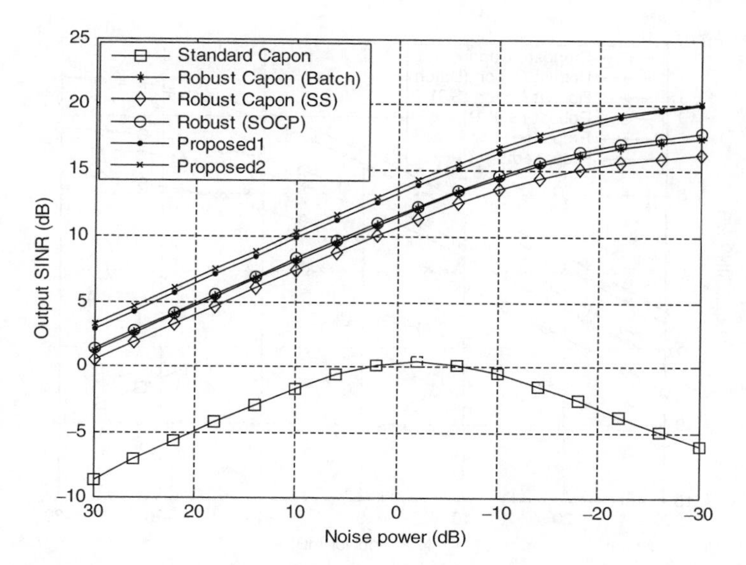

Figure 7.10 Output SINR versus noise power with 0.03π mismatch angle.

recursive adaptation. Otherwise, the ellipsoidal constraint can manage the situation.

It is appealing now to examine the behaviors of the beamformers with noise power variation (SNR). Figures 7.10 and 7.11 compare the six aforementioned beamformers in terms of SINR vs. noise power in the two scenarios. The figures demonstrate that the presented robust beamformers achieve the best SINR over a wide range of the noise power. The joint constraint scheme offers a little improvement over the single constraint scheme, particularly at low SNR.

Finally, we examine the effect of moving interferers on the proposed beamformers as well as the other robust and conventional beamformers. Two moving interferer scenarios are considered here. The first scenario is referred to as the stationary scenario and consists of two sub-scenarios. The trajectories of the interferers' motions versus snapshot index k for the first scenario are given by (refer to the literature for similar scenarios [121–123]):

$$\phi_1 = \pi/6 \pm 0.01\frac{\pi}{180}k \tag{7.106}$$

$$\phi_2 = \pi/4 + 0.1\frac{\pi}{180}k \tag{7.107}$$

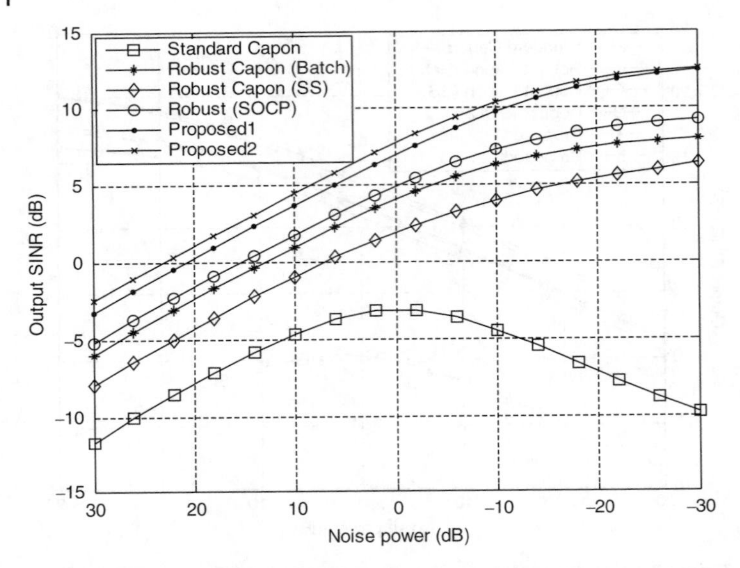

Figure 7.11 Output SINR versus noise power with 0.06π DOA mismatch.

The minus sign in first interferer trajectory simulates a coherent jamming sub-scenario, while the plus sign simulates the non-coherent jamming sub-scenario.

The second scenario is referred to as the non-stationary scenario, and the jammers' trajectories are as follows (again refer to the literature for a similar scenario [66]):

$$\phi_1 = \pi/6 + 10 \times \frac{\pi}{180} \sin(k/15) \tag{7.108}$$

$$\phi_1 = \pi/4 + 10 \times \frac{\pi}{180} \sin(k/15) \tag{7.109}$$

In addition, a 0.01π DOA mismatch is added to build a full dynamic scenario. The steering vectors of the interferers and hence the system matrix will be time varying and hence the SINR will be computed on a time-index basis. The joint constraint approach is excluded from this simulation as it does not appear to offer any performance improvement over the single constraint approach. To validate the previous finding the noise power is set to -20 dB at each antenna element. Figures 7.12–7.14 demonstrate the beamformers' output SINR for the two scenarios.

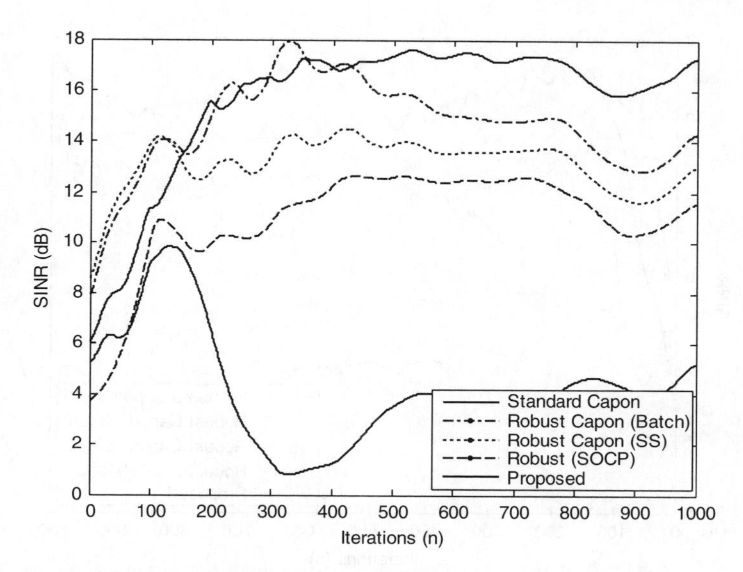

Figure 7.12 Output SINR versus snapshot for non-coherent stationary moving scenario.

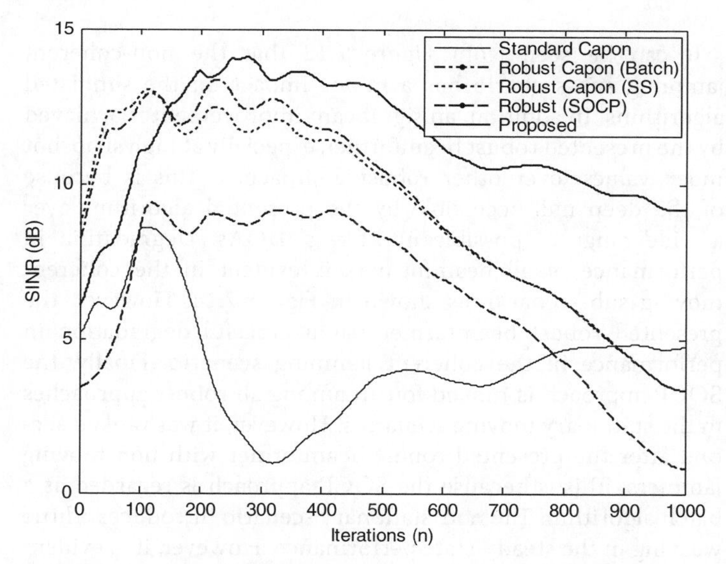

Figure 7.13 Output SINR versus snapshot for coherent stationary moving scenario.

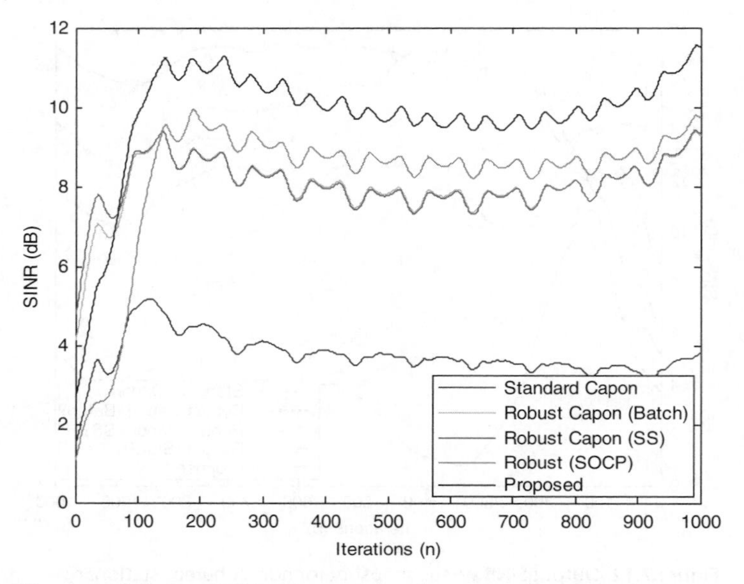

Figure 7.14 Output SINR versus snapshot for non-stationary moving scenario.

It can be seen from Figure 7.12 that the non-coherent jamming sub-scenario has a minor impact on the simulated algorithms. In addition, an significant improvement is achieved by the presented robust beamformer, especially at high snapshot index values, over other robust approaches. This is because of the deep null accessible by the presented algorithm over a wide range of possible interferers' DOAs. Degradation in performance of all beamformers is evident in the coherent moving sub-scenario, as shown in Figure 7.14. However, the presented robust beamformer has a graceful degradation in performance in the coherent jamming scenario. Finally, the SOCP approach is ranked fourth among all robust approaches in the stationary moving scenarios. However, it was ranked second after the presented robust beamformer with non-moving jammers. This is because the SOCP approach is regarded as a batch algorithm. The non-stationary scenario introduces a little weaving in the steady-state performance. However, it is evident from Figure 7.14 that the presented robust beamformer offers an improvement of about 2 dB over SOCP.

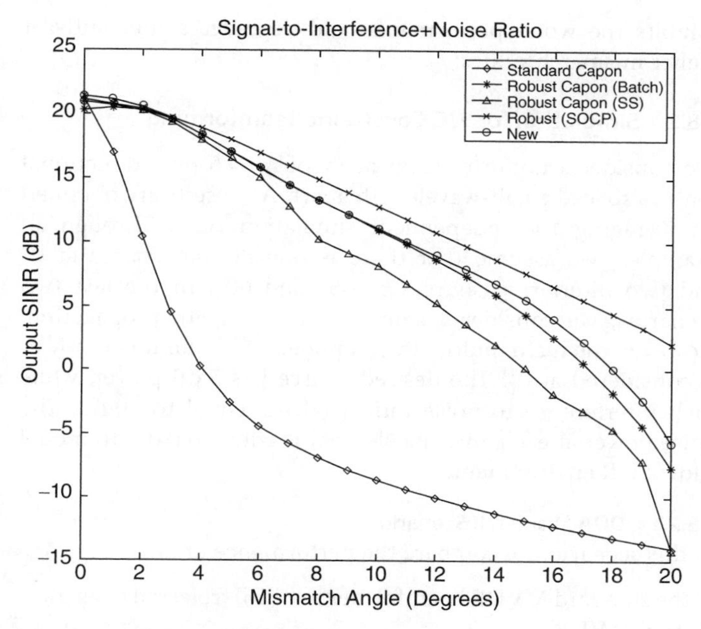

Figure 7.15 SINR versus mismatch angle.

It is worthwhile to analyze the beamforming robustness against the mismatch angle. Figure 7.15 demonstrates the SINR versus signal mismatch in degrees for the five beamformers analyzed in this section. The joint constraint beamformer is not presented in this figure as it is similar to the single constrained beamformer in terms of steady-state performance. The figures demonstrate the ultimate robustness of the algorithm presented in Box 7.1 compared to other robust algorithms. The SOCP is the ranked second after the algorithm presented in Box 7.1 at low mismatch angle i.e., less than 4 degree. However, the SOCP beamformer is the best at higher mismatch angle i.e., more than 4 degrees. All algorithms performed equally well without mismatches, at a level similar to the standard Capon beamformer. The standard Capon beamformer degraded significantly with an increase of the mismatch angle. All robust algorithms degrade with the increase in steering vector mismatch angle. The eigendecomposition beamformer with subspace tracking exhibits the worst performance and degrades significantly at higher mismatch values.

7.8.2 Simulation for WC Constraint Beamforming

We consider a uniform linear array of $M = 5$ omnidirectional sensors spaced a half-wavelength apart. All results are obtained by averaging 100 independent simulation runs. Through all examples, we assume that there is one desired source at 0° and two interfering sources at 45° and 60°. In the last two scenarios, we consider a source with multipath propagation. However, the main multipath component with dominant power is considered at 0°. The desired source has 5 dB power, while each interference-to-noise ratio (INR) is equal to 10 dB. The noise power at each antenna element is equal to 0 dB, to model a low-SNR environment.

7.8.2.1 DOA Mismatch Scenario

In this scenario, we compare the performance of:

- the standard MVDR beamformer in (7.16) (referred to as standard MVDR)
- robust MVDR beamformers with WC constraint, implemented using:
  - SOCP [56], referred to as robust MVDR-WC/SOCP
  - the eigendecomposition [56, 59] approach, referred to as robust MVDR-WC/EigDec.
- the robust adaptive beamformer outlined in this chapter, referred to as robust MVDR-WC/proposed.

These beamformers are simulated using a mismatched steering vector of the desired source where the presumed AOI equals 5°. The robust MVDR-WC/EigDec beamformer is computed using (7.86) and its norm is obtained using a Newton-like algorithm [56, 59]. In addition, we simulate the benchmark MVDR beamformer at (7.16) with the actual steering vector of the desired source. The benchmark MVDR beamformer is implemented using the well-known RLS algorithm. The update of the sample covariance matrix in (7.14) is used with all beamformers, with $\eta = 0.97$. The WC constrained parameter $\zeta = 1.8$ is chosen for both robust MVDR-WC/EigDec and robust MVDR-WC/proposed beamformers, while $\zeta = 3$ is

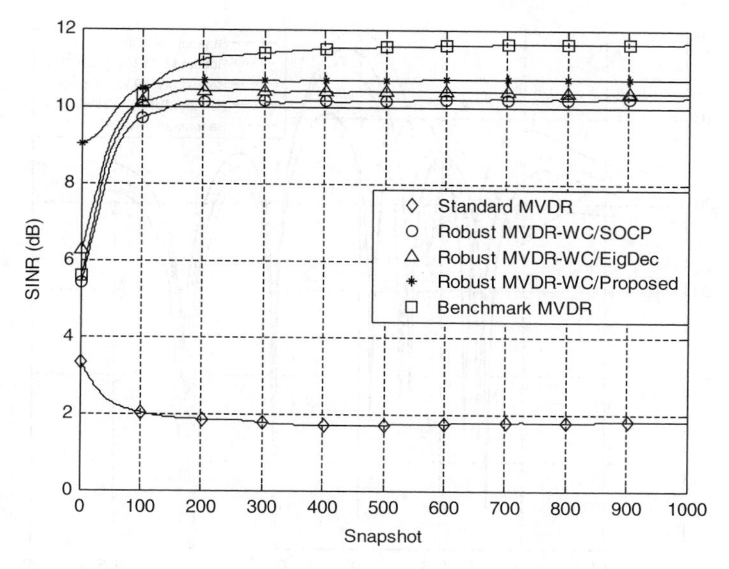

Figure 7.16 Output SINR versus snapshot index for the first scenario.

chosen for the robust MVDR-WC/SOCP beamformer. This is because the SOCP method is initialized with the normalized weight vector. The WC constrained parameter is selected based on the best performance achieved over several simulation runs. In practice, it is selected based on some preliminary (coarse) knowledge about wireless channels or using Monte Carlo simulation. Figure 7.16 shows the output SINR of the beamformers versus snapshot and beam patterns against AOI are illustrated in Figure 7.17. The presented robust WC beamformer has the better SINR than other robust approaches and faster convergence speed than the benchmark MVDR beamformer using the RLS algorithm. The eigendecomposition and SOCP methods are considered as batch algorithms; the weight vector of the robust MVDR-WC/EigDec beamformer is computed using the closed form in (7.86) and the weight vector of the robust MVDR-WC/SOCP beamformer is recomputed completely with each snapshot. Finally, the presented robust WC beamformer is the best at eliminating sidelobes and interference compared to other robust approaches, as evident from Figure 7.17 where it is ranked after the benchmark MVDR beamformer.

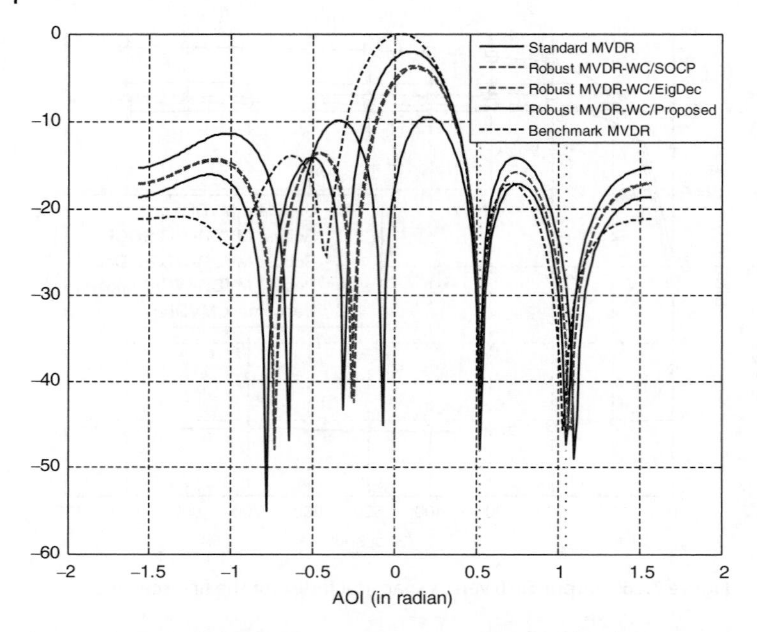

Figure 7.17 Steady state array beam patterns versus AOI (in radians) for the first scenario.

Figure 7.18 shows the output SINR versus noise power using a fixed training sample size of 50 (that is, with low sample support). The figure clearly demonstrates the superiority of the presented robust WC beamformer, especially at low noise power (high SNR). This is thanks to its optimality at low snapshot index, as observed from Figure 7.16.

In order to analyze the Lagrange multiplier in (7.98), we investigate the parameters of (7.97), which are given in (7.99). These parameters and the Lagrange multiplier λ (referred to as the WC parameters) are plotted versus snapshot index in Figure 7.19 at $\zeta = 1.8$. The figure illustrates that $A > 0, B > 0$ and $C < 0$. Therefore, (7.97) has one real positive root, as explained in Section 4.1. It should be noted that the algorithm commenced the WC optimization from the first snapshot, as shown in Figure 7.19.

It is worthwhile to investigate the sensitivity of the presented robust WC beamformer against the WC constrained value ζ. Figure 7.20 demonstrates the performance of the

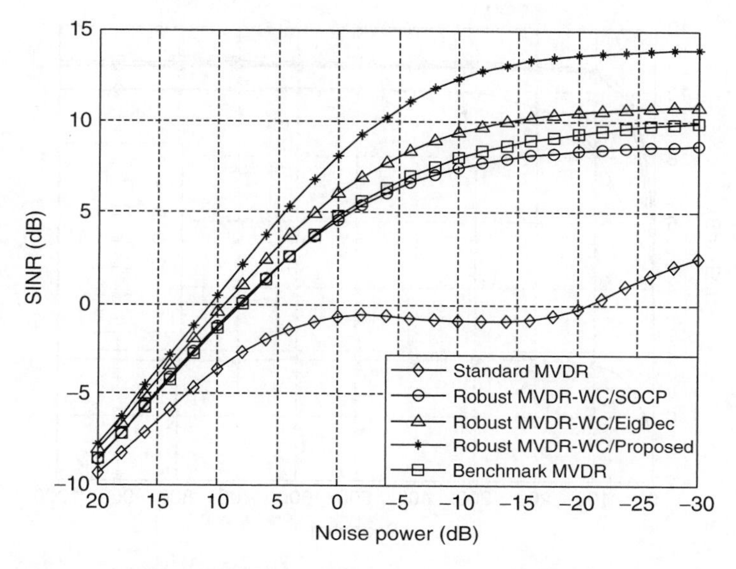

Figure 7.18 Output SINR versus noise power for the first scenario with training data size $N = 50$.

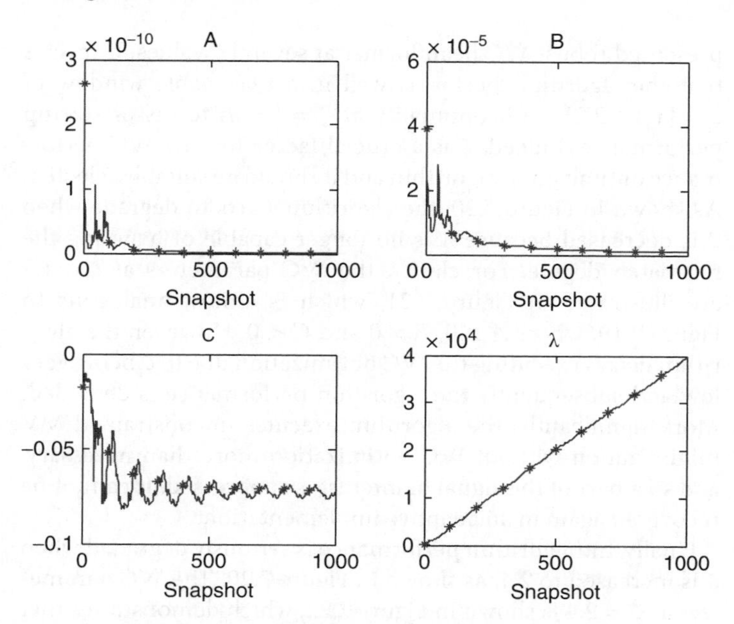

Figure 7.19 WC parameters of the robust MVDR-WC/proposed beamformer at $\zeta = 1.8$.

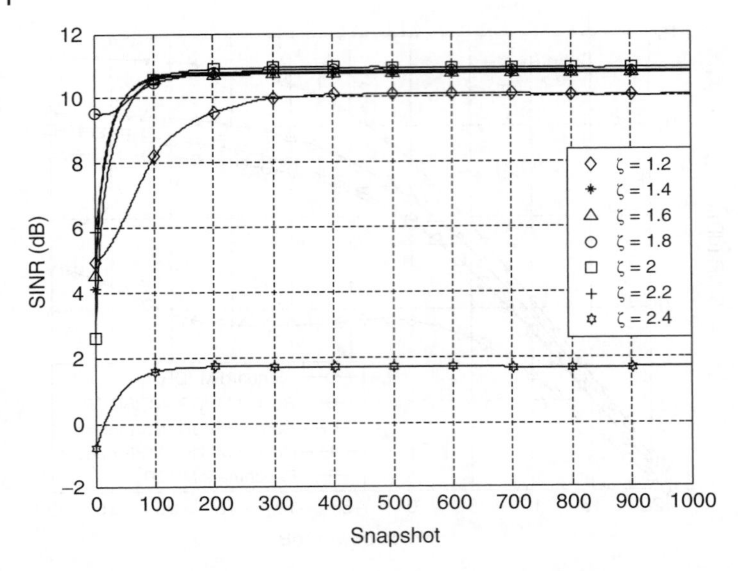

Figure 7.20 Effect of WC parameter ζ on the output SINR of the first scenario.

presented robust WC beamformer at several ζ values. It reveals that the algorithm performs well in a reasonable window of $\zeta = [1.4 : 2.2]$, with optimality at $\zeta = 1.8$ in terms of startup performance. Indeed, ζ is a crucial factor for any WC performance optimization algorithm and it should be suitably selected. As shown in Figure 7.20, the algorithm starts to degrade when ζ is decreased because it is no longer capable of handling the mismatch degree. For clarity, the WC parameters at $\zeta = 1.2$ are illustrated in Figure 7.21, which is almost analogous to Figure 7.19, where $A > 0$, $B > 0$ and $C < 0$. However, the algorithm delays executing the WC optimization due to ζ being very low and subsequently the algorithm performance is degraded. More significantly, the algorithm executes unconstrained MV minimization without WC optimization more than necessary, and so a part of the signal of interest is suppressed. It cannot be recovered again in an adaptive implementation.

Finally, the algorithm performance is seriously degraded when ζ is increased to 2.4, as shown in Figure 7.20. The WC parameters at $\zeta = 2.4$ is shown in Figure 7.22, which demonstrates that

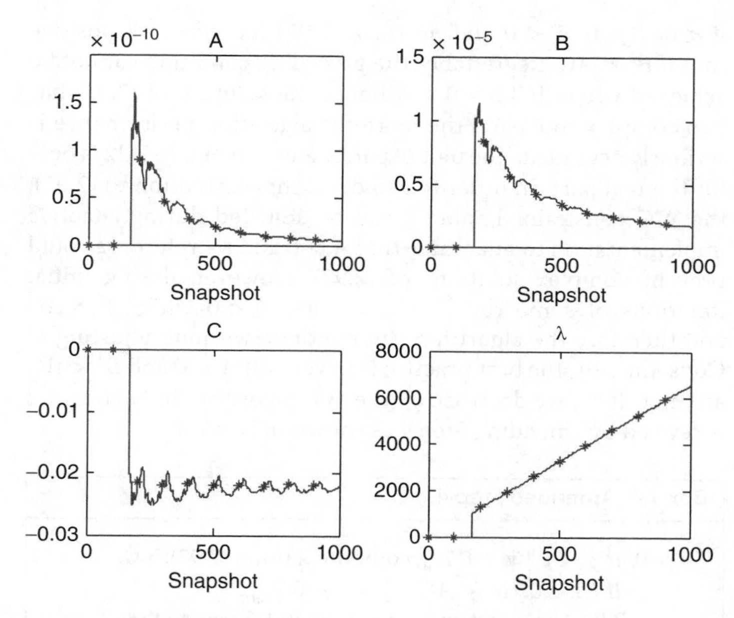

Figure 7.21 WC parameters of the robust MVDR-WC/proposed beamformer at $\zeta = 1.2$.

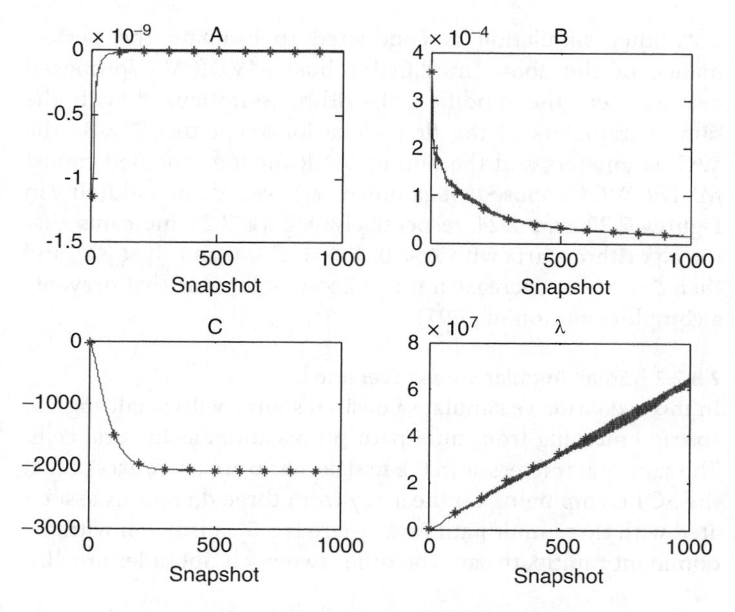

Figure 7.22 WC parameters of the robust MVDR-WC/proposed beamformer at $\zeta = 2.4$.

$A < 0$, $B > 0$, $C < 0$ and, in turn, (7.97) has one real positive root if $B^2 < AC$. Regrettably, the preceding condition cannot be achieved where $|C| >> 0$ and hence the solution of (7.97) has two complex roots and therefore the algorithm performance is seriously degraded. The plot of parameter λ in Figure 7.22 is only for the real part. In order to avoid a complex solution to (7.97), the WC constrained value ζ can be adjusted during adaptive implementation to guarantee that $A > 0$ and therefore we could prevent complex solutions of (7.97). However, during initial iterations of some runs, we may have $A < 0$ while $B^2 > AC$ and therefore the algorithm can continue without adjusting ζ. Consequently, the best practice is to verify if $A < 0$ and $B^2 < AC$ are met. If so, we decrease ζ. The WC algorithm in Section 4.2 is revised by amending Step 4 as shown in Box 7.4.

Box 7.4 Amended Step 4.

Step 4) ***If*** $\chi < \zeta \|\widehat{\mathbf{w}}(k+1)\|$, compute λ using (7.98); $5M$.
If $(A < 0 \varsigma (B^2 < AC))_{,}$ $\zeta = \zeta - 0.1$, *end*
Else $\lambda = 0$ and $\mathbf{w}(k+1) = \widehat{\mathbf{w}}(k+1) \rightarrow$ go to Step 1.

Another simulation is conducted to evaluate the performance of the above modified robust MVDR-WC/proposed beamformer. The modified algorithm is initialized with the same parameters of the first scenario except that $\zeta = 3$. The WC parameters and the output SINR for the modified robust MVDR-WC/proposed beamformer are demonstrated in Figures 7.23 and 7.24 respectively. Figure 7.23 indicates that the algorithm starts with $A < 0$, $B > 0$, $C < 0$ and $B^2 < AC$, and then ζ starts to decrease until an acceptable value that prevents a complex solution of (7.97).

7.8.2.2 Small Angular Spread Scenario

In this scenario, we simulate a desired source with small angular spread emerging from multipath propagation, as in rural cells. The same parameters as in the first scenario are used, except that the SOI is impinging on the array from three directions associated with three multipath rays. There is a 5° mismatch with the dominant multipath ray. The other two ray amplitudes are 40%

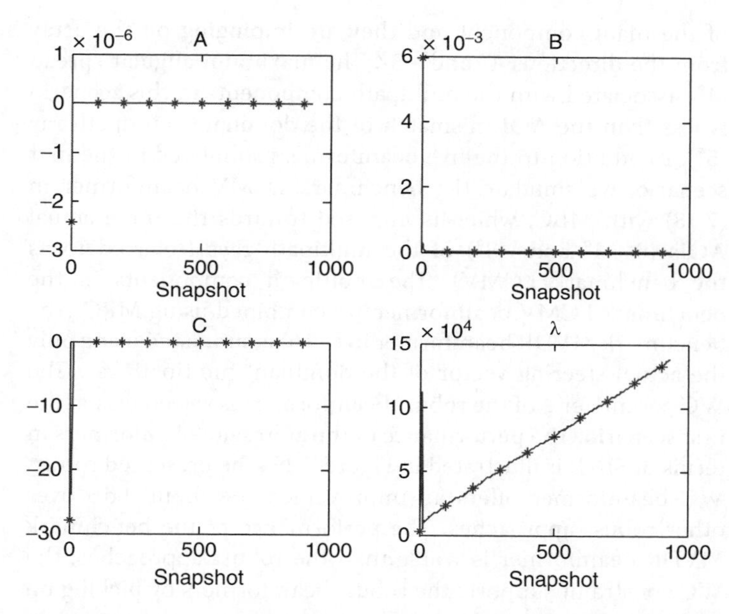

Figure 7.23 WC parameters of the modified robust MVDR-WC/proposed beamformer at $\zeta = 3$.

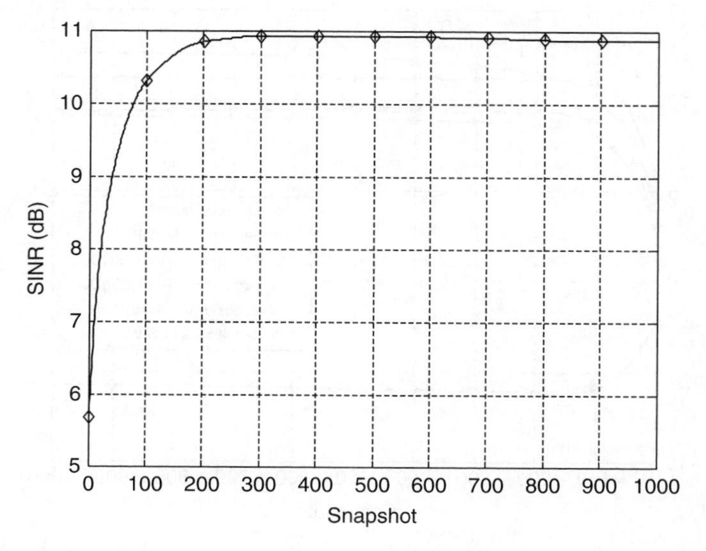

Figure 7.24 Output SINR versus snapshot index for the modified robust MVDR-WC/proposed beamformer with the parameters of the first scenario.

of the main component and they are impinging on the array from the directions 4° and –3°. The maximum angular spread (4°) associated with the multipath components in this scenario is less than the AOI mismatch of the dominant multipath ray (5°). In addition to the five beamformers simulated in the first scenario, we simulate the benchmark LCMV beamformer in (7.18) with MBC, which is imposed towards the three actual AOIs (0°, 4° and –3°) of the multipath rays (referred to as the benchmark LCMV). The multipath components in the benchmark LCMV beamformer are combined using MRC. The benchmark MVDR beamformer in (7.16) is simulated using only the actual steering vector of the dominant multipath ray. The WC parameter ζ of the robust beamformers is selected as in the first scenario. The performance of the aforesaid beamformers in terms of SINR is illustrated in Figure 7.25. The presented robust WC beamformer offers an improvement of about 2 dB over other robust approaches. The performance of the benchmark MVDR beamformer is worse than the robust approaches; the WC constraint supports the robust beamformers by picking up

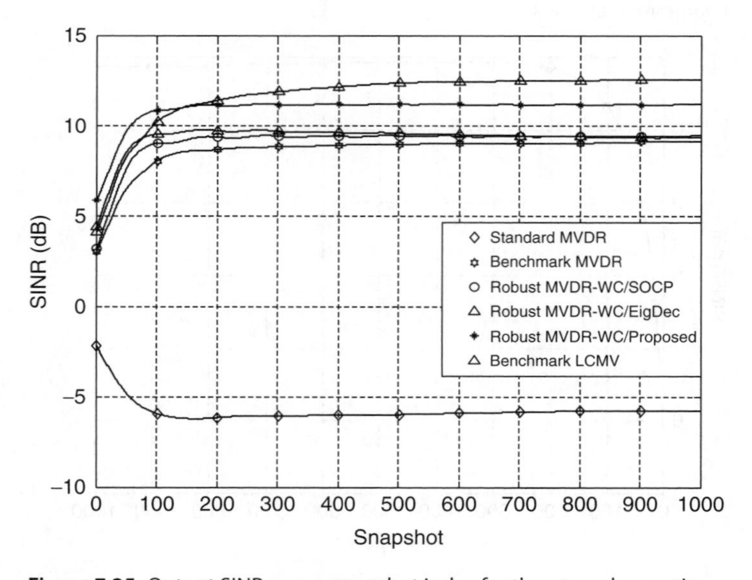

Figure 7.25 Output SINR versus snapshot index for the second scenario.

some signal components from the multipath signals scattered inside the mismatch region bounded by ζ.

7.8.2.3 Large Angular Spread Scenario

In this scenario, we simulate a large angular spread as in the cellular indoor environment. The simulation system is similar to the previous scenario except that the three multipath components are impinging on the array from directions 0°, −30°, and −80°. The dominant ray impinges on the array from the 0° direction and there is a 5° look direction mismatch. We assume that the phases of the multipath rays are independent and uniformly drawn from the interval $[-\pi, \pi]$ in each run. The phases associated with the multipaths vary from run to run while remaining constant during adaptive implementation of each run. In this scenario, we simulate the beamformers in the previous experiments in addition to the presented robust LCMV beamformer with MBWC constraints (referred to as the robust LCMV-MBWC). Moreover, the standard LCMV beamformer with multiple constraints in (7.18) (referred to as the standard LCMV) is simulated using a mismatched steering vector of the dominant multipath ray. There are no mismatches with the other two rays (that is, 5°, −30°, and −80°). The multipath rays of the benchmark LCMV and the standard LCMV beamformers are combined using MRC. The phases of the multipath rays are unknown to all beamformers except the benchmark beamformers. The beamformers of the first scenario are simulated using the same parameters, while the robust LCMV-MBWC beamformer is simulated using $\boldsymbol{\upsilon} = [1.6 \quad 0.2 \quad 0.2]$. The selection of $\boldsymbol{\upsilon}$ is obtained in practice using several simulation runs. It is somehow related to the amplitude distribution of multipath rays. However, in-depth analysis for the tuning of this vector and even optimal estimation are good candidates for future research.

The SINR performance of the seven beamformers is shown in Figure 7.26. First of all, the benchmark MVDR is considerably degraded compared to all beamformers despite tracing the dominant multipath ray. This is because the large angular spread deforms the effective steering vector of the SOI. The performances of the robust beamformers with the single WC constraint resemble those in the first scenario. The robust LCMV-MBWC beamformer offers an improvement of about

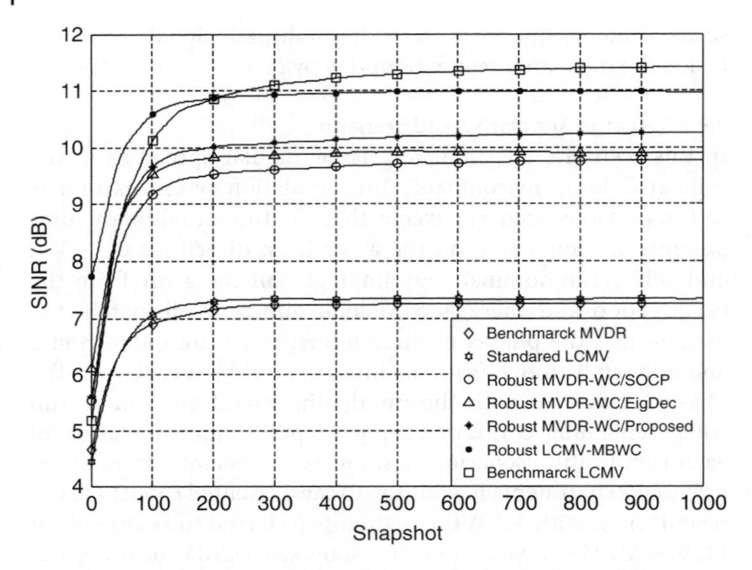

Figure 7.26 Output SINR versus snapshot index for the third scenario.

1 dB over the single WC constraint beamformers due to its efficient multipath handling using multiple WC constraints.

7.9 Summary

We have presented an improved recursive implementation of the robust Capon beamformer including an uncertainty ellipsoidal constraint. It has low computational complexity. Additionally, we developed a joint constraint approach to improve the performance of the single constraint beamformer at low SNR. We verified from computer simulations that the presented adaptive algorithms overall outperform other robust Capon beamformers at high SNR. The single constraint robust beamformer suffers from low convergence speeds in the output SINR at low SNR, while the joint constraints overcome this shortcoming. In addition to the improvements relative to other robust approaches, the presented approaches also have the following merits:

- The diagonal loading term is precisely computed using a simple quadratic equation based on a VL technique.

- The inverse covariance matrix will be only updated. Therefore, the order of computational complexity of this robust approach is comparable with the conventional Capon beamformer.

Robust adaptive beamforming using WC performance optimization is implemented using simple novel approaches. Two efficient implementations are developed using single and multiple WC constraints. The presented implementations are based on iterative gradient minimization with optimum step size. In contrast to existing single-WC robust approaches, the single-WC implementation requires very low computational loads and gives the best performance, especially at low sample support. In addition, the developed algorithm eliminates the need for covariance matrix inversion estimation. The WC performance optimization is generalized to include multiple WC constraints, which produce a robust LCMV beamformer with MBWC constraints. An efficient solution for the LCMV-MBWC beamformer is found by solving a system of nonlinear equations. Simulation results demonstrate the superiority of the presented beamformers over existing robust approaches. Future work may include fine tuning of the constrained vector of the LCMV-MBWC beamformer and developing low-complexity adaptive implementations.

References

1 D. Reynolds, X. Wang, and H.V. Poor, "Blind adaptive space-time multiuser detection with multiple transmitter and receiver antennas," *IEEE Trans. Signal Process.*, vol. 50, no. 6, pp. 1261–1276, 2002.

2 A. Elnashar, S. Elnoubi, and H. Elmikati, "Further study on robust adaptive beamforming with optimum diagonal loading," *IEEE Trans. Antennas Propag.*, vol. 54, no. 12, pp. 3647–3658, 2006.

3 A. Elnashar, "On efficient implementation of robust adaptive beamforming based on worst-case performance optimization," *IET Signal Process.*, vol. 2, no. 4, pp. 381–393, 2008.

4 M. Martone, "Fast adaptive super-exponential multistage beamforming for cellular base-station transceivers with

antenna arrays," *IEEE Trans. Vehic. Tech.*, vol. 48, no. 4, 1999.

5 S. Kapoor, S. Gollamudi, S. Nagaraj, and Y.-F. Huang, "Adaptive multiuser detection and beamforming for interference suppression in CDMA mobile radio," *IEEE Trans. Vehic. Techn.*, vol. 48, no. 5, pp. 1341–1355, 1999.

6 Z. Tian, *Blind Multi-user Detection with Adaptive Space-time Processing for DS-CDMA Wireless Communications*, PhD thesis, George Mason University, 2000.

7 Y.J. Zhang and K.B. Letaief, "An efficient resource-allocation scheme for spatial multiuser access in MIMO/OFDM systems", *IEEE Trans. Commun.*, vol. 53, no. 1, pp. 107–116, 2005.

8 A.J. Paulraj and C.B. Papadias, "Space-time processing for wireless communications," *IEEE Pers. Commun. Mag.*, vol. 14, no. 5, pp. 49–83, 1997.

9 D. Reynolds, X. Wang, and H.V. Poor, "Blind adaptive space-time multiuser detection with multiple transmitter and receiver antennas," *IEEE Trans. Signal Process.*, vol. 50, no. 6, pp. 1261–1276, 2002.

10 H. Dai and H.V. Poor, "Iterative space-time processing for multiuser detection in multipath CDMA channels," *IEEE Trans. Signal Process.*, vol. 50, no. 9, pp. 2116–2127, 2002.

11 X. Wang and H.V. Poor, "Space-time multi-user detection in multipath CDMA channels," *IEEE Trans. Signal Process.*, vol. 47, no. 9, pp. 2356–2374, 1999.

12 X. Wang and H.V. Poor, *Wireless Communication Systems: Advanced Techniques for Signal Reception*, Prentice Hall, 2002.

13 M. Martone, "Fast adaptive super-exponential multistage beamforming for cellular base-station transceivers with antenna arrays," *IEEE Trans. Vehic. Tech.*, vol. 48, no. 4, pp. 1017–1028, 1999.

14 A. Yener, R.D. Yates, and S. Ulukus, "Combined multiuser detection and beamforming for CDMA systems: filter structures," *IEEE Trans. Vehic. Techn.*, vol. 51, no. 5, pp. 1087–1095, 2002.

15 A. Yener, R.D. Yates, and S. Ulukus, "Interference management for CDMA systems through power control,

multiuser, and beamforming," *IEEE Trans. Commun.*, vol. 49, pp. 1227–1239, 2001.
16 B.D. van Veen and K.M. Buckley, "Beamforming: a versatile approach to spatial filtering," *IEEE Acoust. Speech Signal. Process. Mag.*, pp. 4–24, 1988.
17 J. Litva and T.K.-Y. Lo. *Digital Beamforming in Wireless Communications.* Artech House, 1996.
18 B.D. Carlson, "Covariance matrix estimation errors and diagonal loading in adaptive arrays," *IEEE Trans. Aerosp. Electron. Syst.*, vol. 24, no. 4, pp. 397–401, 1988.
19 H. Cox, R. Zeskind, and M. Owen, "Robust adaptive beamforming", *IEEE Trans. Acoust. Speech Signal Process.*, vol. ASSP-35, no. 10, pp. 365–1376, 1987.
20 J. Litva and T.K.-Y. Lo, *Digital Beamforming in Wireless Communications.* Artech House, 1996.
21 L.J. Griffiths and C.W. Jim, "An alternative approach to linearly constrained adaptive beamforming," *IEEE Trans. Antennas Propag.*, vol. AP–30, pp. 27–34, 1982.
22 M. Agrawal and S. Prasad, "Robust adaptive beamforming for wide-band moving, and coherent jammers via uniform linear arrays," *IEEE Trans. Antennas Propag.*, vol. 47, no. 8, pp. 1267–1275, 1999.
23 F. Qian and B.D. van Veen, "Quadratically constrained adaptive beamforming for coherent signals and interference," *IEEE Trans. Signal Process.*, vol. 43, no. 8, pp. 1890–1900, 1995.
24 Q. Wu and K.M. Wong, "Blind adaptive beamforming for cyclostationary signals," *IEEE Trans. Signal Process.*, vol. 44, no. 11, pp. 2757–2767, 1996.
25 M. Martone, "Fast adaptive super-exponential multistage beamforming for cellular base-station transceivers with antenna arrays," *IEEE Trans. Vehic. Tech.*, vol. 48, no. 4, pp. 1017–1028, 1999.
26 L.C. Godara, "Applications of antenna arrays to mobile communications, Part II: beamforming and directional-of-arrival consideration," *Proc. IEEE*, vol. 85, no. 5, pp 1195–1245, 1997.
27 K.L. Bell, Y. Ephraim, and H.L. Van Trees, "A Bayesian approach to robust adaptive beamforming," *IEEE Trans. Signal Process.*, vol. 48, pp. 386–398, 2000.

28 B. Yang, "Projection approximation subspace tracking," *IEEE Trans. Signal Process.*, vol. 43, pp. 95–107, 1995.

29 M.W. Ganz, R.L. Moses, and S.L. Wilson, "Convergence of the SMI and diagonally loaded SMI algorithm with weak interference," *IEEE Trans. Antennas Propag.*, vol. 38, no. 3, pp. 394– 399, 1990.

30 B.D. Carlson, "Covariance matrix estimation errors and diagonal loading in adaptive arrays," *IEEE Trans. Aerosp. Electron. Syst.*, vol. 24, no. 4, pp. 397–401, 1988.

31 A. Elnashar, S. Elnoubi, and H. Elmikati, "Low-complexity robust adaptive generalized sidelobe canceller detector for DS/CDMA systems," *Int. J. Adapt. Control Signal Process.*, vol. 23, no. 3, pp. 293–310, 2008.

32 T.F. Ayoub and A.M. Haimovich, "Modified GLRT signal detection algorithm," *IEEE Trans. Aerosp. Electron. Syst.*, vol. 36, no. 3, pp. 394–399, 2000.

33 J.R. Gurrci and J.S. Bergin, "Principal components, covariance matrix tapers, and the subspace leakage problem," *IEEE Trans. Aerosp. Electron. Syst.*, vol. 38, no. 1, pp. 152–162, 2002.

34 S.Z. Kalson, "Adaptive array CFAR detection," *IEEE Trans. Aerosp. Electron. Syst.*, vol. 31, no. 2., pp. 534–542, 1995.

35 X. Mestre and M.A. Lagunas, "Diagonal loading effect on the asymptotic output SINR of the minimum variance beamformer," Technical report, CTTC/RC/2003–001, CTTC, 2003.

36 D.N. Swingler. "Finite-time performance of a quadratically constrained MVDR beamformer," *IEEE J. Ocean Eng.*, vol. 22, no. 2, pp. 376–384, 1997.

37 Z. He and Y. Chen, "Robust blind beamforming using neural network," *IEEE P. Radar Sonar Nav.*, vol. 147, no. 1, pp. 41–46, 2000.

38 Z. Tian, K.L. Bell, and H.L. Van Trees, "A recursive least squares implementation for LCMP beamforming under quadratic constraint," *IEEE Trans. Signal Process.*, vol. 49, no. 6, pp. 1138–1145, 2001.

39 Z. Tian, K.L. Bell, and H.L. Van Trees, "A RLS implementation for adaptive beam-forming under quadratic constraint." In: *9th IEEE Workshop on Stat. Sig. and Array Processing*, Portland, Oregon, September 1998.

40 Z. Tian, K.L. Bell, and H.L. Van Trees, "Quadratically constrained RLS filtering for adaptive beamforming and DS-CDMA multi-user detection." In: *Adaptive Sensor Array Processing Workshop*, MIT Lincoln Lab, Lexington, MA, March 1999.

41 Z. Tian, K.L. Bell, and H.L. Van Trees, "Robust RLS implementation of the minimum output energy detector with multiple constraints," Technical report, Center of Excellence in C3I, George Mason University, USA, July 15, 2000.

42 Z. Tian, K.L. Bell, and H.L. Van Trees, "Robust constrained linear receivers for CDMA wireless systems," *IEEE Trans. Signal Process.*, vol. 49, no. 7, pp. 1510–1522, 2001.

43 A. Elnashar, S. Elnoubi, and H. Elmikati, "Robust adaptive beamforming with variable diagonal loading." In: *Proceedings of Sixth International Conference on 3G and Beyond*, 07–09 November 2005, London, pp. 489–493.

44 A. Elnashar, S. Elnoubi, and H. Elmikati, "A robust block-Shanno adaptive blind multiuser receiver for DS-CDMA systems." In: *Proceedings of IST Mobile & Wireless Communications Summit 2005*, Dresden, Germany, 19–23 June 2005.

45 A. Elnashar, S. Elnoubi, and H. Elmikati, "A robust linearly constrained CMA for adaptive blind multiuser detection." In: *Proceedings of IEEE WCNC 2005 Conference*, New Orleans, LA, USA, 13–17 March 2005, vol. 1, pp. 233–238.

46 A. Elnashar, S. Elnoubi, and H. Elmikati, "A robust quadratically constrained adaptive blind multiuser receiver for DS/CDMA systems." In: *IEEE International Symposium on Spread Spectrum Techniques and Applications* (ISSSTA 2004), Sydney, Australia, 30 August–3 September 2004.

47 S.A. Vorobyov, A.B. Gershman, and Z.-Q. Luo, "Robust adaptive beamforming using worst-case performance optimization," *IEEE Trans. Signal Process.*, vol. 51, pp. 313–324, 2003.

48 S.A. Vorobyov, A.B. Gershman, and Z.-Q. Luo, "Robust MVDR beamforming using worst-case performance optimization." In: *Proceedings of 10th Workshop on Adaptive Sensor Array Processing*, Lexington, March 2002.

49 R.G. Lorenz and S.P. Boyd: "Robust minimum variance beamforming," *IEEE Trans. Signal Process.*, vol. 53, no. 5, pp. 1684–1696, 2005.

50 R.G. Lorenz and S.P Boyd, "Robust minimum variance beamforming." In: *37th Asilomar Conference on Signals, Systems & Computers*, 9–12 November 2003, vol 2, pp. 1345–1352.

51 P. Stoica, Z. Wang, and J. Li, "Robust Capon beamforming," *IEEE Signal Process. Lett.*, vol. 10, no. 6, pp. 172–175, 2003.

52 J. Li, P. Stoica, and Z. Wang, "On robust Capon beamforming and diagonal loading," *IEEE Trans. Signal Process.*, vol. 51, no. 7, pp. 1702–1715, 2003.

53 J.F. Sturm, "Using SeDuMi 1.02, a MATLAB toolbox for optimization over symmetric cones," *Optim. Meth. Softw.*, vol. 11–12, pp. 625–653, 1999.

54 S. Shahbazpanahi, A.B. Gershman, Z.Q. Luo, and K.M. Wong, "Robust adaptive beamforming for general-rank signal model," *IEEE Trans. Signal Process.*, vol. 51, pp. 2257–2269, 2003.

55 A.B. Gershman, "Robust adaptive beamforming in sensor arrays," *AEU Int. J. Electron. Commun.*, vol. 53, no. 6, pp. 305–314, 1999.

56 J. Li and P. Stoica, *Robust Adaptive Beamforming*, John Wiley & Sons, 2006.

57 S. Cui, M. Kisialiou, Z.-Q. Luo, and Z. Ding, "Robust blind multiuser detection against signature waveform mismatch based on second order cone programming," *IEEE Trans. Wireless Commun.*, vol. 4, no. 4, pp. 1285–1291, 2005.

58 S. Shahbazpanahi and A.B. Gershman, "Robust blind multiuser detection for synchronous CDMA system using worst-case performance," *IEEE Trans. Wireless Commun.*, vol. 3, no. 6, pp. 2232–2245, 2004.

59 K. Zarifi, S. Shahbazpanahi, A.B. Gershman, and Z.-Q. Luo, "Robust blind multiuser detection based on the worst-case performance optimization of the MMSE receiver," *IEEE Trans. Signal Process.*, vol. 53, no. 1, pp. 295–205, 2005.

60 O. Besson and F. Vincent: "Performance analysis of beamformers using generalized loading of the covariance matrix in the presence of random steering vector errors," *IEEE Trans. Signal Process.*, vol. 53, no. 2, pp. 452–459, 2005.

61 F. Vincent and O. Besson: "Steering vector errors and diagonal loading," *IEEE P. Radar Son. Nav.*, vol. 151, no. 6, pp. 337–343, 2004.
62 C.C. Gaudes, I. Santamaría, J. Vía, E.M. Gómez, and T.S. Paules, "Robust array beamforming with sidelobe control using support vector machines," *IEEE Trans. Signal Process.*, vol. 55, no. 2, pp. 574–584, 2007
63 S. Kikuchi, H. Tsuji, and A. Sano, "Autocalibration algorithm for robust Capon beamforming," *IEEE Antennas Wireless Propag. Lett.*, vol. 5, p. 251, 2006.
64 S.-J. Kim, A. Magnani, A. Mutapcic, S.P. Boyd, and Z.-Q. Luo, "Robust beamforming via worst-case SINR maximization," *IEEE Trans. Signal Process.*, vol. 56, no. 4, pp. 1539–1547, 2008.
65 Y.C. Eldar, A. Nehorai, and P.S. La Rosa, "A competitive mean-squared error approach to beamforming," *IEEE Trans. Signal Process.*, vol. 55, no. 11, pp. 5143–5150, 2007.
66 S.A. Vorobyov, A.B. Gershman, Z.-Q. Luo, and N. Ma," Adaptive beamforming with joint robustness against mismatched signal steering vector and interference nonstationarity," *IEEE Signal Process. Lett.*, vol. 11, no. 2, pp. 108–111, 2004.
67 C.-Y. Chen and P.P. Vaidyanathan, "Quadratically constrained beamforming robust against direction-of-arrival mismatch," *IEEE Trans. Signal Process.*, vol. 55, no. 8, pp. 4139–4150, 2007.
68 W. Liu and S. Ding, "An efficient method to determine the diagonal loading factor using the constant modulus feature," *IEEE Trans. Signal Process.*, vol. 56, no. 12, pp. 6102–6106, 2008.
69 A. van der Veen, "Analytical method for blind binary signal separation," *IEEE Trans. Signal Process.*, vol. 45, no. 1, pp. 1078–1082, 1997.
70 S.-J. Kim, A. Magnani, A. Mutapcic, S.P. Boyd and Z.-Q. Luo, "Robust beamforming via worst-case SINR maximization", *IEEE Trans. Signal Process.*, vol. 56, no. 4, pp. 1539–1547, 2008.
71 S.A. Vorobyov, A.B. Gershman, and Y. Rong, "On the relationship between the worst-case optimization-based and probability-constrained approaches to robust adaptive

beamforming." In: *Proceedings of IEEE ICASSP*, Honolulu, HI, April 2007, vol. 2, pp. 977–980.

72 Y. Selén, R. Abrahamsson, and P. Stoica, "Automatic robust adaptive beamforming via ridge regression," *Signal Process.*, vol. 88, no. 1, pp. 33–49, 2008.

73 J. Yang, X. Ma, C. Hou, Y. Liu, and W. Li, "Fully automatic robust adaptive beamforming via principal component regression." In: *Proceedings of International Conference on Signal Processing*, October 2008, pp. 358–361.

74 J. Yang, X. Ma, C. Hou, Y. Liu, and W. Li, "Robust adaptive beamforming using partial least squares." In: *Proceedings of International Conference on Signal Processing*, October 2008, pp. 362–365.

75 L. Du, J. Li and P. Stoica, "Fully automatic computation of diagonal loading levels for robust adaptive beamforming", *IEEE Trans. Aerosp. Electron. Syst.*, vol. 46, no. 1, pp. 449–458, 2010.

76 J. Yang, X. Ma, C. Hou, and Y. Liu, "Automatic generalized loading for robust adaptive beamforming," *IEEE Signal Process. Lett.*, vol. 16, no. 3, pp. 219–222, 2009.

77 J. Yang, X. Ma, C. Hou, Y. Liu, and Z. Yao "Automatic robust linear receiver for multi-access space-time block coded MIMO systems", *IEEE Signal Process. Lett.*, vol. 16, no. 8, p. 687, 2009.

78 C.W. Xing, S.D. Ma and Y.C. Wu, "On low complexity robust beamforming with positive semidefinite constraints", *IEEE Trans. Signal Process.*, vol. 57, pp. 4942–4945, 2009.

79 L. Lei, J.P. Lie, A.B. Gershman and C.M.S. See, "Robust adaptive beamforming in partly calibrated sparse sensor arrays", *IEEE Trans. Signal Process.*, vol. 58, no. 3, pp. 1661–1667, 2010.

80 C.J. Lam and A.C. Singer, "Bayesian beamforming for DOA uncertainty: theory and implementation," *IEEE Trans. Signal Process.*, vol. 54, no. 11, Nov. 2006.

81 Y. Lu, J. An and X. Bu, "Adaptive Bayesian beamforming with sidelobe constraint", *IEEE Commun. Lett.*, vol. 14, no. 5, pp. 369–371, 2010.

82 O. Ledoit, and M. Wolf, "A well-conditioned estimator for large-dimensional covariance matrices," *J. Multivar. Anal.*, vol. 88, pp. 365–411, 2004.

83 P. Stoica, J. Li, X. Zhu, and J.R. Guerci, "On using a priori knowledge in space-time adaptive processing," *IEEE Trans. Signal Process.*, vol. 56, pp. 2598–2602, 2008.
84 X. Zhu, J. Li, and P. Stoica, "Knowledge-aided adaptive beamforming," *IET Signal Process.*, vol. 2, pp. 335–345, 2008.
85 B. Liao, K.M. Tsui, and S.C. Chan, "Robust beamforming with magnitude response constraints Using iterative second-order cone programming," *IEEE Trans. Antennas Propag.*, vol. 59, no. 9, pp. 3477–3482, 2011.
86 Y. Zhao and W. Liu, "Robust wideband beamforming with frequencyresponse variation constraint subject to arbitrary norm-bounded error", *IEEE Trans. Antennas Propag.*, vol. 60, no. 5, pp. 2566–2571, 2012.
87 A. Khabbazibasmenj, S.A. Vorobyov, and A. Hassanien, "Robust adaptive beamforming based on steering vector estimation with as little as possible prior information," *IEEE Trans. Signal Process.*, vol. 60, no. 6, pp. 2974–2987, 2012.
88 A. Beck and Y.C. Eldar, "Doubly constrained robust Capon beamformer with ellipsoidal uncertainty sets," *IEEE Trans. Signal Process.*, vol. 55, pp. 753–758, 2007.
89 Y. Huang and D.P. Palomar, "Rank-constrained separable semidefinite programming with applications to optimal beamforming," *IEEE Trans. Signal Process.*, vol. 58, pp. 664–678, 2010.
90 Z.-Q. Luo, W.-K. Ma, A.M.-C. So, Y. Ye, and S. Zhang, "Semidefinite relaxation of quadratic optimization problems," *IEEE Signal Process. Mag.*, vol. 27, no. 3, pp. 20–34, 2010.
91 M. Rahmani, M.H. Bastani, and S. Shahraini, "Two layers beamforming robust against direction-of-arrival mismatch," *IET Signal Process.*, vol. 8, no. 1, pp. 49–58, 2014.
92 H.L. van Trees, *Optimum Array Processing: Part IV of Detection, Estimation, and Modulation Theory*. Wiley, 2002.
93 K.C. Toh, M.J. Todd, and R.H. Tutuncu, "SDPT3 – a Matlab software package for semidefinite programming," *Opt. Methods Soft.*, vol. 11, pp. 545--581, 1999.
94 R.H Tutuncu, K.C. Toh, and M.J. Todd, "Solving semidefinite-quadratic-linear programs using SDPT3," *Math. Prog. B*, vol. 95, pp. 189–217, 2003.

95 http://cvxr.com/cvx/

96 http://www.ece.ualberta.ca/~vorobyov/robustbeam.m

97 S.A. Vorobyov, A.B. Gershman, and Z.-Q. Luo, "Robust adaptive beamforming using worst-case performance optimization: a solution to the signal mismatch problem" *IEEE Trans. Signal Process.*, vol. 51, no. 2, pp. 313–324, 2003.

98 R.B. Ertel, P. Cardieri, K.W. Sowerby, T.S. Rappaport, and J.H. Reed: "Overview of spatial channel models for antenna communication systems," *IEEE Personal Commun.*, vol. 5, no. 1, pp. 10–22, 1998.

99 Van der Veen A.-J.: "Algebraic methods for deterministic blind beamforming," *Proc. IEEE*, vol. 86, no. 10, pp. 1987–2008, 1998.

100 M.K. Tsatsanis and Z. Xu, "Performance analysis of minimum variance CDMA receivers," *IEEE Trans. Signal Process.*, vol. 46, no. 11, pp. 3014–3022, 1998.

101 S.-J. Yu and J.-H. Lee, "The statistical performance of eigenspace-based adaptive array processors," *IEEE Trans. Antennas Propag.*, vol. 44, no. 5, pp. 665–667, 1996.

102 P.S. Chang and A.N. Willson Jr.,, "Analysis of conjugate gradient algorithms for adaptive filtering," *IEEE Trans. Signal Process.*, vol. 48, no. 2, pp. 409–418, 2000.

103 S. Choi, D. Shim, and T.K. Sarkar, "A comparison of tracking-beam arrays and switching-beam arrays operating in a CDMA mobile communications," *IEEE Antennas Propag. Mag.*, vol. 41, no. 6, pp. 10–22, 1999.

104 E.W. Swokowski, M. Olinick, and D. Pence, *Calculus*, 6th edn. PWS, 1994.

105 Z. Xu, "Further study on MOE-based multiuser detection in unknown multipath," *EURASIP J. Appl. Signal Process.*, vol. 12, pp. 1377–1386, 2002.

106 Z. Xu, "Improved constraint for multipath mitigation in constrained MOE multiuser detection," *J. Commun. Netw.*, vol. 3. no. 3, pp. 249–256, 2001.

107 Z. Xu and M.K. Tsatsanis, "Blind adaptive algorithms for minimum variance CDMA receivers", *IEEE Trans. Signal Process.*, vol. 49, no. 1, pp. 180–194, 2001.

108 Z. Xu and L. Ping, "Constrained CMA-based multiuser detection under unknown multipath." In: *2001 12th IEEE International Symposium on Personal, Indoor and Mobile*

Radio Communications, 30 September–3 October 2001, vol. 1, pp. A-21–A-25.

109 Z. Xu, *Blind Channel Estimation and Multiuser Detection for CDMA Communications*, PhD thesis, Steven Institute of Technology, 1999.

110 Z. Xu, L. Ping, and X. Wang, "Blind multiuser detection: from MOE to subspace methods," *IEEE Trans. Signal Process.*, vol. 52, no. 2, pp. 510–524, 2004.

111 S. Haykin, *Adaptive Filter Theory*, 4th edn, Prentice-Hall, 2002.

112 L. Fertig and J. McClellan, "Dual forms for constrained adaptive filtering," *IEEE Trans. Signal Process.*, vol. 42, no. 1, pp. 11–23, 1994.

113 S. Attallah and K. Abed-Meraim, "Fast algorithms for subspace tracking," *IEEE Trans. Signal Process. Lett.*, vol. 8, no. 7, pp. 203–206, 2001.

114 A. Elnashar, S. Elnoubi, and El-Makati H., "Performance analysis of blind adaptive MOE multiuser receivers using inverse QRD-RLS algorithm," *IEEE Trans. Circuits Syst. I*, vol. 55, no. 1, pp. 398–411, 2008.

115 M.J.D. Powell, "A Fortran subroutine for solving systems of nonlinear algebraic equations." In: P. Rabinowitz (ed.) *Numerical Methods for Nonlinear Algebraic Equations*, Gordon and Breach, 1988.

116 N.R. Conn, N.I.M Gould, and P.L Toint, *Trust-Region Methods*. MPS/SIAM Series on Optimization, SIAM and MPS, 2000.

117 S. Choi and D. Shim: "A novel adaptive beamforming algorithm for smart antenna system in a CDMA mobile communication environment," *IEEE Trans. Vehic. Techn.*, vol. 49, no. 5, pp. 1793–1806, 2000.

118 E. Oja, "A simplified neuron model as a principal component analyzer," *J. Math. Biol.*, vol. 15, pp. 267–273, 1982.

119 E. Oja, "Principal components, minor components, and linear networks," *Neural Networks*, vol. 5, pp. 927–935, 1992.

120 K. Abed-Meraim, S. Attallah, A. Chkeif, and Y. Hua, "Orthogonal Oja algorithm," *IEEE Signal Process. Lett.*, vol. 7, no. 5, pp. 116–119, 2000.

121 A.B. Bershman, G.V. Serebryakov, and J.F. Bohme, "Constrained Hung-Turner adaptive beam-forming

algorithm with additional robustness to wideband and moving jammers," *IEEE Trans. Antennas Propag.*, vol. 44, no. 3, pp. 361–367, 1996.

122 M. Agrawal and S. Prasad, "Robust adaptive beamforming for wide-band moving, and coherent jammers via uniform linear arrays," *IEEE Trans. Antennas Propag.*, vol. 47, no. 8, pp. 1267–1275, 1999.

123 S.-J. Chern and C.-Y. Chang, "Adaptive linearly constrained inverse QRD-RLS beamforming algorithm for moving jammers suppression", *IEEE Trans. Antennas Propag*, vol. 50, no. 8, pp. 1138–1150, 2002.

124 S.E. Nai, W. Ser, Z.L. Yu, and H.W. Chen, "Iterative robust minimum variance beamforming." *IEEE Trans. Signal Process.*, vol. 59, no. 4, pp. 1601–1611, 2011.

125 C.C. Gaudes, J. Via, I. Santamaria, E.M.M. Gomez, and T.S. Paules, "Robust array beamforming with sidelobe control using support vector machines," *IEEE Trans. Signal Process.*, vol. 55, no. 2, pp. 574–584, 2007.

8

Minimum BER Adaptive Detection and Beamforming

8.1 Introduction

The traditional minimum mean square error (MMSE) detector is the most popular for beamforming. An adaptive implementation of the MMSE can be achieved by minimizing the MSE between the desired output and the actual array output. The LCMV and MVDR beamformers presented in Chapter 7 are different forms of MMSE detectors. For a practical communication system, it is the achievable bit error rate (BER) or block BER (BLER), not the MSE performance, which really matter. Ideally, the system design should be based directly on minimizing the BER, rather than the MSE. For applications to single-user channel equalization, multi-user detection, and beamforming, it has been shown that the MMSE solution can, in certain situations, be distinctly inferior to the minimum BER (MBER) solution. However, the BER cost function is not a linear function of the detector or the beamformer, which makes it difficult to minimize. Several adaptive MBER beamformer/detector implementations have been developed [1–29].

It must be stated here that the cost function of the MMSE criterion has a circular shape. This means that we have one global minimum and convergence can be easily achieved. In contrast, the cost function of the BER is highly nonlinear. This means that during minimization we may converge to a local minimum. This can be seen from Figure 8.1 [5]. The MMSE and MBER solutions choose the detector's weight vector very differently. Figure 8.1 illustrates the full conditional PDFs, marginal conditional PDFs, and the corresponding signal subsets at SNR = −2 dB [5].

Simplified Robust Adaptive Detection and Beamforming for Wireless Communications, First Edition. Ayman Elnashar.

Companion website: www.wiley.com/go/elnashar49

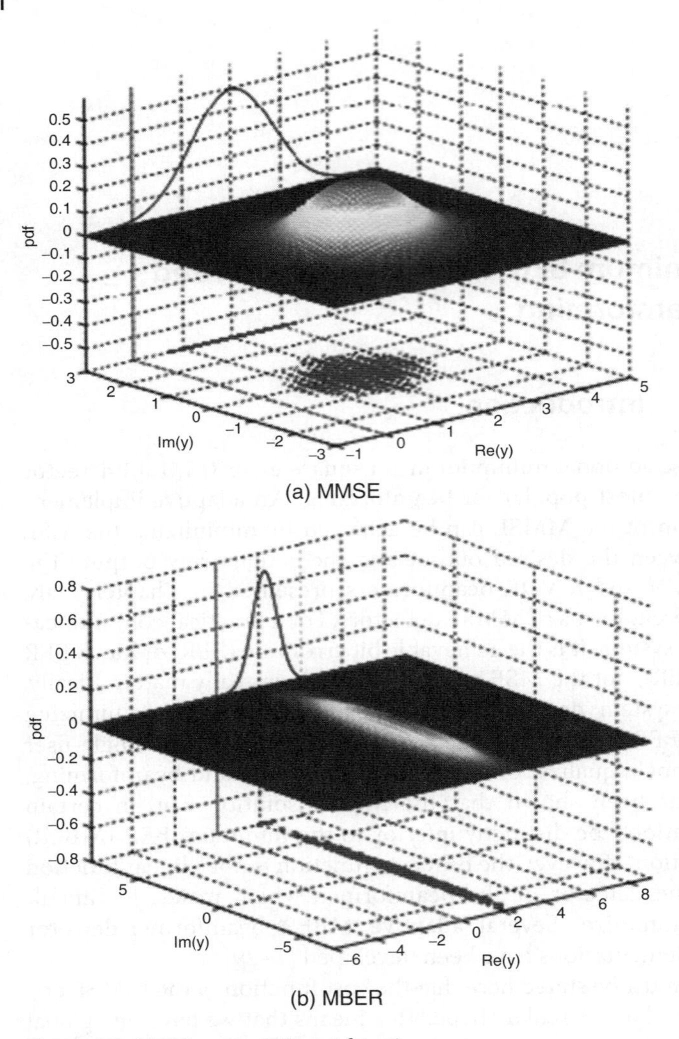

Figure 8.1 MMSE versus BER cost function.

Clearly, the MBER design is more intelligent in utilizing the detector's resources. However, special attention is mandatory during minimization in order to avoid convergence to a local minimum. If this happens, the algorithm will diverge rather than converge.

Beamforming is a key technology in smart antenna systems. It can increase capacity and coverage and mitigate multipath propagation in mobile radio communication systems. The most popular criterion for linear beamforming MMSE. However, the mean square error (MSE) cost function is not optimal in terms of the bit error probability performance of the system. Here we present a class of adaptive beamforming algorithms based on minimizing the BER cost function directly. Unfortunately, the popular least minimum BER (LMBER) stochastic beamforming algorithm suffers from low convergence speed. Gradient Newton algorithms will be presented to speed up the convergence rate and enhance performance but at the expense of complexity. In this chapter, a block processing objective function for the minimum BER (MBER) is formulated, and a nonlinear optimization strategy, which produces the so-called block-Shanno MBER (BSMBER), is developed. There is a complete discussion of the complexity calculations of the proposed algorithm. Simulation scenarios are carried out in a multipath Rayleigh-fading direct-sequence code-division multiple access (DS-CDMA) system to explore the performance of the proposed algorithm. Simulation results show that the proposed algorithm offers good performance in terms of convergence speed, steady-state performance, and even system capacity compared to other MBER- and MSE-based algorithms.

Also in this chapter, we will extend the adaptive filtering algorithms using the concept of spatial MUD in a MIMO/OFDM system model rather than beamforming in a DS-CDMA model. As stated above, a fundamental goal in any digital communications system is to directly minimize the BER. Wiener solution based algorithms indirectly minimize the BER by optimizing other cost functions (e.g. SNR, SINR, or MSE), which may result in sub-optimal BER performance [30–34].

8.2 MBER Beamformer

In order to develop a minimum BER beamformer for a binary system such as the DS/CDMA system with antenna array presented in section 2.5.2, we start by forming its BER cost function.

To derive the BER cost function for a linear detector with weight vector $\boldsymbol{w}$, let us define the following signed variable:

$$y_s(n) = \text{sgn}(b(n))y_R(n) \tag{8.1}$$

where $y_s(n)$ is an error indicator for the binary decision. If it is positive then we have a correct decision, if it is negative then there is an error. Let $N_b = 2^K$ be the number of possible transmitted bit sequences b_v of $b(n)$, where $1 \leq v \leq N_b$. The first element of b_v is therefore the desired user data. The noiseless array output signal $\overline{\boldsymbol{x}}(n)$ takes values from the signal set $\chi \triangleq \{\overline{\boldsymbol{x}}_v = b_v\boldsymbol{h}, 1 \leq v \leq N_b\}$. Similarly, the noiseless beamformer's output, $\overline{y}(n)$, takes values from the scalar set $Y(\boldsymbol{w}) \triangleq \{\overline{y}_v(\boldsymbol{w}) = \boldsymbol{w}^H\overline{\boldsymbol{x}}_v, 1 \leq v \leq N_b\}$. Thus the real part of the beamformer's output $\overline{y}_R(n)$ can only take values from the set $Y_R(\boldsymbol{w}) \triangleq \{\overline{y}_{R,v}(\boldsymbol{w}) = \text{Re}\{\overline{y}_v(\boldsymbol{w})\}, 1 \leq v \leq N_b\}$. Therefore, the PDF of the error indicator, $y_{sign}(n)$, is a mixed sum of Gaussian distributions [13]:

$$P(y_s) = \frac{1}{N_b\sqrt{2\pi\sigma_\eta^2\boldsymbol{w}^H\boldsymbol{w}}} \cdot \sum_{v=1}^{N_b} \exp\left(-\frac{(y_s - \text{sgn}(b(n))y_{R,v}(\boldsymbol{w}))^2}{2\sigma_\eta^2\boldsymbol{w}^H\boldsymbol{w}}\right) \tag{8.2}$$

So the error probability of the beamformer $\boldsymbol{w}$ – the BER cost function – is given by:

$$\begin{aligned} P_E(\boldsymbol{w}) &= \frac{1}{N_b\sqrt{2\pi\sigma_\eta^2\boldsymbol{w}^H\boldsymbol{w}}}\sum_{v=1}^{N_b}\int_{q(\boldsymbol{w})}^{\infty}\exp\left(-\frac{u^2}{2}\right)du \\ &= \frac{1}{N_b}\sum_{v=1}^{N_b} Q(q(\boldsymbol{w})) \end{aligned} \tag{8.3}$$

where $Q(.)$ is the Gaussian error function, given by:

$$Q(x) = \frac{1}{\sqrt{2\pi}}\int_x^{\infty}\exp\left(-\frac{y^2}{2}\right)dy \tag{8.4}$$

and

$$q(\boldsymbol{w}) = \frac{\text{sgn}(b(n))y(n)}{\sigma_\eta\sqrt{\boldsymbol{w}^H\boldsymbol{w}}} = \frac{\text{sgn}(b(n))\text{Re}\{\boldsymbol{w}^H\boldsymbol{x}(n)\}}{\sigma_\eta\sqrt{\boldsymbol{w}^H\boldsymbol{w}}} \tag{8.5}$$

In practice, the set of Y is not available. A widely used approach for approximating the PDF is the kernel density or Parzen window-based estimate [6–35]. A kernel density estimate is known to produce a reliable PDF estimate with short data records. Given a block of Z training samples $\{\boldsymbol{x}(Z), b(Z)\}$, a kernel density estimate of the PDF is given by:

$$\widehat{P}(y_s) = \frac{1}{Z\sqrt{2\pi\rho^2 \boldsymbol{w}^H\boldsymbol{w}}} \cdot \sum_{i=1}^{Z} \exp\left(-\frac{(y_s - \text{sgn}(b(i))y_R(i))^2}{2\rho^2 \boldsymbol{w}^H\boldsymbol{w}}\right) \tag{8.6}$$

where the radius parameter ρ is related to the noise standard deviation σ_η [6–35]. Therefore, the block BER cost function can be derived from the kernel density estimate of the PDF as follows:

$$P_E(\boldsymbol{w}) = \frac{1}{Z}\sum_{z=1}^{Z} Q(q_Z(\boldsymbol{w})) \tag{8.7}$$

where

$$q_Z(\boldsymbol{w}) = \sum_{i=1}^{Z} \frac{\text{sgn}(b(i))\text{Re}\{\boldsymbol{w}^H\boldsymbol{x}(i)\}}{\sigma_\eta\sqrt{\boldsymbol{w}^H\boldsymbol{w}}} \tag{8.8}$$

After deriving the BER cost function, we now define the optimization problem. Our objective is to minimize the BER of the system. The MBER beamforming solution is defined as:

$$\boldsymbol{w}_{MBER} = \arg\min_{\boldsymbol{w}} P_E(\boldsymbol{w}) \tag{8.9}$$

In order to develop a minimum BER beamformer for a binary system, we start by forming its BER cost function. The BER cost function for a linear detector with weight vector $\boldsymbol{w}$ is defined in Chapter 2. After deriving the BER cost function, we now define the optimization problem. The objective is to minimize the BER of the system. Hence, the MBER beamforming solution is defined as:

$$w_{MBER} = \arg\min_{w} P_E(w) \tag{8.10}$$

$P_E(w)$ is given in Chapter 2. In order to solve the optimization problem in (8.10), a gradient estimate of the BER cost function stated in (2.54) with respect to $\boldsymbol{w}$ should be estimated. An iterative gradient optimization algorithm can be used. As a result, the block gradient of the BER cost function, $P_E(\boldsymbol{w})$, is given by:

$$\nabla P_E(\boldsymbol{w}) = \frac{\partial P_E(\boldsymbol{w})}{\partial \boldsymbol{w}} = \frac{1}{2Z\sqrt{2\pi}\rho\sqrt{\boldsymbol{w}^H\boldsymbol{w}}} \times \sum_{i=1}^{Z} \exp\left(\frac{-(\boldsymbol{x}(i))^2}{2\rho^2\boldsymbol{w}^H\boldsymbol{w}}\right) \operatorname{sgn}(\boldsymbol{x}(i))\boldsymbol{x}(i) \tag{8.11}$$

where ρ is the radius parameter of the kernel density estimate [36, 37]. Alternatively, a sample-by-sample estimate for the BER cost function can be given as [27]:

$$P_E(\boldsymbol{w}, z) = \frac{1}{\sqrt{2\pi\rho^2\boldsymbol{w}^H\boldsymbol{w}}} \exp\left(-\frac{(y_s - \operatorname{sgn}(b(i))y(i))^2}{2\rho^2\boldsymbol{w}^H\boldsymbol{w}}\right) \tag{8.12}$$

Therefore, the instantaneous gradient of $P_E(\boldsymbol{w})$ with respect to $\boldsymbol{w}$ is given by:

$$\nabla P_E(\boldsymbol{w}) = \frac{1}{2\sqrt{2\pi}\rho\sqrt{\boldsymbol{w}^H\boldsymbol{w}}} \exp\left(\frac{-(\boldsymbol{x}(i))^2}{2\rho^2\boldsymbol{w}^H\boldsymbol{w}}\right) \operatorname{sgn}(\boldsymbol{x}(i))\boldsymbol{x}(i) \tag{8.13}$$

The general update equation for the stochastic gradient is given by:

$$\boldsymbol{w}_k(i+1) = \boldsymbol{w}_k(i) + \mu g(\boldsymbol{w}_k(i)) \tag{8.14}$$

where μ is the step size and $g(w_k(i))$ is a function to approximate an expression for a coefficient vector $\boldsymbol{w}_k(i)$ that achieves a MBER performance with linear receiver structures [6]. It has different forms depending on the algorithm itself. We have replaced the sampled index n by the iteration index i in a sample-by-sample iteration.

We will discuss two of the most successful and suitable algorithms for adaptive implementations. Then we will introduce Newton modifications of them, which increase the convergence rate and enhance the performance. We will apply normalization to the MBER cost function in order to produce more robustness of the MBER algorithms. After that, we will combine both modifications, Newton and normalization, and produce two new complex MBER algorithms with superior performance. Ultimately, we produce a new linear complexity MBER algorithm with good performance.

The following algorithms will be discussed in this chapter based on the above system model [29]:

- approximate MBER (AMBER)
- least MBER (LMBER)
- Newton-AMBER
- Newton-LMBER
- normalized-AMBER
- normalized-LMBER
- Newton-normalized-AMBER
- Newton-normalized-LMBER
- Block-Shanno-MBER.

8.2.1 AMBER

AMBER is a stochastic gradient algorithm, which attempts to approximate the exact MBER performance [18]. The algorithm is appealing due to its very low complexity, simplicity and straightforward extension to the complex signaling case. Given the desired user's transmitted training sequence d, the bit error probability, $P(\varepsilon|d)$, is expressed by [38, 60]:

$$\begin{aligned} P(\varepsilon|d) = P_E &= P(d(n)\text{sgn}(x(n)) = -1) \\ P_E &= P(\text{sgn}(d(n)x(n)) = -1) \\ &= P(d(n)x(n) < 0) \end{aligned} \tag{8.15}$$

$x(n)$ is the estimated symbol for the desired user at symbol "n" (assuming user number one and user number is removed for simplicity). The beamforming solution that minimizes the BER criterion via the AMBER algorithm employs the vector function:

$$g(\boldsymbol{w}(n)) = E\left[Q\left(\frac{d(n)w^H\overline{\boldsymbol{x}}(n)}{\|w(n)\|\sigma_\eta}\right)d(n)\overline{\boldsymbol{x}}(n)\right] \tag{8.16}$$

$\overline{\boldsymbol{x}}(n)$ is defined in equation 2.98. Note that the quantity $Q\left(\frac{d(n)w^H\overline{\boldsymbol{x}}(n)}{\|w(n)\|\sigma_\eta}\right)$ inside the expected value operator in (8.16) corresponds to the conditional bit error probability given the product $d(n)\overline{\boldsymbol{x}}(n)$. This quantity can be replaced by an error indicator function $I_d(n)$ given by:

$$I_d(n) = \frac{1}{2}\left(1 - \text{sgn}(d(n)\overline{\boldsymbol{x}}(n))\right) \tag{8.17}$$

By replacing the sampled index n by the iteration index i in a sample-by-sample iteration (one shot receiver), the AMBER algorithm [18] is described by the following equalities:

$$w(i+1) = w(i) + \mu E\left[Q\left(\frac{d(i)w^H\overline{x}(i)}{\|w(i)\|\sigma_\eta}\right)d(i)\overline{x}(i)\right] \quad (8.18)$$

$$w(i+1) = w(i) + \mu E[E[I_d(i)|d(i)\overline{x}(i)]d(i)\overline{x}(i)] \quad (8.19)$$

$$w(i+1) = w(i) + \mu E[I_d(i)d(i)\overline{x}(i)] \quad (8.20)$$

Since $\overline{x}(i) = x(i) - \eta(i)$, and $I_d(i)$ and $d(i)$ are statistically independent, we have:

$$E[I_d(i)d(i)\overline{x}(i)] = E[d(i)]E[I_d(i)\eta(i)] = 0 \quad (8.21)$$

thus:

$$w(i+1) = w(i) + \mu E[I_d(i)d(i)x(i)] \quad (8.22)$$

This algorithm updates when an error is made and when an error is almost made, making it a good choice for updating the filter coefficients.

8.2.2 LMBER

Chen *et al.* developed the LMBER algorithm [19], which is one of the most popular stochastic gradient algorithms for minimizing the BER cost function. The LMBER algorithm is preferred due to its simplicity, low complexity, and good performance. LMBER beamforming seeks the minimization of the cost function in (8.12). The beamforming solution that minimizes the BER criterion via the LMBER algorithm employs the vector function:

$$g(w_k(n)) = \frac{\partial P_k}{\partial w_k} = \nabla P \quad (8.23)$$

Consequently, its update equation is given by:

$$\begin{aligned} w(i+1) &= w(i) + \mu \nabla P_E(w) \\ &= w(i) + \mu \frac{\operatorname{sgn}(x(i))}{\sqrt{2\pi}\rho\sqrt{w^H w}} \exp\left(\frac{-(x(i))^2}{2\rho^2 w^H w}\right) x(i) \end{aligned} \quad (8.24)$$

By normalizing the weight vector to unit length – $w^H w = 1$ – the update equation reduces to:

$$w(i+1) = w(i) + \mu \frac{\text{sgn}(x(i))}{\sqrt{2\pi}\rho} \exp\left(\frac{-(x(i))^2}{2\rho^2}\right) x(i) \quad (8.25)$$

The adaptive gain (step size) μ and the kernel width ρ are two algorithm parameters that have to be set appropriately to ensure a fast convergence rate and a small steady-state BER maladjustment.

8.2.3 Gradient Newton Algorithms

A direct approach to ensure convergence irrespective of the signal input energy is to calculate a suitable step-size value using the function and the gradient. This usually requires finding estimates of the Hessian of the objective function. The gradient-Newton algorithm [6] incorporates second-order statistics of the input signals. It usually has a faster convergence rate than the standard gradient technique, but it requires a higher computational complexity. In practice, only estimates of the covariance matrix and the gradient vector are available. These estimates can be applied to Newton's formula to provide an update equation for the gradient-Newton algorithm as follows:

$$w(i+1) = w(i) + \mu R_{xx}^{-1}(i) g(w(i)) \quad (8.26)$$

where $R_{xx} = \frac{1}{N}\sum_{n=1}^{N} x(n)x^H(n)$ is the autocorrelation matrix of the received signal. Since matrix inversion requires a lot of computation power, the inverse can be computed according to the following rank-1 update:

$$R_{xx}^{-1}(i) = R_{xx}^{-1}(i-1) - \frac{R_{xx}^{-1}(i-1)x(i)x^H(i)R_{xx}^{-1}(i-1)}{1 + x^H(i)R_{xx}^{-1}(i-1)x(i)}$$

$$R_{xx}^{-1}(0) = \frac{1}{\varepsilon} I, \quad \varepsilon > 0 \quad (8.27)$$

The convergence factor μ is introduced to protect the algorithm from divergence due to the use of the noisy estimates of the covariance matrix and the gradient vector.

8.2.3.1 Newton-AMBER

Applying the Newton modification to the AMBER algorithm results in the Newton-AMBER algorithm. The AMBER weight vector update (8.22) can be modified to be the Newton-AMBER as:

$$\boldsymbol{w}(i+1) = \boldsymbol{w}(i) + \mu \boldsymbol{R}_{xx}^{-1}(i) I_d(i) d(i) \boldsymbol{x}(i) \tag{8.28}$$

8.2.3.2 Newton-LMBER

Applying the Newton modification to the LMBER algorithm results in the Newton-LMBER algorithm. The LMBER weight vector update (8.25) can be modified to be the Newton-LMBER as:

$$\boldsymbol{w}(i+1) = \boldsymbol{w}(i) + \mu \boldsymbol{R}_{xx}^{-1}(i) \frac{\mathrm{sgn}(\boldsymbol{x}(i))}{\sqrt{2\pi}\rho} \exp\left(\frac{-(\boldsymbol{x}(i))^2}{2\rho^2}\right) \boldsymbol{x}(i) \tag{8.29}$$

8.2.4 Normalized Gradient Algorithms

Close examination of the gradient function above reveals that the L2-norm $\|\nabla P_E(\boldsymbol{w})\|$ changes with respect to the array input energy $\|\boldsymbol{x}\|^2$ [17]. Since $y = w^H x$, we find that:

$$\|\nabla P_E(\boldsymbol{w})\| \propto \sum_{p=1}^{P} \exp\left(-\frac{\|\boldsymbol{x}\|^2}{2\sigma_\eta^2}\right) \|\boldsymbol{x}\|^2 \tag{8.30}$$

The exponential decrease of $\|\nabla P_E(\boldsymbol{w})\|$ with an increase in input signal energy potentially alters weight adaptation speeds with changes in SNR. The variation of the terminal beamformer weight values due to variation in convergence speed consequently affects the performance of beamformers that do not compensate for changes in signal input energy. In flat-fading environments, where array input energy varies, the convergence property and BER performance of non-normalized beamformers, such as MBER stochastic gradient, vary. A direct approach to ensure convergence irrespective of signal input energy is to calculate suitable step-size values using the function and the gradient. This usually requires finding estimates of the Hessian, which makes this approach computationally expensive and unsuitable for real-time implementations. An alternative approach to ensure convergence and provide robustness is to reshape (or normalize) the cost function into one that is easily searchable irrespective of signal energy, yet would yield minima that are near the minima of the original [17].

8.2.4.1 Normalized-AMBER

Applying the normalization modification to the AMBER algorithm results in the normalized-AMBER (NAMBER) algorithm. The AMBER weight vector update (8.22) can be modified to be the NAMBER by normalizing the received signal as:

$$\boldsymbol{w}(i+1) = \boldsymbol{w}(i) + \mu(i) Id(i) d(i) \boldsymbol{x}(i) \big/ \|x(i)\| \tag{8.31}$$

8.2.4.2 Normalized-LMBER

Applying the normalization modification to the LMBER algorithm results in the normalized-LMBER (NLMBER) algorithm. The LMBER weight vector update (8.25) can be modified to be the NLMBER by normalizing the received signal as:

$$\boldsymbol{w}(i+1) = \boldsymbol{w}(i) + \mu \frac{\operatorname{sgn}(d(i))}{\sqrt{2\pi}\rho} \exp\left(\frac{-(\boldsymbol{x}(i))^2}{2\rho^2} \right) \frac{(\boldsymbol{x}(i))}{\|(x(i))\|} \tag{8.32}$$

8.2.5 Normalized Newton Gradient Algorithms

We can take advantage of the normalized modification and the Newton algorithms together to enhance the performance and the convergence rate of stochastic gradient algorithms but at the cost of complexity.

8.2.5.1 Normalized-Newton-AMBER

Applying both the normalization and the Newton modifications to the AMBER algorithm results in the normalized-Newton-AMBER (NNAMBER) algorithm. The weight update equation for NNAMBER can be driven from equations (8.22), (8.28), and (8.31) as:

$$\boldsymbol{w}(i+1) = \boldsymbol{w}(i) + \mu \boldsymbol{R}_{xx}^{-1}(i) Id(i) d(i) \boldsymbol{x}(i) / \|x(i)\| \tag{8.33}$$

8.2.5.2 Normalized-Newton-LMBER

Applying both the normalization and the Newton modifications to the LMBER algorithm results in the normalized-Newton-LMBER (NNLMBER) algorithm. The weight vector update equation for NNLMBER can be driven from equations (8.25), (8.29), and (8.32) as:

$$\boldsymbol{w}(i+1) = \boldsymbol{w}(i) + \mu \boldsymbol{R}_{xx}^{-1}(i)\frac{\text{sgn}(d(i))}{\sqrt{2\pi}\rho} \times \exp\left(\frac{-(x(i))^2}{2\rho^2}\right)\frac{(\boldsymbol{x}(i))}{\|(x(i))\|} \tag{8.34}$$

8.2.6 Block-Shanno MBER

Stochastic gradient algorithms choose a constant step size. The constant step size depends on the interference as well as the channel coefficients. In order to guarantee convergence of the AMBER and LMBER algorithms, the step-size choice is the bottleneck for the algorithm to converge. The Newton algorithm optimizes the step size in order to guarantee convergence, so it requires computing, updating and storing of the inverse of the Hessian matrix. These operations on the Hessian matrix are the most costly part of the Newton algorithm. As a result, there is a need to implement a new algorithm to take advantage of the Newton approach while maintaining linearity in complexity. The Shanno algorithm is a memoryless modified Newton algorithm, which conducts the conjugate gradient-type search without fully optimizing the step size; it is chosen to be within a range specified to ensure that convergence is guaranteed. The higher bound of the step size must satisfy the following inequality [39]:

$$P_E(\boldsymbol{w}(i)) < P_E(\boldsymbol{w}(i-1)) + \alpha\mu\nabla P_E(\boldsymbol{w}(i-1))^T \boldsymbol{d}(\boldsymbol{w}(i-1)) \tag{8.35}$$

In addition, the lower bound of the step size must satisfy the following inequality [39]:

$$\nabla P_E(\boldsymbol{w}(i))^T \boldsymbol{d}(\boldsymbol{w}(i)) > \beta\nabla P_E(\boldsymbol{w}(i-1))^T \boldsymbol{d}(\boldsymbol{w}(i-1)) \tag{8.36}$$

where α and β are constants and $\boldsymbol{d}(\boldsymbol{w}(i))$ is the search direction vector. The search direction vector is a linear combination of the negative gradient, the gradient difference between the current gradient and the previous gradient, and previous search direction. It is defined as:

$$\boldsymbol{d}(i) = -\nabla P_E(\boldsymbol{w}(i)) + a(i)\boldsymbol{u}(i) + [b(i) - c(i)a(i)]\boldsymbol{d}(i-1) \tag{8.37}$$

where $\boldsymbol{u}(i)$ is the gradient difference between the current gradient and the previous gradient and is defined as:

$$\boldsymbol{u}(i) = \nabla P_E(\boldsymbol{w}(i)) - \nabla P_E(\boldsymbol{w}(i-1)) \tag{8.38}$$

and

$$a(i) = \frac{\boldsymbol{d}^T(i-1)\nabla P_E(\boldsymbol{w}(i))}{\boldsymbol{d}^T(i-1)\boldsymbol{u}(i)} \tag{8.39}$$

$$b(i) = \frac{\boldsymbol{u}^T(i)\nabla P_E(\boldsymbol{w}(i))}{\boldsymbol{d}^T(i-1)\boldsymbol{u}(i)} \tag{8.40}$$

$$c(i) = \mu(i-1) + \frac{|\boldsymbol{u}(i)|^2}{\boldsymbol{d}^T(i-1)\boldsymbol{u}(i)} \tag{8.41}$$

The search direction vector in the Shanno algorithm, which is linear in complexity, will replace the Hessian matrix, which is quadratic in complexity, in the Newton algorithm. This feature reduces the number of calculations as it involves only an implicit calculation of the Hessian matrix [40]. As a result, the weight update equation for the Shanno algorithm is given by:

$$\boldsymbol{w}(i) = \boldsymbol{w}(i-1) + \mu\boldsymbol{d}(i) \tag{8.42}$$

Another advantage of Shanno algorithm is its quick and efficient ability to minimize nonlinear objective functions. The BER cost function is nonlinear, so any minimization process can lead to a local minimum rather than the global minimum [28]. As a result, the Shanno algorithm is the best algorithm for optimizing the BER cost function.

The proposed BSMBER algorithm processes the data on a block-by-block basis; that is, it takes in a block of data and iterates until the convergence tolerance criterion is matched. In order to use the block-Shanno algorithm we should convert the problem from a complex data to a real data format. First, we define:

$$\boldsymbol{w}(i) = \boldsymbol{w}_r(i) + j\boldsymbol{w}_i(i) \tag{8.43}$$

$$\boldsymbol{x}(i) = \boldsymbol{x}_r(i) + j\boldsymbol{x}_i(i) \tag{8.44}$$

where $j = \sqrt{-1}$. Then we define the following new vectors:

$$\boldsymbol{w}_c(i) = \begin{bmatrix} \boldsymbol{w}_r(i) \\ \boldsymbol{w}_i(i) \end{bmatrix}, \ \boldsymbol{x}_c(i) = \begin{bmatrix} \boldsymbol{x}_r(i) \\ \boldsymbol{x}_i(i) \end{bmatrix} \tag{8.45}$$

Now, the detected signal y_c is given by:

$$y_c = \boldsymbol{w}_c^T \boldsymbol{x}_c \tag{8.46}$$

The proposed BSMBER algorithm is summarized in Box 8.1. First, we initialize the main algorithm parameters. Then we perform the matched filter operation for the received signal. The algorithm consists of two main loops. The outer loop is for each block of data and the inner loop is repeated over the same block of data until:

- a certain number of iterations is reached (for example 25 iterations)
- the norm of the gradient vector is sufficiently small
- the gradient difference becomes zero (to prevent dividing by zero).

Box 8.1 Summary of BSMBER algorithm in DS-CDMA system.

Initialization

- $i = 1$, $\mu = 0.1$, $\Delta\mu = 0.1\mu$, $\varepsilon = 0.1$, $\alpha = 0.25$, $\beta = 0.5$, $\rho = 30\sigma_\eta$, $Z = 100$, $\boldsymbol{w}(0) = \dfrac{\boldsymbol{x}(0)}{\|\boldsymbol{x}(0)\|^2}$

Matched filter process; $x^{(m)} = \boldsymbol{r}^{(m)}\boldsymbol{c}_1$

Outer loop (for $i = 1 : \lfloor N/Z \rfloor$**, where** $\lfloor . \rfloor$ **is the floor)**

- Form a block of data from the received signals.
- Convert the complex data into real, as in (8.43)–(8.45).

- Initialize $o = 1$, $P_E = 1$, $\nabla P_E = \mathbf{1}_M$, where $\mathbf{1}_M$ is a vector of size $M \times 1$ with all elements equal to 1.

  Inner loop (while $o \leq 25$ or $\|\nabla P_E\| < \varepsilon$ or $u = 0$)

  - Calculate the cost function BER and the gradient vector over the block from (2.54) and (8.11), respectively, ($2MZ$).
  - If $o = 1$ then $\boldsymbol{d} = -\nabla P_E$ else calculate $\boldsymbol{d}$ from (8.37), ($10M$).
  - Check the direction vector $\boldsymbol{d}$; if $\dfrac{|\boldsymbol{d}^T(o)\nabla P_E(o)|}{\|\boldsymbol{d}^T(o)\| \cdot \|\nabla P_E(o)\|} < \varepsilon$ then reset $\boldsymbol{d}$ to $-\nabla P_E$, ($2M$).
  - Check the step-size μ, as in (8.35) and (8.36). If it falls outside the boundaries, then increase or decrease the step-size as μ, ($2M$ and $4M$).
  - Update the weight vector as in (8.42).
  - Transfer $\boldsymbol{w}$ to the complex form again.
  - Determine the detected signals for all the blocks of data, as in (8.46), in order to use for calculating the BER cost function, ($2MZ$).
  - Increment the iteration number $o = o + 1$
  - end of inner loop
- end of outer loop

In the main loop, we formulate a block of data (100 bits) from the output of the matched filter operation, and convert it to real format, as in (8.43)–(8.45). Then we initialize the inner loop iteration index, o, the BER cost function, and the gradient vector to unity. In the inner loop, we compute the BER cost function and the gradient vector from (2.54) and (8.11), respectively. Then we compute the search direction vector from (8.37), unless we are at the beginning of the inner loop, in which case we set it to be equal to the negative of the gradient. After that, we check the search direction vector; if it is in the wrong direction, we reset it to the negative of the gradient. Then, we check the step size and adjust it according to its boundary. When the step size is above the upper bound in (8.35), we decrease the step size as follows:

$$\mu(i) = \mu(i-1) - \Delta\mu \tag{8.47}$$

where $\Delta\mu$ is a fraction of μ. When the step size is lower than the lower bound in (8.36), we increase the step size as follows:

$$\mu(i) = \mu(i-1) + \Delta\mu \tag{8.48}$$

Otherwise, the algorithm resumes without changing the step size. Then, we compute the weight update vector from (8.42) and convert it back to the complex form. At the end of the inner loop, we determine the detected signal by multiplying the computed weight vector by the received signal. This is used for calculating the BER cost function again. We then increment the inner loop iteration index. The inner loop iterates until any of the stop criteria is met. After that, we go back to the main loop and form another block of data, and so on. These processes iterate until we finish all the incoming data. As we can see, the algorithm keeps track of the received signal with different channel and interference parameters rather than having a constant step size that might force the algorithm to diverge when the channel and interference parameters change.

A detailed calculation of the BSMBER algorithm's complexity is shown in Box 8.1. Given that multiplication operations are more complex than addition operations, we base our complexity calculations on multiplication operations only. We assume that all vectors have the same length M. From a mathematical point of view, multiplying two vectors results in M complexity. Back to our algorithm, converting the vectors from complex format to real format results in a doubling of the vectors' length, $2M$. Following the algorithm complexity in Box 8.1, we will find that the total number of multiplications is $4M.Z + 18M$. This number of multiplications is required to determine Z bits. Hence the BSMBER algorithm needs $(4MZ + 18M)/Z$ multiplication operations in order to detect one bit.

8.3 MBER Simulation Results

In this section, we present simulation results that illustrate the performance of different MBER beamforming techniques, as well as MMSE beamforming. We employ the elliptical channel model described in Chapter 2. The simulation parameters are described in Table 8.1. The DOA angles were randomly generated between $\pm 60°$, three sectors site, and the multipaths for the same user have DOA angles spread in a range $\pm 10°$ from the main path. One of the interferers is separated by only $10°$ from the desired user. The DOA angles are unknown to the receiver.

Table 8.1 MBER simulation parameters in DS-CDMA system.

Parameter	Value
SNR	30 dB
Frame length	1000 bits
Spreading factor	31 chip Gold code
Pulse shape	RAISED cosine ($\alpha = 0.22$)
Modulation	BPSK
Antenna elements	5-element ULA with half-wavelength spacing between omni elements
Number of users	5
User power distribution	Scenario 1: equal power Scenario 2: desired user −10 dB from interfering
Maximum channel length	10 chips
Number of multi-rays	3
Maximum multi-ray delay	5 chips
Standard deviation of multipaths	0.3
Channel	Synchronous Rayleigh fading
Noise	Complex AWGN
Receiver type	Beamforming matched filtered
Beamforming technique	MBER

The simulations were done in two scenarios. The first scenario assumed perfect power control (equal power distribution). The second scenario assumed the desired user's power is less than the interferers' power by 10 dB to model the near-far effect.

8.3.1 BER Performance versus SNR

In this section, BER performance is assessed against the SNR of the desired user. For SNRs between 1 and 30 dB, the algorithms are run for 500 bits (5 blocks in the BSMBER) in the first scenario and for 1000 bits (10 blocks in the BSMBER) in the second scenario. The average BER based on the Q function is determined after the steady state is reached. We can make various conclusions from the simulation results.

From Figures 8.2–8.5, we can see the enhancement in BER from using the Newton and the normalized algorithms, over the normal algorithms. In addition, we can see that using both the normalization and the Newton modifications together gives lower BER.

Figures 8.6 and 8.7 compare the BER performance against the SNR of the AMBER family against the LMS-MMSE and the LMBER family against SD-MMSE algorithms, respectively, at equal power gain. We can see the enhancement in BER of using the MBER techniques over the MMSE techniques. In addition, we can see the enhancement in performance when using Newton algorithms over the normal algorithms. Again, using the normalization modification gives robustness to the MBER algorithms. Finally, combining both modifications gives the best BER performance.

Figures 8.8 and 8.9 assess the BER performance against SNR for MMSE, LMBER, and Newton-LMBER beamformers and the BSMBER algorithm, in the first and second scenarios, respectively. It is evident that the BSMBER algorithm performs

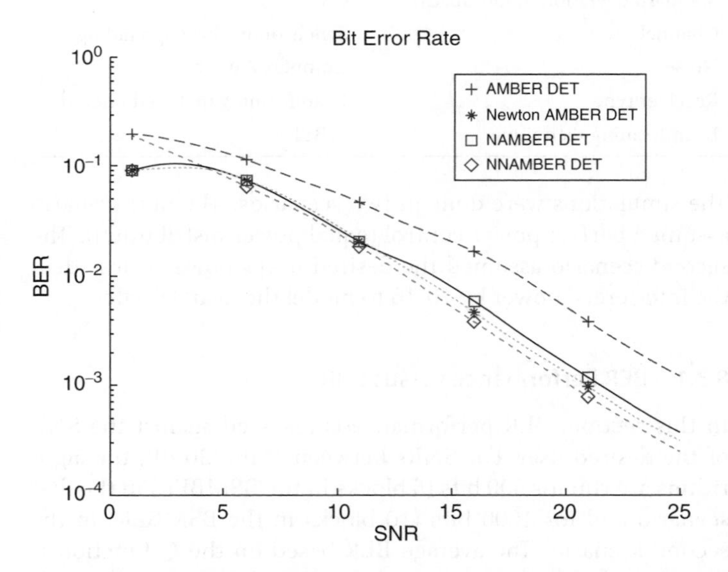

Figure 8.2 BER vs SNR for selected algorithms minimizing the BER; equal power distribution.

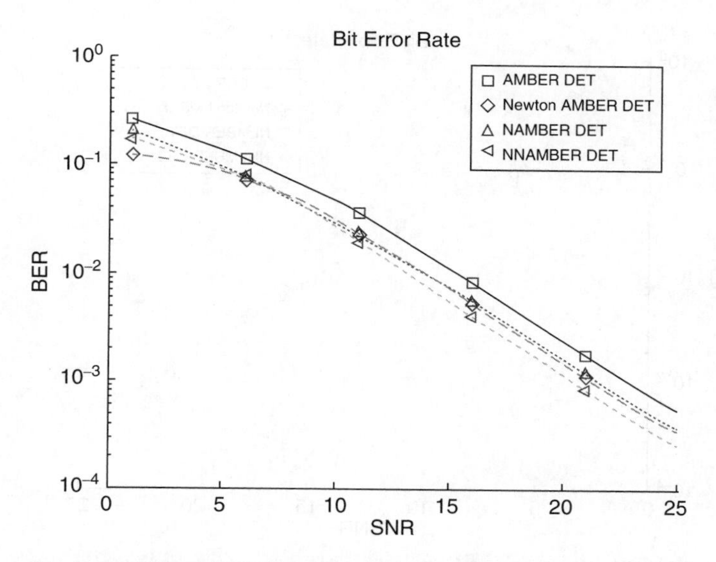

Figure 8.3 BER vs SNR for selected algorithms minimizing the BER; desired user power 10 dB below interferers.

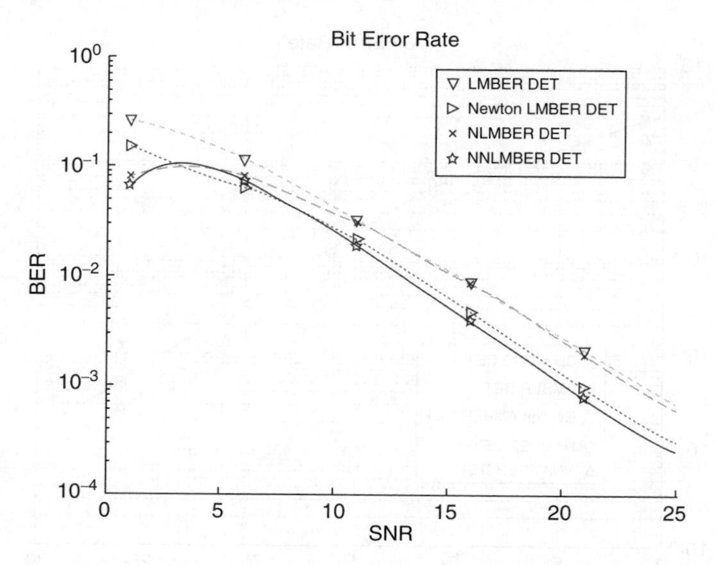

Figure 8.4 BER vs SNR for selected algorithms minimizing the BER; equal power distribution.

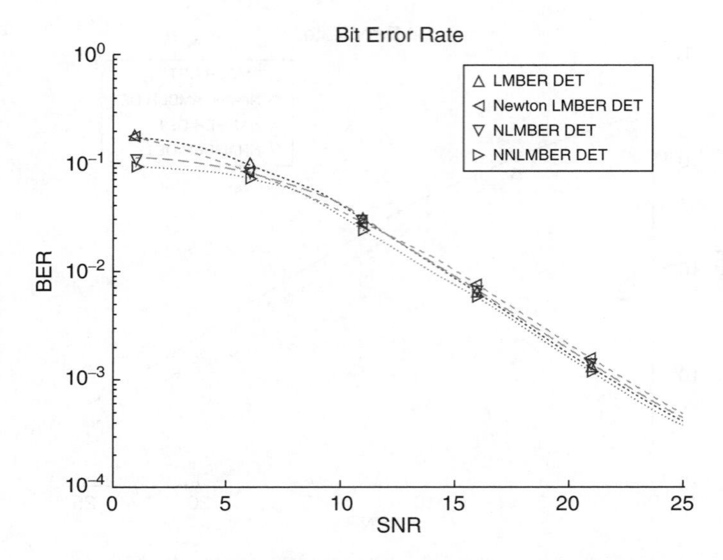

Figure 8.5 BER vs SNR for selected algorithms minimizing the BER; desired user power 10 dB below interferers.

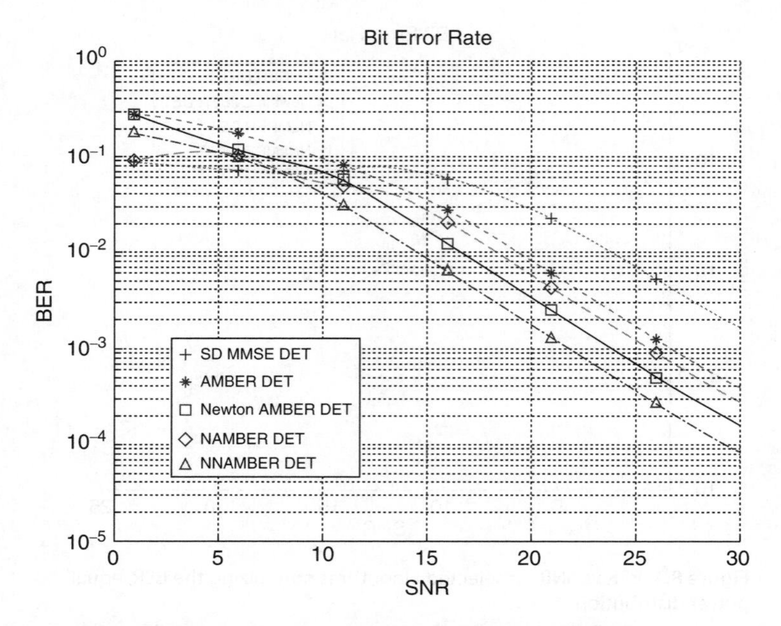

Figure 8.6 BER vs SNR for AMBER family algorithms and SD-MMSE algorithm; equal power distribution.

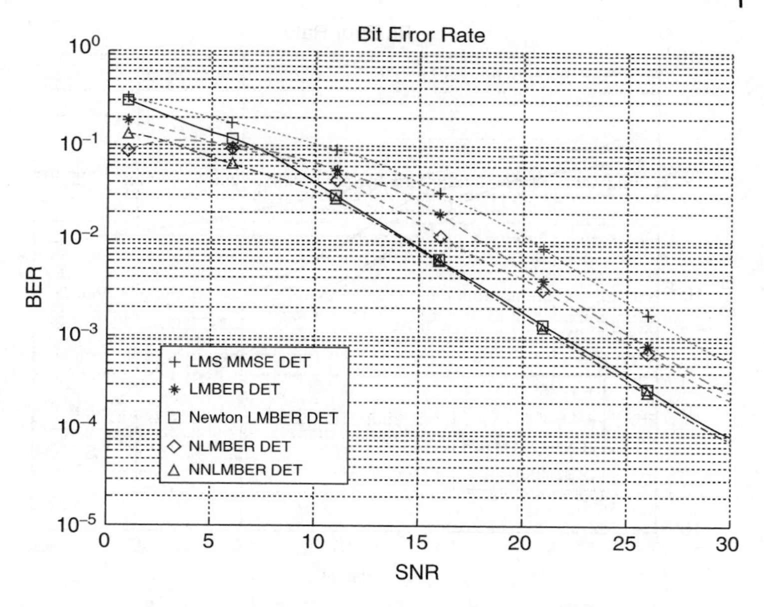

Figure 8.7 BER vs SNR for LMBER family algorithms and LMS-MMSE algorithm; equal power distribution.

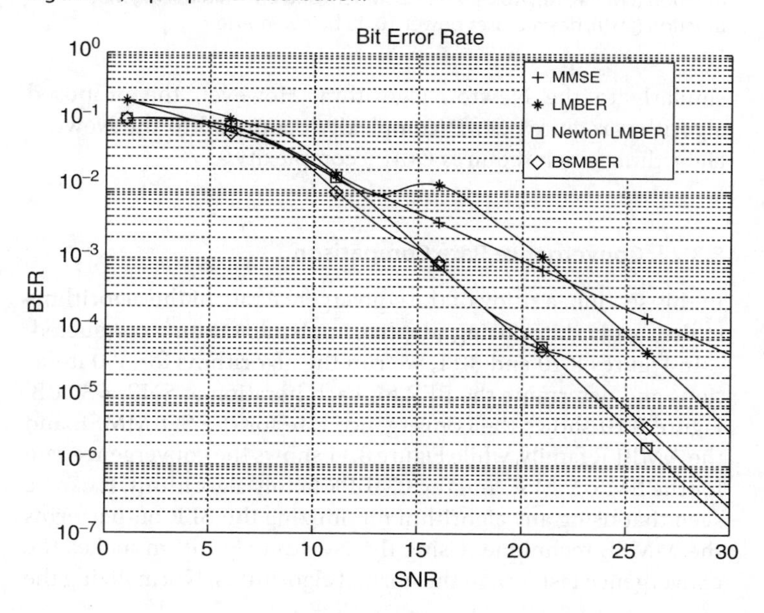

Figure 8.8 BER vs SNR for three algorithms minimizing BER cost function (LMBER, Newton-LMBER, BSMBER) and one minimizing MSE cost function (DMI); equal power distribution.

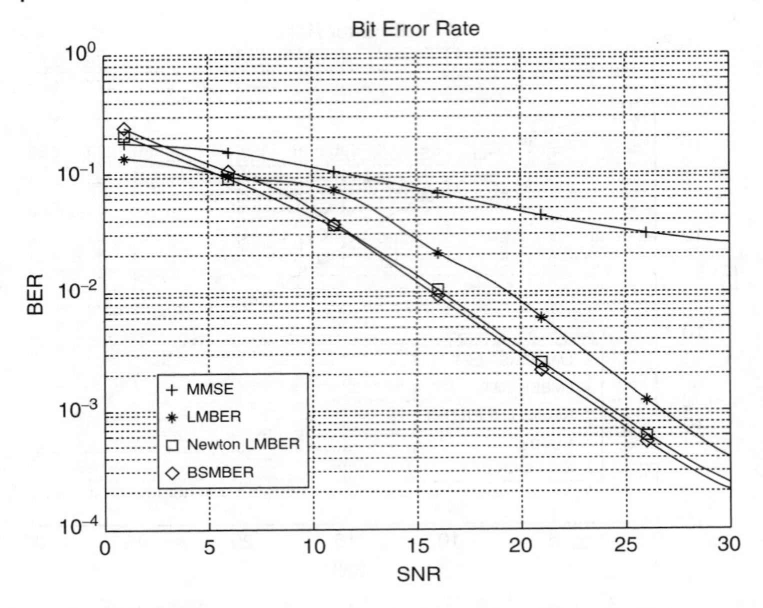

Figure 8.9 BER versus SNR for three algorithms minimizing BER cost function (LMBER, Newton-LMBER, BSMBER) and one minimizing MSE cost function (DMI); desired user power 10 dB below interferers.

similarly to the Newton algorithm. However, the proposed algorithm requires $O(M)$ complexity compared to the Newton algorithm, which requires $O(M^2)$ complexity.

8.3.2 Convergence Rate Comparison

In this section, a study of the convergence rate of the algorithms to the steady state during adaptive implementation is conducted.

In Figures 8.10 and 8.11, we run the algorithms for 350 iterations and determine the BER for each iteration at SNR = 30 dB. Figure 8.10 shows the convergence rate for the SD-MMSE and the AMBER family, while Figure 8.11 shows the convergence rate for the LMS-MMSE and the LMBER family algorithms. It can be seen that using any algorithm minimizing the BER outperforms the MMSE technique. Using the Newton algorithm makes the convergence faster than the normal algorithms. Normalizing the

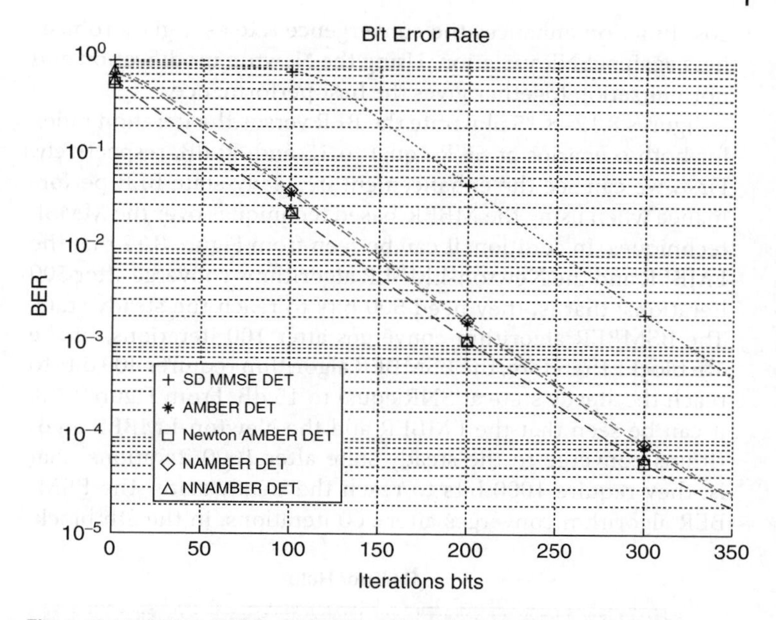

Figure 8.10 BER vs SNR for AMBER family and SD-MMSE algorithms; equal power distribution at SNR = 30 dB.

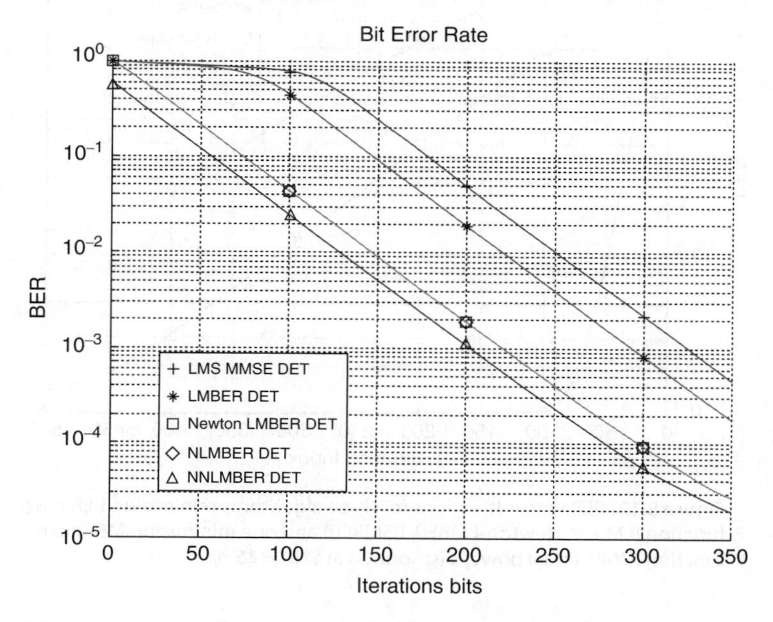

Figure 8.11 BER vs iterations for LMBER family and LMS-MMSE algorithms; equal power distribution at SNR = 30 dB.

cost function enhances the convergence rate as it gives robustness against SNR variation. Using the Newton modification and normalization together gives the best performance.

Figures 8.12–8.15 illustrate the BER versus the iteration index for both scenarios at SNR equal to 15 and 30 dB, respectively. First, we can see the enhancement in steady-state BER performance when using the MBER-based techniques over the MMSE techniques. In addition, it can be seen from Figure 8.12 that the LMBER and the Newton-LMBER algorithms converge after 500 iterations; that is, they need 500 bits to reach the steady state. The BSMBER algorithm converges after 100 iterations, in the 4th block. Therefore, the BSMBER algorithm requires 400 bits to reach the steady state at SNR equal to 15 dB. From Figure 8.13, it can be seen that the LMBER and the Newton-LMBER algorithms converge to the steady state after 1000 iterations; that is, they require 1000 bits to reach the steady state. The BSMBER algorithm converges after 200 iterations, in the 8th block.

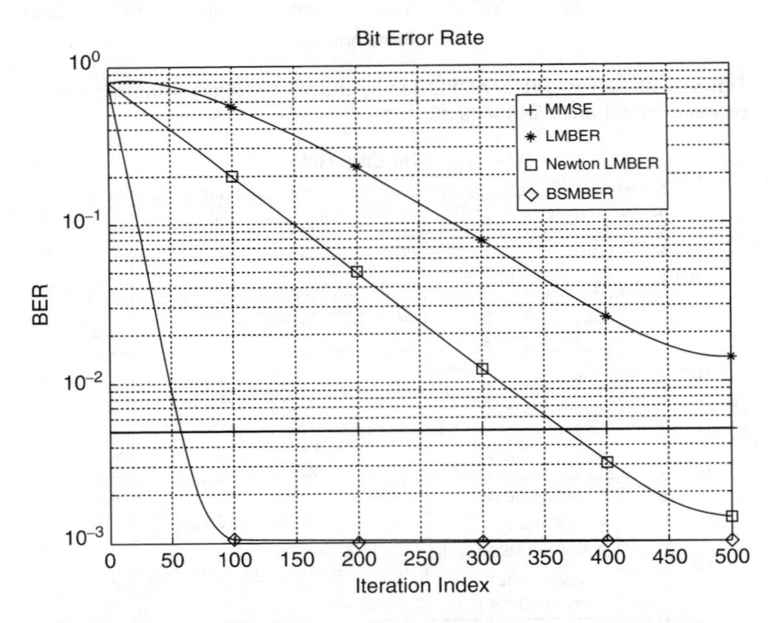

Figure 8.12 BER versus Iterations for three algorithms minimizing BER cost function (LMBER, Newton-LMBER, BSMBER) and one minimizing MSE cost function (DMI); equal power distribution at SNR = 15 dB.

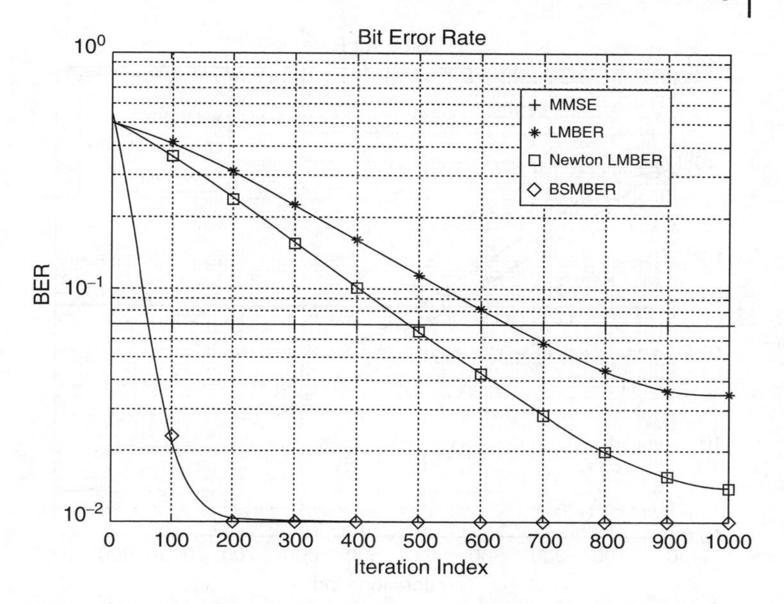

Figure 8.13 BER versus iterations for three algorithms minimizing BER cost function (LMBER, Newton-LMBER, BSMBER) and one minimizing MSE cost function (DMI); desired user power 10 dB below interferers at SNR = 15 dB.

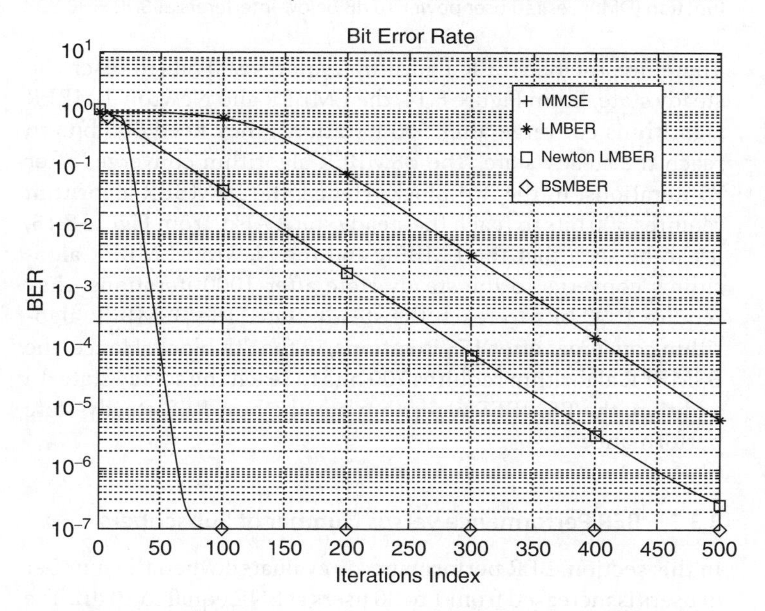

Figure 8.14 BER versus iterations for three algorithms minimizing BER cost function (LMBER, Newton-LMBER, BSMBER) and one minimizing MSE cost function (DMI); equal power distribution at SNR = 30 dB.

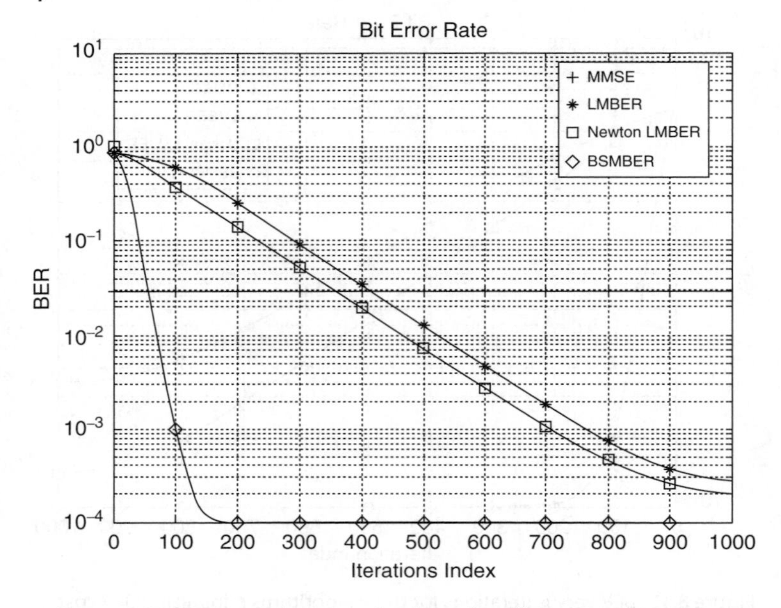

Figure 8.15 BER versus iterations for three algorithms minimizing BER cost function (LMBER, Newton-LMBER, BSMBER) and one minimizing MSE cost function (DMI); desired user power 10 dB below interferers at SNR = 30 dB.

Therefore, the BSMBER algorithm requires 800 bits to reach the steady state. From Figure 8.14, the LMBER and Newton-LMBER algorithms converge after 500 iterations; they need 500 bits to reach the steady state. The BSMBER algorithm converges after 75 iterations, in the 3rd block. Hence, the BSMBER algorithm requires 300 bits to reach the steady state. Also, from Figure 8.15, it can be seen that the LMBER and the Newton-LMBER algorithms converge to the steady state after 1000 iterations; they require 1000 bits to reach the steady state. The BSMBER algorithm converges after 175 iterations, in the 7th block. Hence, the BSMBER algorithm requires 700 bits to reach the steady state. In addition, the BSMBER algorithm has the best BER steady-state performance.

8.3.3 BER Performance versus Number of Subscribers

In this section, BER performance is evaluated when the number of users is increased from 1 to 30 users at SNR equal to 30 dB. The algorithms are run for 500 bits (5 blocks with the BSMBER) in

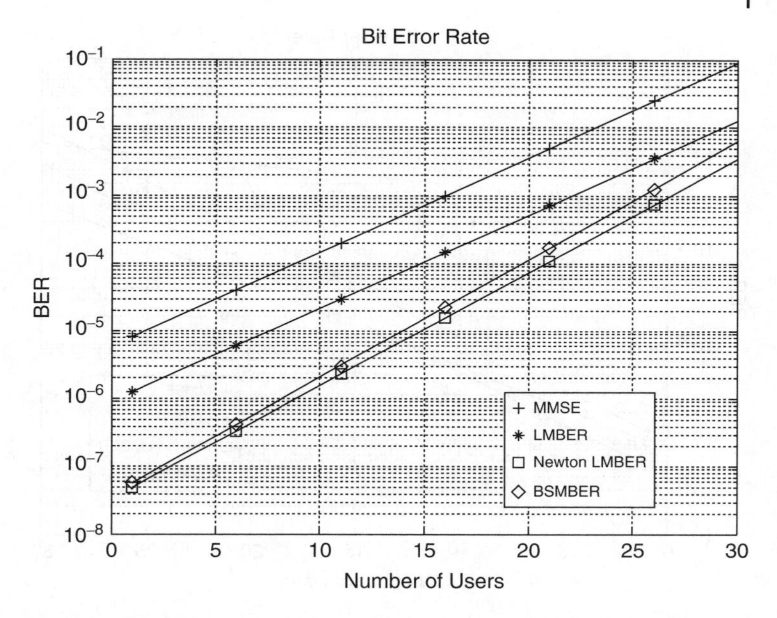

Figure 8.16 BER versus number of users for three algorithms minimizing the BER cost function (LMBER, Newton-LMBER, BSMBER) and one minimizing the MSE cost function (DMI) with equal power distribution at SNR = 30 dB.

the first scenario and for 1000 bits (10 blocks with the BSMBER) in the second scenario. The average BER based on the *Q* function is determined after the algorithms converge.

Figures 8.16 and 8.17 show the BER performance for the first and second scenarios, respectively. The figures demonstrate that the BSMBER algorithm exhibits the best performance in terms of low BER at high numbers of subscribers. The MMSE beamformer is dramatically affected in the second scenario; MMSE algorithms are not robust against the near–far effect. Meanwhile, the MBER algorithms retain their good performance when the system suffers from the near–far effect and this agrees with the results of Chen *et al.* [19].

8.3.4 Computational Complexity

In this section, we compare all the algorithms in the terms of computational complexity [29]. Since addition is much easier than multiplication, we will focus on multiplication in order

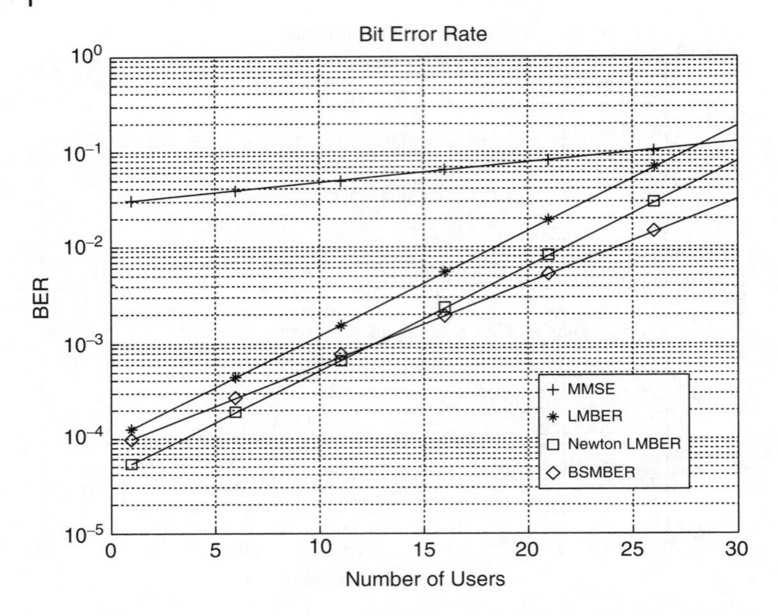

Figure 8.17 BER versus number of users for three algorithms minimizing BER cost function (LMBER, Newton-LMBER, BSMBER) and one minimizing MSE cost function (DMI); desired user power 10 dB below interferers at SNR = 30 dB.

to compare the computational complexity. Table 8.2 illustrates the number of multiplications required to complete a single iteration, i.e., detecting one bit. It can be seen that the normal algorithms have linear complexity in the number of antenna elements. Using the normalization does not affect the complexity. When using the Newton modification, the complexity becomes quadrate in the sense of number of antenna elements. The worst complexity is seen when we combine both the Newton and Normalization modifications together. The BSMBER maintain the linearity in complexity however, its performance is at least similar to the Newton algorithm with quadratic complexity [29].

8.4 MBER Spatial MUD in MIMO/OFDM Systems

In order to develop a minimum BER spatial MUD algorithm for a binary system, we start by forming the BER cost function.

Table 8.2 Computational complexity comparison.

Algorithm	Computational complexity
LMS	$O(N)$
SD	$O(N^2+N/2)$
LMBER	$O(N)$
AMBER	$O(N)$
Newton-LMBER	$O(3N^2+(5/2)N)$
Newton-AMBER	$O(3N^2+(5/2)N)$
NLMBER	$O(2N)$
NAMBER	$O((3/2)N)$
NNLMBER	$O(3N^2+(7/2)N)$
NNAMBER	$O(3N^2+3N)$
BSMBER	$O(4M+18M/Z)$

The BER cost function for a linear detector with weight vector $\boldsymbol{W}$ is defined in Chapter 2. After deriving it, we define the optimization problem. Our objective is to minimize the BER of the system. The MBER spatial MUD solution is defined as:

$$\boldsymbol{W}_{MBER} = \arg\min_{\boldsymbol{W}} P_E(\boldsymbol{W}) \tag{8.49}$$

In order to solve the optimization problem in (8.49), a gradient estimate of the BER cost function in (2.80) with respect to $\boldsymbol{W}$ should be estimated. An iterative gradient optimization algorithm can be used. As a result, the block gradient of the BER cost function, $P_E(\boldsymbol{W})$, is given by:

$$\nabla P_E(\boldsymbol{W}) = \frac{\partial P_E(\boldsymbol{W})}{\partial \boldsymbol{W}} = \frac{1}{2Z\sqrt{2\pi}\rho} \times \sum_{i=1}^{Z} \exp\left(\frac{-(\boldsymbol{x}(i))^2}{2\rho^2}\right) \text{sgn}(\boldsymbol{x}(i))\boldsymbol{x}(i) \tag{8.50}$$

Alternatively, a sample-by-sample estimate for the BER cost function can be given as [27]:

$$P_E(\boldsymbol{W}, z) = \frac{1}{\sqrt{2\pi\rho^2}} \exp\left(-\frac{(\boldsymbol{y}_s - \text{sgn}(\boldsymbol{s}(i))\boldsymbol{y}(i))^2}{2\rho^2}\right) \tag{8.51}$$

Therefore, the gradient of $P_E(\boldsymbol{W})$ with respect to $\boldsymbol{W}$ is given by:

$$\nabla P_E(\boldsymbol{W}) = \frac{1}{2\sqrt{2\pi}\rho} \exp\left(\frac{-(\boldsymbol{x}(i))^2}{2\rho^2}\right) \text{sgn}(\boldsymbol{x}(i))\boldsymbol{x}(i) \quad (8.52)$$

The general update equation for the stochastic gradient is:

$$\boldsymbol{W}(i+1) = \boldsymbol{W}(i) + \mu z(\boldsymbol{W}(i)) \quad (8.53)$$

where μ is the step size and $z(\boldsymbol{W}(i))$ is a function to approximate an expression for a coefficient matrix $\boldsymbol{W}(i)$ that achieves a MBER performance with linear receiver structures. It has different forms depending on the algorithm itself. We have replaced the sampled index n by the iteration index i in sample-by-sample iteration (one shot receiver). Extension of the BER cost function from BPSK modulation to QPSK modulation scenarios can be found in the literature [16].

Similar to the beamforming algorithms developed earlier in this chapter, we will discuss two of the most successful and suitable algorithms for adaptive implementation. Then we will introduce the Newton modifications to those two algorithms in order to increase the convergence rate and enhance the performance. Then we will apply the normalization to the MBER cost function in order to produce more robustness in the MBER algorithms. After that, we will combine both modifications – Newton and normalization – and propose two new complex MBER algorithms with superior performance. Ultimately, we will introduce new linear complexity MBER algorithms with good performance [29, 41–43]:

- approximate MBER (AMBER)
- least MBER (LMBER)
- Newton-AMBER
- Newton-LMBER
- normalized-AMBER
- normalized-LMBER
- Newton-normalized-AMBER
- Newton-normalized-LMBER
- block-Shanno-MBER.[endbl]

8.4.1 AMBER

AMBER is a stochastic gradient algorithm that attempts to approximate the exact MBER performance [18]. The algorithm is appealing due to its very low complexity, simplicity and straightforward extension to the complex signaling case. Given the desired user's transmitted training sequence $\boldsymbol{b}$, the bit error probability is expressed by:

$$P_E = P\left(\text{sgn}(\boldsymbol{b}(n)\boldsymbol{x}(n)) = -\mathbf{1}_{M_t}\right) = P\left(\boldsymbol{b}(n)\boldsymbol{x}(n) < \mathbf{0}_{M_t}\right) \tag{8.54}$$

where $\mathbf{1}_{M_t}$ and $\mathbf{0}_{M_t}$ are vectors of size $1 \times M_t$ consisting of ones and zeros, respectively. The linear MUD (LMUD) solution that minimizes the BER criterion via the AMBER algorithm employs the matrix function:

$$z(\boldsymbol{W}(n)) = E\left[Q\left(\frac{\boldsymbol{b}(n)\boldsymbol{W}\overline{\boldsymbol{x}}(n)}{\sigma_\eta}\right)\boldsymbol{b}(n)\overline{\boldsymbol{x}}(n)\right] \tag{8.55}$$

where σ_η is the noise variance. Note that the quantity $Q\left(\frac{\boldsymbol{b}(n)\boldsymbol{W}\overline{\boldsymbol{x}}(n)}{\sigma_\eta}\right)$ inside the expected value operator in (8.55) corresponds to the conditional bit error probability given the product $\boldsymbol{b}(n)\overline{\boldsymbol{x}}(n)$. This quantity can be replaced by an error indicator function $\boldsymbol{I}_b(n)$ given by:

$$\boldsymbol{I}_b(n) = \frac{1}{2}\left(\mathbf{1}_{M_t} - \text{sgn}(\boldsymbol{b}(n)\overline{\boldsymbol{x}}(n))\right) \tag{8.56}$$

Replacing the sampled index n with the iteration index i in sample-by-sample iterations, the AMBER algorithm [44] is described by the following equalities:

$$\boldsymbol{W}(i+1) = \boldsymbol{W}(i) + \mu E\left[Q\left(\frac{\boldsymbol{b}(i)\boldsymbol{W}\overline{\boldsymbol{x}}(i)}{\sigma_\eta}\right)\boldsymbol{b}(i)\overline{\boldsymbol{x}}(i)\right] \tag{8.57}$$

$$\boldsymbol{W}(i+1) = \boldsymbol{W}(i) + \mu E[E[\boldsymbol{I}_b(i)|\boldsymbol{b}(i)\overline{\boldsymbol{x}}(i)]\boldsymbol{b}(i)\overline{\boldsymbol{x}}(i)] \tag{8.58}$$

$$\boldsymbol{W}(i+1) = \boldsymbol{W}(i) + \mu E[\boldsymbol{I}_b(i)\boldsymbol{b}(i)\overline{\boldsymbol{x}}(i)] \tag{8.59}$$

Since $\boldsymbol{I}_b(\boldsymbol{i})$ and $\boldsymbol{b}(i)$ are statistically independent, we have:

$$E[\boldsymbol{I}_b(i)\boldsymbol{b}(i)\overline{\boldsymbol{x}}(i)] = E[\boldsymbol{b}(i)]E[\boldsymbol{I}_b(i)\overline{\boldsymbol{x}}(i)] = \mathbf{0}_{M_t} \tag{8.60}$$

thus:

$$\boldsymbol{W}(i+1) = \boldsymbol{W}(i) + \mu E[\boldsymbol{I}_b(i)\boldsymbol{b}(i)\boldsymbol{x}(i)] \tag{8.61}$$

This algorithm updates the filter coefficients when an error is made or almost made.

8.4.2 LMBER

The LMUD-BER depends on the distribution of the decision variable $\boldsymbol{x}(n)$, which is a function of the filter weights. The gradient terms of the probability of error are:

$$\frac{\partial P}{\partial \boldsymbol{W}} = \frac{1}{\sqrt{2\pi}\rho} \exp\left(\frac{-(\boldsymbol{x}(n))^2}{2\rho^2}\right) \mathrm{sgn}(\boldsymbol{b}(n))(\boldsymbol{x}(n)) \tag{8.62}$$

From [29] we can have:

$$z(\boldsymbol{W}(n)) = \frac{\partial P}{\partial \boldsymbol{W}} = \nabla P \tag{8.63}$$

Hence, by replacing the sampled index n by the iteration index i in a sample-by-sample iteration, the LMBER weight updated equation [29] becomes:

$$\boldsymbol{W}(i+1) = \boldsymbol{W}(i) + \mu \nabla P = \boldsymbol{W}(i) + \mu \frac{\mathrm{sgn}(\boldsymbol{b}(i))}{\sqrt{2\pi}\rho} \exp\left(\frac{-(\boldsymbol{x}(i))^2}{2\rho^2}\right)(\boldsymbol{x}(i)) \tag{8.64}$$

The adaptive gain (step size) μ and the kernel width ρ are two algorithm parameters that have to be set appropriately to ensure fast convergence and small steady-state BER maladjustment.

8.4.3 Gradient Newton Algorithms

Gradient-Newton algorithms incorporate the second order statistics of input signals, increasing their convergence rate [6]. They usually have faster convergence rates than gradient techniques, but at the cost of computational complexity. In practice, estimates of the covariance and gradient matrices are required to converge to the desired solution. The weight update of Newton's method is given by:

$$\boldsymbol{W}(i+1) = \boldsymbol{W}(i) + \mu \boldsymbol{R}_{xx}^{-1}(i) z(\boldsymbol{W}(i)) \tag{8.65}$$

The convergence factor μ is introduced to protect the algorithm from divergence due to the use of the noisy estimates of the covariance matrix and the gradient vector. Since matrix inversion requires many computations, the inverse is computed using the following rank-1 update:

$$R_{xx}^{-1}(i) = R_{xx}^{-1}(i-1) - \frac{R_{xx}^{-1}(i-1)x(i)x^H(i)R_{xx}^{-1}(i-1)}{\mathbf{1}_{M_r \times M_t} + x^H(i)R_{xx}^{-1}(i-1)x(i)}$$

$$R_{xx}^{-1}(0) = \frac{1}{\varepsilon}I, \quad \varepsilon > 0 \tag{8.66}$$

8.4.3.1 Newton-AMBER

Applying the Newton modification to the AMBER algorithm results in the Newton-AMBER algorithm. The AMBER weight vector update (8.61) can be modified to be the Newton-AMBER as:

$$W(i+1) = W(i) + \mu R_{xx}^{-1}(i)I_b(i)b(i)x(i) \tag{8.67}$$

8.4.3.2 Newton-LMBER

Applying the Newton modification to the LMBER algorithm results in the Newton-LMBER algorithm. The LMBER weight vector update (8.64) can be modified to be the Newton-LMBER as:

$$W(i+1) = W(i) + \mu R_{xx}^{-1}(i)\frac{\text{sgn}(b(i))}{\sqrt{2\pi}\rho}\exp\left(\frac{-(x(i))^2}{2\rho^2}\right)(x(i)) \tag{8.68}$$

8.4.4 Normalized Gradient Algorithms

Close examination of the gradient function reveals that the $L_2 - norm\|\nabla P_E(W)\|$ changes with respect to the array input energy $\|x\|^2$ [37]. Since $y = Wx$, we find that:

$$\|\nabla P_E(w)\| \propto \sum_{p=1}^{P} \exp\left(-\frac{\|x\|^2}{2\sigma_\eta^2}\right)\|x\|^2 \tag{8.69}$$

The exponential decrease of $\|\nabla P_E(W)\|$ with an increase in input signal energy potentially alters the weight adaptation with changes in SNR. This variation of the filter weights affects the performance of LMUD algorithms that do not compensate for changes in signal input energy. In flat-fading environments, where array input energy varies, the convergence property and

BER performance of no-normalizing LMUD algorithms, such as the MBER stochastic gradient, vary. A direct approach to ensure convergence, irrespective of signal input energy, is to calculate a suitable step size using the function and the gradient. This usually requires finding estimates of the Hessian, which makes it computationally expensive and unsuitable for real-time implementation. An alternative approach to ensure convergence is to reshape (or normalize) the cost function into one that is easily searchable, irrespective of signal energy, yet would yield minima that are near the original's minima [31].

8.4.4.1 Normalized-AMBER

Applying the normalization modification to the AMBER algorithm results in the normalized-AMBER (NAMBER) algorithm. The AMBER weight vector update (8.61) can be modified to be the NAMBER by normalizing the received signal as:

$$\boldsymbol{W}(i+1) = \boldsymbol{W}(i) + \mu(i)\boldsymbol{I}_b(i)\boldsymbol{b}(i)\boldsymbol{x}(i)/\|\boldsymbol{x}(i)\| \tag{8.70}$$

8.4.4.2 Normalized-LMBER

Applying the Normalization modification to the LMBER algorithm results in the Normalized-LMBER (NLMBER) algorithm. The LMBER weight vector update (8.64) can be modified to be the NLMBER by normalizing the received signal as:

$$\boldsymbol{W}(i+1) = \boldsymbol{W}(i) + \mu\frac{\operatorname{sgn}(\boldsymbol{b}(i))}{\sqrt{2\pi}\rho}\exp\left(\frac{-(\boldsymbol{x}(i))^2}{2\rho^2}\right)\frac{(\boldsymbol{x}(i))}{\|(\boldsymbol{x}(i))\|} \tag{8.71}$$

8.4.5 Normalized Newton Gradient Algorithms

Combining both the normalization and the Newton algorithms enhances the performance and the convergence rates of stochastic gradient algorithms.

8.4.5.1 Normalized-Newton-AMBER

Applying both the normalization and the Newton modification to the AMBER algorithm results in the normalized-Newton-AMBER (NNAMBER) algorithm. The weight update equation for NNAMBER can be driven from (8.61), (8.67), and (8.70) as:

$$\boldsymbol{W}(i+1) = \boldsymbol{W}(i) + \mu\boldsymbol{R}_{xx}^{-1}(i)\boldsymbol{I}_b(i)\boldsymbol{b}(i)\boldsymbol{x}(i)/\|\boldsymbol{x}(i)\| \tag{8.72}$$

8.4.5.2 Normalized-Newton-LMBER

Applying both, the Normalization and the Newton modification to the LMBER algorithm results in the Normalized-Newton-LMBER (NNLMBER) algorithm. The weight update equation for NNLMBER can be driven from equations (8.64), (8.68), and (8.71) as:

$$\boldsymbol{W}(i+1) = \boldsymbol{W}(i) + \mu \boldsymbol{R}_{\boldsymbol{xx}}^{-1}(i) \frac{\mathrm{sgn}(\boldsymbol{b}(i))}{\sqrt{2\pi}\rho} \times \exp\left(\frac{-(\boldsymbol{x}(i))^2}{2\rho^2}\right) \frac{(\boldsymbol{x}(i))}{\|(\boldsymbol{x}(i))\|} \tag{8.73}$$

8.4.6 Block-Shanno MBER

The Shanno algorithm is a memoryless modified Newton algorithm [45]. Like the gradient algorithm, the Shanno algorithm involves an implicit computation of the Hessian. However, it does the conjugate gradient type search without fully optimizing the step size. The step size is chosen to be within a specified range, such that convergence is guaranteed. The higher and lower bounds of the step size must satisfy the following inequalities:

$$P_E(\boldsymbol{W}(i)) < P_E(\boldsymbol{W}(i-1)) + \alpha\mu\nabla P_E(\boldsymbol{W}(i-1))^T \boldsymbol{D}(\boldsymbol{W}(i-1)) \tag{8.74}$$

$$\nabla P_E(\boldsymbol{W}(i))^T \boldsymbol{D}(\boldsymbol{W}(i)) > \beta\nabla P_E(\boldsymbol{W}(i-1))^T \boldsymbol{D}(\boldsymbol{W}(i-1)) \tag{8.75}$$

where $\boldsymbol{D}(\boldsymbol{W}(i))$ is the search direction matrix, which will be defined later, and α and β are constants. This feature saves on computation, as it involves only an implicit computation of the Hessian matrix. There is no need to compute, update and store the inverse of the Hessian, the most costly part of the Newton algorithm. The Shanno algorithm is also known for its fast convergence rate, and it is designed to quickly and efficiently minimize nonlinear objective functions. The weight update equation for the Shanno algorithm is given by:

$$\boldsymbol{W}(i) = \boldsymbol{W}(i-1) + \mu \boldsymbol{D}(i) \tag{8.76}$$

BSMBER processes the data on a block-by-block basis; it takes in a block of data and iterates until the convergence tolerance criterion is met. The search direction matrix is a linear combination of the negative gradient, the gradient difference between the current gradient and the previous gradient, and previous search direction. In order to calculate the search direction matrix, we first define the gradient difference, $\boldsymbol{G}(i)$, between the current gradient and the previous gradient as:

$$\boldsymbol{G}(i) = \nabla P_E(\boldsymbol{W}(i)) - \nabla P_E(\boldsymbol{W}(i-1)) \tag{8.77}$$

In addition, we define the following matrices:

$$\boldsymbol{A}(i) = \frac{\boldsymbol{D}(i-1)\nabla P_E(\boldsymbol{W}(i))}{\boldsymbol{D}(i-1)\boldsymbol{G}(i)} \tag{8.78}$$

$$\boldsymbol{B}(i) = \frac{\boldsymbol{G}(i)\nabla P_E(\boldsymbol{W}(i))}{\boldsymbol{D}(i-1)\boldsymbol{G}(i)} \tag{8.79}$$

$$\boldsymbol{C}(i) = \mu(i-1)\boldsymbol{I}_{M_r} + \frac{|\boldsymbol{G}(i)|^2}{\boldsymbol{D}(i-1)\boldsymbol{G}(i)} \tag{8.80}$$

The search direction matrix is given by:

$$\boldsymbol{D}(i) = -\nabla P_E(\boldsymbol{W}(i)) + \boldsymbol{A}(i)\boldsymbol{G}(i) + (\boldsymbol{B}(i) - \boldsymbol{C}(i)\boldsymbol{A}(i))\boldsymbol{D}(i-1) \tag{8.81}$$

In order to use the Block shanno algorithm we must convert the complex data into real. We define:

$$\boldsymbol{W}(i) = \boldsymbol{W}_r(i) + j\boldsymbol{W}_i(i) \tag{8.82}$$

$$\boldsymbol{x}(i) = \boldsymbol{x}_r(i) + j\boldsymbol{x}_i(i) \tag{8.83}$$

where $j = \sqrt{-1}$. Then we define the following new vectors:

$$\boldsymbol{W}_c(i) = \begin{bmatrix} \boldsymbol{W}_r(i) \\ \boldsymbol{W}_i(i) \end{bmatrix} \quad \boldsymbol{x}_c(i) = \begin{bmatrix} \boldsymbol{x}_r(i) \\ \boldsymbol{x}_i(i) \end{bmatrix} \tag{8.84}$$

The detected signal $\boldsymbol{y}$ is given by:

$$\boldsymbol{y} = \boldsymbol{W}_c \boldsymbol{x}_c \tag{8.85}$$

The proposed BSMBER algorithm has two main loops. The outer loop is for each block of data and the inner loop is repeated over the same block of data until:

- a certain number of iterations reached
- the norm of the gradient matrix is sufficiently small
- the gradient difference becomes zero (to prevent dividing by zero).

The BSMBER algorithm is shown in Box 8.2 [41–43].

Box 8.2 Summary of BSMBER algorithm in MIMO/OFDM system.

Initialization

$i = 1$, $\mu = 0.1$, $\varepsilon = 0.1$, $\alpha = 0.25$, $\beta = 0.5$, $\rho = 10\sigma_\eta$, Block = 10,
$\boldsymbol{W}(0) = \boldsymbol{R}_{xx}(0)/\|\boldsymbol{R}_{xx}(0)\|^2$.

Outer loop (1: floor (all bits/block))

- Form a block of data from the received signals.
- Initialize $P_E(\boldsymbol{W}) = \mathbf{1}_{M_r}$, $\nabla P_E(\boldsymbol{W}) = \mathbf{1}_{M_r \times M_r}$, $j = 1$.

Inner loop (while $j < 11$ *or* $\|\nabla P_E\| < \varepsilon$ *or* $\boldsymbol{D} = \mathbf{0}$

- Calculate the cost function BER and the gradient matrix over the block from equations (2.75) and (8.50).
- If $j = 1$ then $\boldsymbol{D} = -\nabla P_E$ else calculate $\boldsymbol{D}$ from equation (8.81).
- Check the direction matrix $\boldsymbol{D}$, if $\frac{|\boldsymbol{D}(j)\nabla Pe(j)|}{\|\boldsymbol{D}(j)\| \cdot \|\nabla Pe(j)\|} < \varepsilon$ then reset $\boldsymbol{D}$ to be equal to $-\nabla P_E$.
- Check the step size μ as in equations (8.74) and (8.75). If it falls outside the boundaries, then increase or decrease the step size as $\mu(i) = \mu(i-1) \pm \Delta\mu$.
- Update the weight matrix as $\boldsymbol{W}_c(i) = \boldsymbol{W}_c(i-1) + \mu\boldsymbol{D}(i)$.
- Increment the iteration number $i = i + 1$

End of inner loop

End of outer loop

8.5 MBER Simulation Results

In this section, we present simulation results that illustrate the performance of different MBER spatial MUD techniques in addition to the MMSE spatial MUD. We employ the elliptical channel model described in Chapter 2. The simulation parameters are described in Table 8.3. The DOA angles were randomly generated between ±60°, three sectors site, and the multipaths for the same user have DOA angles spread in a range ±10° from the main path. One of the interferers is separated by only 10° from the desired user. The DOA angles are unknown to the receiver. The simulations were done in a perfect power control environment (equal power distribution).

Table 8.3 MBER simulation parameters in MIMO/OFDM system.

Parameter	Value
Noise variance (σ_η)	30 dB
Step size (μ)	0.01
Kernel radius (ρ)	10 σ_η
Number of subcarriers	64
OFDM symbol time	64 symbol periods
Guard time	16 symbol periods
Coherence time	50 symbol periods
Number of users	4
User transmit antennas	4 for each user
BS receive antennas	4
Antenna spacing	$d = \lambda/2$
Power delay profile	[1 e^{-1} e^{-2}]
Modulation	BPSK
Angle of arrival	Unknown random [−60, 60]
Channel	Synchronous Block Rayleigh fading
Noise	Complex AWGN

8.5.1 Convergence Rate Comparison

In this section, we run the algorithms for 100 iterations (10 iterations for each block, which consists of 10 bits) and determine the BER for each iteration. Figures 8.18 and 8.19 show the convergence rates for the AMBER and LMBER families, respectively, compared with the conventional matched filter (MF), MMSE, and single user (SU) detectors. LMUD outperforms the conventional MF in terms of the steady-state BER. In addition, using any algorithm minimizing the BER outperforms the MMSE technique. Using the Newton algorithm makes the convergence faster than the normal algorithms. Normalizing the cost function enhances the convergence rate as it gives robustness against SNR variation. Using the Newton modification and normalization together gives the best performance. Figure 8.20 shows BER versus the iteration index for the MMSE, LMBER, Newton-LMBER, and BSMBER algorithms, with SU and MF

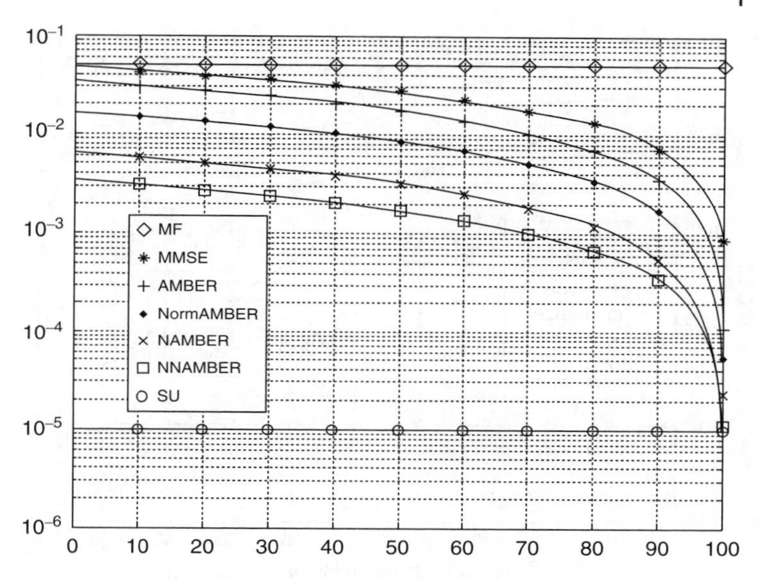

Figure 8.18 BER vs iterations for AMBER family and MMSE algorithms; SU and MF as higher and lower steady-state limits, respectively; equal power distribution at SNR = 30 dB.

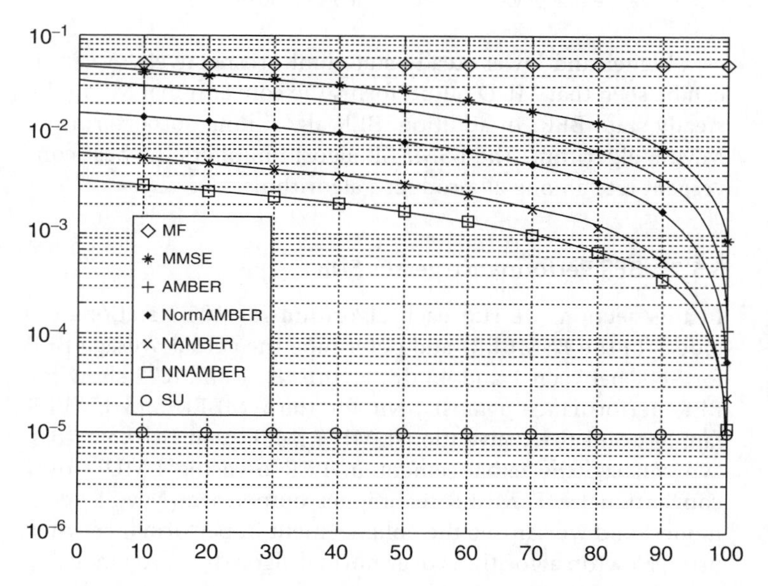

Figure 8.19 BER vs iterations for LMBER family and MMSE algorithms; SU and MF as higher and lower steady-state limits, respectively; equal power distribution at SNR = 30 dB.

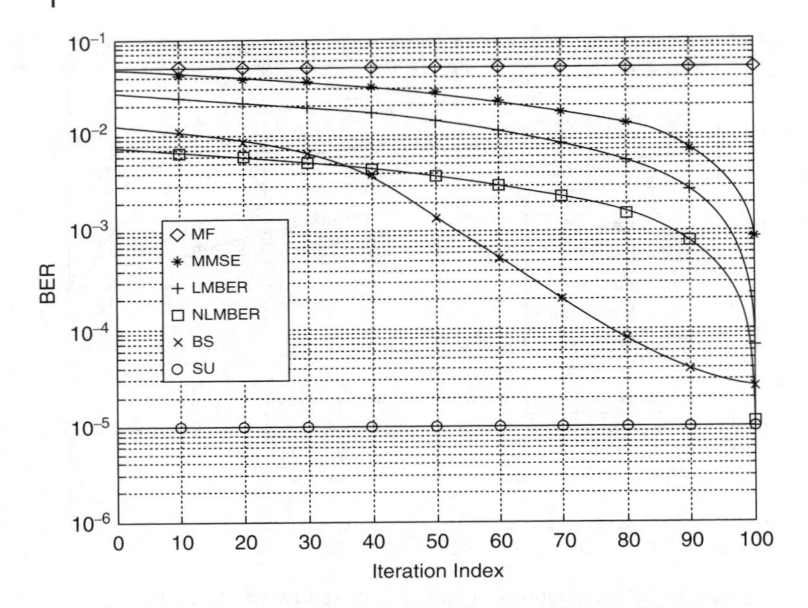

Figure 8.20 BER vs iterations for LMBER, Newton-LMBER, BSMBER and MMSE algorithms; SU and MF as higher and lower steady-state limits, respectively; equal power distribution at SNR = 30 dB.

as a higher and lower steady-state limit, respectively. Again, it can be seen that MUD outperforms the conventional MF in the steady state BER. In addition, BER algorithms outperform the MMSE algorithm. The BSMBER algorithm offers the best convergent rate among all produced algorithms.

8.5.2 BER Performance versus SNR

In this section, we run each algorithm for 100 iterations (10 blocks in the BSMBER) and determine the actual average BER for SNR between 1 and 45 dB. Figures 8.21 and 8.22 show the BER performance against SNR for the AMBER and LMBER families and the conventional MF, MMSE, and SU detectors. We can see the enhancement in BER of using LMUD over conventional MF. MBER techniques outperform MMSE techniques, and we can see the enhancement in performance when using Newton algorithms over normal algorithms. Again, using

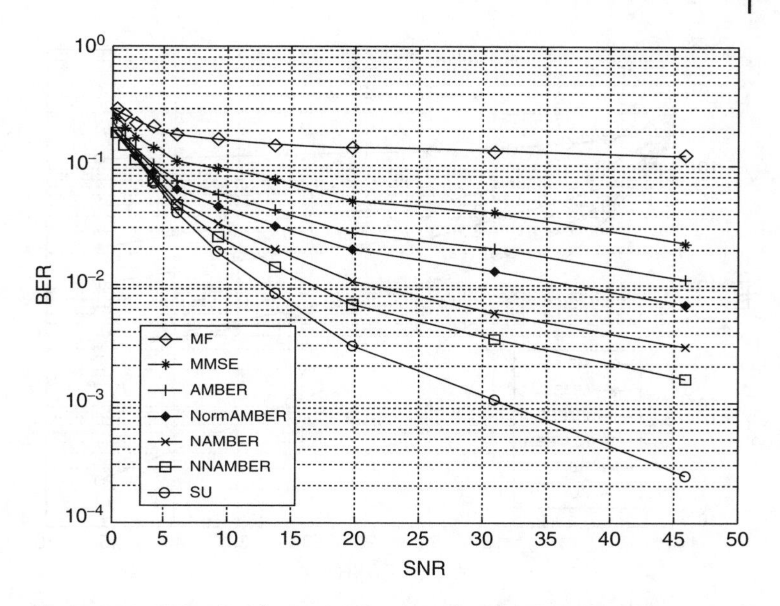

Figure 8.21 BER vs SNR for AMBER family and MMSE algorithms; SU and MF as higher and lower steady-state limits, respectively; equal power distribution at SNR = 30 dB.

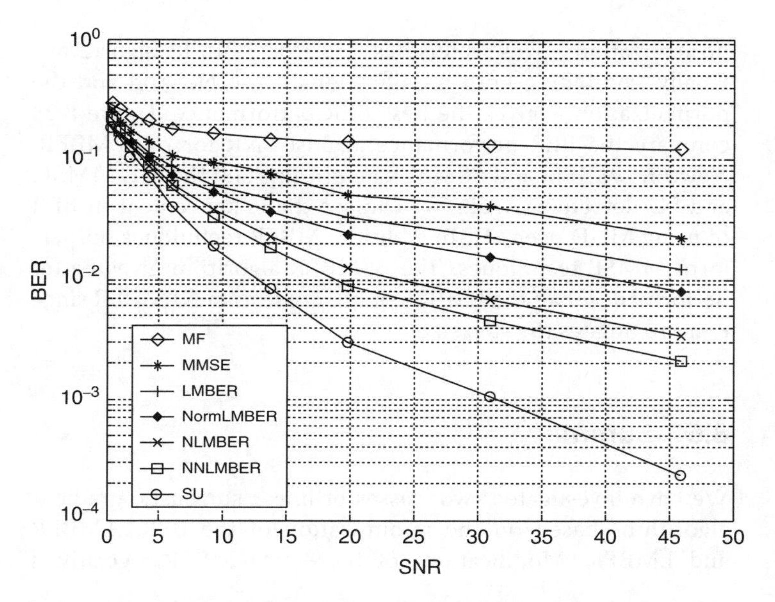

Figure 8.22 BER vs SNR for LMBER family and MMSE algorithms; SU and MF as higher and lower steady-state limits, respectively; equal power distribution at SNR = 30 dB.

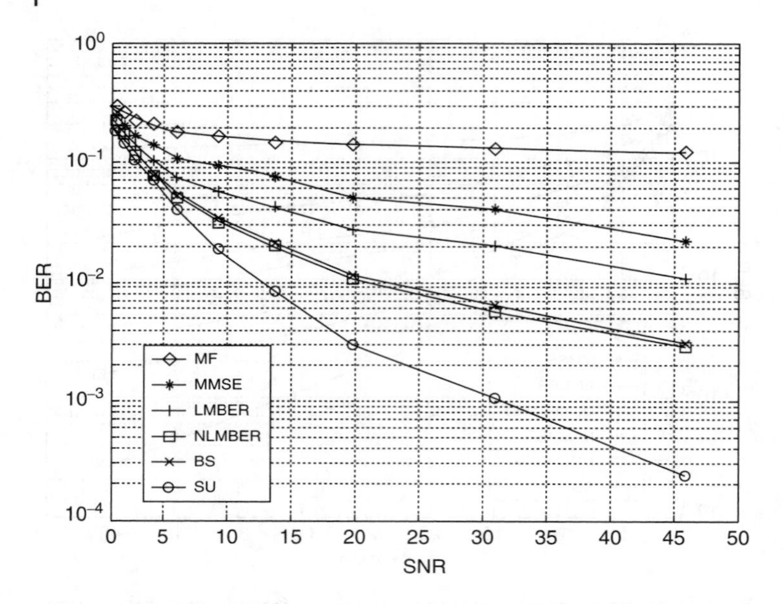

Figure 8.23 BER vs SNR for LMBER, Newton-LMBER, BSMBER and MMSE algorithms; SU and MF as higher and lower steady-state limits, respectively; equal power distribution at SNR = 30 dB.

normalization gives robustness to the MBER algorithms. Finally, combining both modifications – the Newton and the normalization – gives the best BER performance. Figure 8.23 compare the BER performance against SNR for the LMBER, Newton LMBER, and BSMBER algorithms and the MF, MMSE, and SU detectors. Again, we can see the enhancement in BER of using MUD over MF. In addition, MBER techniques outperform MMSE techniques. The BSMBER algorithm gives better performance than the LMBEE but is worse than NLMBER since it approximates the step size.

8.6 Summary

We have investigated two classes of linear stochastic gradient algorithms based on the minimization of the BER: AMBER and LMBER. Modifications of these two MBER algorithms give rise to two families of algorithms: the AMBER family and the LMBER family. These modifications are the Newton modification and the normalization modification of the BER cost function. The modifications were made in order to enhance the performance, increase the convergence rate and give robustness against channel variations. The Newton modification led to the Newton-AMBER and Newton-LMBER algorithms. The normalization modification led to the normalized-AMBER (NAMBER) and the normalized-LMBER (NAMBER) combined modifications. The Newton and the normalization modifications together led to the normalized-Newton-AMBER (NNAMBER) and the normalized-Newton-LMBER (NNLMBER) algorithms. The combined modifications were made in order to get the advantages of both Newton and normalized modifications together, but this came at the cost of complexity. The block-Shanno MBER (BSMBER) algorithm was also presented. It has similar performance to the Newton algorithms but with linear complexity. We have compared these algorithms with the most popular MMSE-based algorithms, assessing their performance in terms of convergence rate, BER, system capacity, and computational complexity. The literature gives more details of MBER algorithms [29, 41–43], including recent examples [46–60].

References

1 S. Chen, "Adaptive minimum bit-error-rate filtering", *IEE Proc. Vis. Image Signal Process.*, vol. 151, no. 1, 5 Feb. pp. 76-85, 2004.

2 S. Chen, B. Mulgrew, E.S. Chng, and G.J Gibson, "Space translation properties and the minimum-BER linear-combiner DFE", *IEE Proc. Commun.*, vol. 145, no. 5, pp. 316–322, 1998.

3 B. Mulgrew and S. Chen, "Stochastic gradient minimum-BER decision feedback equalizers." In *The IEEE Adaptive Systems for Signal Processing, Communications, and Control Symposium 2000*, 1–4 October 2000, pp. 93–98.

4 S. Chen, B. Mulgrew, and L. Hanzo, "Least bit error rate adaptive nonlinear equalizers for binary signaling," *IEE Proc. Commun.*, vol. 150, no. 1, pp. 29–36, 2003.

5 S. Chen, X.C. Yang, and L. Hanzo, "Space-time equalization assisted minimum bit-error ratio multiuser detection for SDMA systems." In: *IEEE 61st Vehicular Technology Conference, 2005*, 30 May–1 June 2005, vol. 2, pp. 1220–1224.

6 R.C. de Lamare and R. Sampaio-Neto, "Adaptive MBER decision feedback multiuser receivers in frequency selective fading channels," *IEEE Commun. Lett.*, vol. 7, no. 2, pp. 73–75, 2003.

7 Y. Liu and Y. Yang, "Adaptive minimum bit error rate multitarget array algorithm." In: *Proceedings of the IEEE 6th Circuits and Systems Symposium on Emerging Technologies: Frontiers of Mobile and Wireless Communication, 2004*, 31 May–2 June 2004, vol. 2, pp. 745–748.

8 Q. Meng, J.-K. Zhang, and K.M. Wong, "Block data transmission: a comparison of performance for MBER precoder designs." In: *Proceedings of Sensor Array and Multichannel Signal Processing Workshop Proceedings, 2004*, 18-21 July 2004, pp. 682–686.

9 S. Chen, A.K. Samingan, B. Mulgrew, and L. Hanzo, "Adaptive minimum-BER linear multiuser detection for DS-CDMA signals in multipath channels," *IEEE Trans. Signal Process.*, vol. 49, no. 6, pp. 1240–1247, 2001.

10 S. Chen, A.K. Samingan, B. Mulgrew, and L. Hanzo, "Adaptive minimum-BER linear multiuser detection." In: *Proceedings of IEEE International Conference on Acoustics, Speech, and Signal Processing, 2001*, 7–11 May 2001, vol. 4, pp. 2253–2256.

11 M.Y. Alias, S. Chen, and L. Hanzo, "Genetic algorithm assisted minimum bit error rate multiuser detection in multiple antenna aided OFDM." In: *IEEE 60th Vehicular Technology Conference*, 26–29 September 2004. vol. 1, pp. 548–552.

12 M.Y. Alias, A.K. Samingan, S. Chen, and L. Hanzo, "Multiple antenna aided OFDM employing minimum bit error rate multiuser detection," *Electron. Lett.*, vol. 39, no. 24, pp. 1769–1770, 2003.

13 A. Wolfgang, N.N. Ahmad, S. Chen, and L. Hanzo, "Genetic algorithm assisted minimum bit error rate beamforming." In: *IEEE 59th Vehicular Technology Conference*, 17-19 May 2004, vol. 1, pp. 142–146.
14 M.Y. Alias, S. Chen, and L. Hanzo, "Multiple-antenna-aided OFDM employing genetic-algorithm-assisted minimum bit error rate multiuser detection," *IEEE Trans. Vehic. Techn.*, vol. 54, no. 5, pp. 1713–1721, 2005.
15 S. Chen, L. Hanzo, and N.N. Ahmad, "Adaptive minimum bit error rate beamforming assisted receiver for wireless communications." In: *IEEE International Conference on Acoustics, Speech, and Signal Processing*, 6–10 April 2003, vol. 4, pp. IV–640-3.
16 S. Chen, L. Hanzo, N.N. Ahmad, and A. Wolfgang, "Adaptive minimum bit error rate beamforming assisted QPSK receiver." In: *2004 IEEE International Conference on Communications*, 20–24 June 2004, vol. 6, pp. 3389–3393.
17 I.D.S. Garcia, J.J.S. Marciano Jr., and R.D. Cajote, "Normalized adaptive minimum bit-error-rate beamformers." In: *2004 IEEE Region 10 Conference*, 21–24 November 2004, vol. B, pp. 625–628.
18 C. Yeb, R. Lopez, and J.R. Barry, "Approximate minimum bit-error rate multiuser detection." In: *Global Communications Conference*, 1998.
19 S. Chen, N.N. Ahmad, and L. Hanzo, "Adaptive minimum bit-error rate beamforming," *IEEE Trans. Wireless Commun.*, vol. 4, no. 2, pp. 341–348, 2005.
20 S. Chen, A.K. Samingan, and L. Hanzo, "Adaptive minimum error rate training for neural networks with application to multiuser detection in CDMA communication system," *IEEE Trans. Signal Process.*, vol. 49, pp. 1240–1247, 2001.
21 S. Chen, A.K. Samingan, and L. Hanzo, "Support vector machine multiuser receiver for DSCDMA signals in multipath channels," *IEEE Trans. Neural Netw.*, vol. 12, pp. 604–611, 2001.
22 S. Chen, B. Mulgrew, and L. Hanzo, "Adaptive least error rate algorithm for neural network classifiers." In: *Proceedings of the 2001 IEEE Signal Processing Society Workshop Neural Networks for Signal Processing XI*, 10–12 September 2001, pp. 223–232.

23 I.N. Psaromiligkos, S.N. Batalama, and D.A. Pados, "On adaptive minimum probability of error linear filter receivers for DS-CDMA channels," *IEEE Trans. Commun.*, vol. 47, no. 7, pp. 1092–1102, 1999.

24 C.C. Yeh and J.R. Barry, "Adaptive minimum bit-error rate equalization for binary signaling," *IEEE Trans. Commun.*, vol. 48, no. 7, pp. 1226–1235, 2000.

25 S.J. Yi, C.C. Tsimenidis, O.R. Hinton, and B.S. Sharif, "Adaptive minimum bit error rate multiuser detection for asynchronous MC-CDMA systems frequency selective Rayleigh fading channels," In: *14th IEEE Proceedings on Personal, Indoor and Mobile Radio Communications, 2003,* 7–10 September 2003, vol. 2, pp. 1269–1273.

26 I.S. Garcia, "Performance of adaptive minimum bit- error-rate beamforming algorithms for diversity combining and interference suppression under frequency- flat directional fading channels," MSEE thesis, University of the Philippines, 2004.

27 B. Mulgrew and S. Chen, "Adaptive minimum-BER decision feedback equalizers for binary signaling," *Signal Process.*, vol. 81, no. 7, pp. 1479–1489, 2001.

28 J. Li, F. Chen, and G. Wei, "Local minimum-BER linear multiuser detector," *IEEE International Symposium on Communications and Information Technology,* 2005, 12–14 October 2005, vol. 1, pp. 511–514.

29 T. Samir, S. Elnoubi, and A. Elnashar, "Block-Shanno minimum BER beamforming," *IEEE Transactions Vehic. Techn.*, vol. 57, no. 5, pp. 2981-2990, 2008.

30 A. Bahai and B. Saltzberg, *Multicarrier Digital Communications: Theory and Applications of OFDM*. Kluwer Academic, 1999.

31 Y.J. Zhang and K.B. Letaief, "An efficient resource-allocation scheme for spatial multiuser access in MIMO/OFDM systems," *IEEE Trans. Commun.*, vol. 53, no. 1, pp. 107–116, 2005.

32 M.Y. Alias, S. Chen, and L. Hanzo, "Genetic algorithm assisted minimum bit error rate multiuser detection in multiple antenna aided OFDM." In: *2004 IEEE 60th Vehicular Technology Conference,* 2004, 26–29 September 2004, vol. 1, pp. 548–552.

33 M.Y. Alias, A.K. Samingan, S. Chen, and L. Hanzo, "Multiple antenna aided OFDM employing minimum bit error rate multiuser detection," *Electron. Lett.*, vol. 39, no. 24, pp. 1769–1770, 2003.

34 M.Y. Alias, S. Chen, and L. Hanzo, "Multiple-antenna-aided OFDM employing genetic-algorithm-assisted minimum bit error rate multiuser detection," *IEEE Trans. Vehic. Techn.*, vol. 54, no. 5, pp. 1713–1721, 2005.

35 Chen S. and C.J. Harris, "Design of the optimal separating hyperplane for the decision Feedback equalizer using support vector machines," In: *IEEE International Conference on Acoustics, Speech, and Signal Processing, 2000*, 5–9 June 2000, vol. 5, pp. 2701–2704.

36 E. Parzen, "On estimation of a probability density function and mode," *Ann. Math. Stat.*, vol. 33, pp. 1066–1076, 1962.

37 A.W. Bowman and A. Azzalini, *Applied Smoothing Techniques for Data Analysis*, Oxford University Press, 1997.

38 J. Litva and T.K. Lo, *Digital Beamforming in Wireless Communications*, Artech House, 1996.

39 D.F. Shanno, "Conjugate gradient method with inexact searches," *Math. Oper. Res.*, vol. 3, pp. 244–256, 1978.

40 U.G. Jani, E.M. Dowling, R.M. Golden, and Z. Wang, "Multiuser interference suppression using block Shanno constant modulus algorithm," *IEEE Trans. Signal Process.*, vol. 48, no. 5, pp. 1503–1506, 2000.

41 T. Samir, S. Elnoubi, and A. Elnashar "Block-Shanno MBER algorithm in a spatial multiuser MIMO/OFDM." In: *Proceedings of 14th European Wireless Conference*, 22-25 June 2008.

42 T. Samir, S. Elnoubi, and A. Elnashar "Class of minimum bit error rate algorithms." In: *Proceedings of ICACT 2007*, Korea, February 2007, pp. 168-173.

43 T. Samir, S. Elnoubi, and A. Elnashar, "Block-Shanno minimum BER beamforming." In: *Proceedings of ISSPA 2007*, UAE, February 2007.

44 A. Hjørungnes and P.S.R. Diniz, "Minimum BER prefilter transform for communications systems with binary signaling and known FIR MIMO channel," *IEEE Signal Process. Lett.*, vol. 12, no. 3, pp. 234–237, 2005.

45 C.-C. Yeh and J. Barry, “Adaptive minimum bit-error rate equalization for binary signaling,” *IEEE Trans. Commun.*, vol. 48, no. 7, pp. 1226–1235, 2000.

46 S. Chen, W. Yao, and L. Hanzo, “Semi-blind adaptive spatial equalization for MIMO systems with high-order QAM signalling,” *IEEE Trans. Wireless Commun.*, vol. 7, no. 11, pp. 4486–4491, 2008.

47 W. Yao, S. Chen, and L. Hanzo, “Improved MMSE vector precoding based on MBER criterion.” In: *Proceedings of VTC Spring*, Barcelona, Spain, 26–29 April 2009.

48 S. Chen, A. Livingstone, H.-Q. Du, and L. Hanzo, “Adaptive minimum symbol error rate beamforming assisted detection for quadrature amplitude modulation,” *IEEE Trans. Wireless Commun.*, vol. 7, no. 4, pp. 1140–1145, 2008.

49 S. Tan, *Minimum Error Rate Beamforming Transceivers*, PhD thesis, University of Southampton, 2008.

50 W. Yao, S. Chen, S. Tan, and L. Hanzo, “Particle swarm optimisation aided minimum bit error rate multiuser transmission.” In: *Proceedings of ICC*, Dresden, Germany, 14–18 June 2009, pp. 1–5.

51 W. Yao, S. Chen, S. Tan, and L. Hanzo, “Minimum bit error rate multiuser transmission designs using particle swarm optimisation,” *IEEE Trans. Wireless Commun.*, vol. 8, no. 10, pp. 5012–5017, 2009.

52 A.K. Dutta, K.V.S. Hari, and L. Hanzo, “Minimum-error-probability CFO estimation for multiuser MIMO-OFDM systems,” *IEEE Trans. Vehic. Techn.*, vol. 64, pp. 2804–2818, 2015.

53 S. Chen, A. Livingstone, and L. Hanzo, “Minimum bit-error rate design for space-time equalization-based multiuser detection,” *IEEE Trans. Commun.*, vol. 54, no. 5, pp. 824–832, 2006.

54 W. Yao, S. Chen, and L. Hanzo,“Generalized MBER-based vector precoding design for multiuser transmission,” *IEEE Trans. Vehic. Techn.*, vol. 60, no. 2, pp. 739–745, 2011.

55 W. Yao, S. Chen, and L. Hanzo, “A transceiver design based on uniform channel decomposition and MBER vector perturbation,” *IEEE Trans. Vehic. Techn.*, vol. 59, no. 6, pp. 3153–3159, 2010.

56 M. Alias, S. Chen, and L. Hanzo, "Multiple-antenna-aided OFDM employing genetic-algorithm-assisted minimum bit error rate multiuser detection," *IEEE Trans. Vehic. Techn.*, vol. 54, no. 5, pp. 1713–1721, 2005.

57 S. Sugiura, N. Wu, and L. Hanzo, "Improved Markov chain MBER detection for steered linear dispersion coded MIMO systems." In: *Proceedings of VTC Spring* 2009, pp. 1–5.

58 A. Dutta, K.V. S. Hari, and L. Hanzo, "Channel estimation relying on the minimum bit error ratio criterion for BPSK and QPSK signals," *IET Commun.*, vol. 8, no. 1, pp. 69–76, 2014.

59 P. Weeraddana, N. Rajatheva, and H. Minn, "Probability of error analysis of BPSK OFDM systems with random residual frequency offset," *IEEE Trans. Commun.*, vol. 57, no. 1, pp. 106–116, 2009.

60 R.C. de Lamare and R. Sampaio-Neto, "Adaptive multiuser receivers for DS-CDMA using minimum BER gradient-Newton algorithms." In: *13th IEEE International Symposium on Personal, Indoor and Mobile Radio Communications*, vol. 3, pp. 1290–1294, 2002.

Index

Simplified Robust Adaptive Detection and Beamforming for Wireless Communications, First Edition. Ayman Elnashar.

Companion website: www.wiley.com/go/elnashar49

d

e

n

q

r

S